Genetic Errors of Glycoprotein Metabolism

Genetic Errors of Glycoprotein Metabolism

Edited by
Paolo Durand and John S. O'Brien

edi·ermes
Milano
Springer-Verlag
Berlin Heidelberg New York
1982

Sole distribution rights
in Italy:
Edi.Ermes
Via Timavo, 12
20124 Milano
Italy

Outside Italy:
Springer-Verlag
Berlin Heidelberg New York
Tiergartenstrasse 17
6900 Heidelberg 1
F.R. Germany

ISBN 978-3-642-51584-2 ISBN 978-3-642-51582-8 (eBook)
DOI 10.1007/978-3-642-51582-8

Edi.Ermes Milano

Softcover reprint of the hardcover 1st edition 1982

Contributors

Pertti AULA
Laboratory of Prenatal Genetics, Department of Obstetrics and Gynecology, University of Helsinki, 00290 Helsinki - 29 Finland

Seppo AUTIO
The Children's Castle Hospital, SF-00250 Helsinki 25 - Finland

Carla BORRONE
Third Department of Pediatrics, G. Gaslini Institute - 16148 Genova-Quarto - Italy

M. Alan CHESTER
Department of Clinical Chemistry, University Hospital, S-221 85 Lund - Sweden

Paolo DURAND
Third Department of Pediatrics, G. Gaslini Institute - 16148 Genova-Quarto - Italy

Rosanna GATTI
Third Department of Pediatrics, G. Gaslini Institute - 16148 Genova-Quarto - Italy

Arne LUNDBLAD
Department of Clinical Chemistry, University Hospital, S-221 85 Lund, Sweden

John S. O'BRIEN
Department of Neurosciences, School of Medicine, University of California, San Diego, La Jolla, California 92093 - USA

Per-Arne OCKERMAN
Department of Clinical Chemistry, University Hospital, S-221 85 Lund - Sweden

Aldo PELIZZA
Department of Radiology, G. Gaslini Institute - 16148 Genova-Quarto - Italy

Kari O. RAIVIO
The Children's Castle Hospital, SF-00250 Helsinki 25 - Finland

Juhani RAPOLA
The Children's Castle Hospital, SF-00250 Helsinki 25 - Finland

Martin RENLUND
The Children's Castle Hospital, SF-00250 Helsinki 25 - Finland

Thomas G. WARNER
Department of Neurosciences, School of Medicine, University of California, San Diego, La Jolla, California 92093 - USA

Preface

On starting to write the preface it occured to the editors that it might be a good idea to incorporate some organizational thoughts.

First of all, why have we organized this book on «Glycoprotein storage diseases", an unfamiliar area of pathology unlikely to be encountered in practice?

In preparing such a volume we considered that in the past ten years there has been a large amount of intensive research activity on glycoproteinoses and no detailed review has been entirely devoted to this subject. This research has led to new information about hereditary mental retardation and progressive neurologic degeneration, and as clinicians we are impressed with its clinical importance.

This volume is intended to represent for the uninitiated as well as for the expert a summary of studies on disorders of glycoprotein catabolism written by experts in each area.

Each author has written with freedom about his particular experience, describing the establishment and development of his investigations, his methods, the utility and limitations of his work and eventual sources of problems.

We are grateful for their many fundamental studies and for their having generously accepted to partecipate in this book.

The Editors

Acknowledgements

We wish to express our gratitude to our friends and colleagues G. Romeo, W.S. Sly and G. Tettamanti whose criticism and suggestions have significantly improved the quality of particular chapters or sections.

We are also indebted to Miss Jenny Cottam for her expert assistance and constant pleasantness in the revision of some chapters of this volume.

We furthermore wish to acknowledge the many contributions made by various members of the professional, technical and secretarial staffs of the Department of the Authors.

This book has also been made possible by the help and encouragement of the "Associazione Italiana Prevenzione Cura Handicaps" (A.I.R.H.).

The Authors are grateful to the publishers Edi.Ermes for their cooperation in making this volume possible.

Contents

1 Introduction to Glycoprotein Storage Diseases

Paolo Durand
John S. O'Brien

1.1 GENERAL INTRODUCTION

During the last decade there have been impressive advances in the elucidation of a group of diseases due to defects in the enzyme activities that are required for the catabolism of the oligosaccharide portion of glycoproteins.

At present four inherited disorders: sialidosis, fucosidosis, mannosidosis and aspartylglucosaminuria, can be defined as glycoproteinoses. In each, the metabolic defect predominantly involves faulty catabolism of the oligosaccharide side chain of glycoproteins.

In this book other diseases such as I cell disease (mucolipidosis II), pseudo-Hurler polydystrophy (mucolipidosis III) and Salla disease are mentioned. For these diseases it is probable that future elucidation of the exact nature of the type of gene mutation involved will be an additional reason to confirm or to exclude the inclusion in the category of glycoproteinoses. Other diseases like G_{M1} gangliosidosis and G_{M2} gangliosidosis type 2 or Sandhoff disease are also included since absence of β-galactosidase and β-N-acetylhexosaminidase A and B results in accumulation not only of gangliosides but also of glycopeptides and oligosaccharides.

Delineation of each of these distinct diseases on the basis of clinical features alone is difficult in many instances but in general, gargoyle-like facies, skeletal changes of the dysostosis multiplex type, mental and motor retardation and visual loss are present. In a number of these diseases, however, like some variants of sialidosis, fucosidosis and mannosidosis, gargoyle like facies, skeletal deformities and even psychomotor retardation may be absent. The major goal when studying a specific genetic disease is to find the primary defect which provides an understanding of the biochemical basis of the disorder.

As in other storage diseases, an inherited lysosomal enzyme defect leads to an accumulation of complex carbohydrates with a specific pattern of urinary excretion.

In practice, a relatively simple approach for screening these diseases is to look for an abnormal oligosaccharide excretion in the urine. This procedure together with the clinical features is useful in the differential diagnosis of glycoproteinoses and may be used as a basis for selection of the analysis of the enzyme responsible for the metabolic defect.

Affected individuals have very low or undetectable amounts of the enzyme activity in question and heterozygotes have activity levels between those of diseased and normal individuals, which allows detection of the carrier state.

Knowledge of these disorders at the biochemical level is therefore a prerequisite for the prevention procedures such as heterozygote detection and prenatal diagnosis.

The ability to recognize heterozygotes in relatives of a proband in a family at risk in small inbred communities or in defined populations, is helpful in regard to marriage, reproductive options and/or prospective genetic counselling.

Undoubtedly future studies will reveal an increasing number of variants of glycoproteinoses which will lead to a better definition of some existing disorders. Other than mannosidosis due to α-D-mannosidase deficiency in humans, in cattle and in cats, a caprine storage disorder was recently described, due to an inherited β-D-mannosidase deficiency with an accumulation of mannosylchitobiose.

Although no humans with the condition have been found as yet, β-D-mannosidosis is likely to exist as a human disorder as well.

The glycoproteinoses as "èxperiment of nature" have provided the Rosetta stone for understanding not only the normal pathway of glycoprotein degradation but also the process of recognition, packaging and transport of lysosomal enzymes, as well as pathophysiology of somatic dysmorphism and cerebral degeneration which result from impaired catabolism of glycoproteins, glycosaminoglycans and glycolipids.

1.2 HISTORICAL REVIEW

Between 1966 and 1969 a group of hereditary disorders manifesting some features of mucopolysaccharidosis without mucopolysacchariduria were published. The clinical descriptions were soon followed by biochemical studies and within a short time the molecular nature of these rare afflictions was recognized.

A. Fucosidosis

The first cases of fucosidosis were described by Durand et al., in 1966 as "A new mucopolysaccharide lipid-storage disease" [1]. Two young siblings suffering from this disease developed normally at first, but later showed gradual mental deterioration progressing to dementia, spastic quadriplegia and decerebrate rigidity. Death came at 4 to 5 years of age.

Pathologic studies revealed lysosomal storage in nearly all cell types in the body.

Fucose-rich storage material was found in the nervous system as well as in the liver, heart, pancreas, spleen and kidneys [2-4].

In 1968 Van Hoof and Hers identified the enzyme defect as absence of activity of lysosomal α-L-fucosidase [5] and at that time the term "fucosidosis" was proposed [6].

B. Mannosidosis

The first case of mannosidosis was described by Ockerman in 1967 as a "generalized storage disorder resembling Hurler's syndrome" [7]. The boy had psychomotor retardation, coarse facial features, dysostosis multiplex, abnormal susceptibility to infections and accelerated growth in infancy, followed by practically no increase in height after 2 years of age.

The patient died at 4 years of age. Pathologic studies revealed storage inclusions in cells of many tissues.

Two years later biochemical investigation disclosed a deficiency of α-D-mannosidase activity in the liver with storage of oligosaccharides containing mannose. The abnormal storage of mannosides suggested the name "Mannosidosis" [8,9].

C. Aspartylglucosaminuria

In 1968 - 1969 Pollit and Jenner were the first to draw attention to the clinical characteristics and biochemical abnormalities of "Aspartylglucosaminuria" [10, 11]. They observed two severely mentally retarded siblings, a thirty-one-year old woman and her twenty year old brother, who excreted an unusual urinary compound. The sister suffered also from hypomanic periods and both sibs had deep sagging skin of the cheeks, a finding not present in normal members of the family.

The degradation of aspartylglucosylamine, a compound which had not been found free in nature before, is carried out by a specific enzyme, aspartamido-β-N-acetylglucosamine amidohydrolase; activation of this enzyme was deficient in the two affected individuals [11], Aspartylglucosaminuria is characterized also by slowly coarsening features, slowly progressive mental retardation and mild bone changes. It occurs mainly in Finland and Aula, Autio, Palo and co-workers have performed extensive clinical and biochemical studies on this disease and chapter 5 [12-14].

D. Sialidosis

In 1968 Spranger et al. described two patients with a hereditary disorder beginning in early childhood and characterized by Hurler-like features, skeletal dysplasia, a cherry-red macular spot, progressive ataxia, myoclonic epilepsy, speech impairment, mental retardation and death at the end of childhood. Cytoplasmatic inclusions and membrane-bound vacuoles mainly in Kupffer's cells were found in the electron microscopic studies of the liver [15].

These patients previously described as having a "Lipomucopolysaccharidosis" [15], have been subsequently classified as "Mucopolidosis I" [16] or "Sialidosis, type 2 juvenile" [22]. The primary defect in Mucolipidosis I remained unknown until 1976 - 1977, when Strecker et al. discovered that milder variants of this disease were characterized by an increased urinary excretion of sialyl-oligosaccharides and storage of sialylated compounds [17, 18].

Cantz et al., [19] reported a sialidase deficiency in one of the original cases of mucolipidosis I and Durand et al. [20] and O'Brien et al. [21] reported the same enzymatic defect in incomplete and milder forms of this disease, known as the "cherry red spot-myoclonus syndrome".

A number of variants of this disease have since been recognized by Lowden and O'Brien [22]. Severe infantile and juvenile forms of sialidosis [23-26] with myoclonus epilepsy and renal failure and patients with combined β-galactosidase deficiency in a variety of tissues [27], have been identified.

These variants, collectively called sialidosis [20, 22] extend well beyond the confines of the original description of Mucolipidosis I.

E. Glycoprotein storage in other lysosomal diseases

Mainly through further refinement in biochemical studies it became apparent that in another group of lysosomal inborn errors there is an alteration of glycoprotein catabolism. G_{M1}-gangliosidosis, an inherited storage disease caused by a deficiency of G_{M1}-ganglioside-β-D-galactosidase which leads to an accumulation of a specific monosialoganglioside in neural tissues and visceral organs, is classified as a disorder of glycosphingolipid metabolism, although glycoprotein metabolism is also altered [28-30] since galactosyl-oligosaccharides accumulate and are excreted in excess.

Oligosaccharides and glycopeptides also accumulate in various tissues [31-33] and are excreted in the urine [34-36] of patients with the variant 0 of G_{M2} gangliosidosis (Sandhoff disease). The disease is characterized by a deficiency of activity of isoenzymes A and B of β-hexosaminidase and by a storage of globoside in extraneuronal tissues in addition to a storage of glycolipids G_{M2} and G_{A2} [37].

Some authors [17, 18, 38, 39] showed that sialyloligosaccharides are excreted in considerable amounts in the urine of patients of the two related disorders I-cell disease (Mucolipidosis type II) and pseudo-Hurler polydystrophy (Mucolipidosis type III) first described by Leroy and De Mars [40] and Maroteaux and Lamy [41], respectively.

In I-cell disease there is severe dwarfism, coarse features, severe dysostosis multiplex, dementia, thick skin, hernias, heart disease and hypertrophy of the gingiva. Symptoms are milder in Mucopolidosis type III with later onset and survival to adult life. Both disorders are characterized by multiple lysosomal deficiencies in cultured fibroblasts and a concomitant increase of several lysosomal enzyme activities in the culture medium [42, 43]. The mutant gene product in these disorders has recently been shown to be a phosphotransferase which phosphorylates mannosyl residue present in lysosomal enzymes.

This phosphorylated mannose residue apparently serves as the recognition marker for lysosomal enzyme uptake and packaging within lysosomes; failure to phosphorylate it leads to leakage of lysosomal enzymes from cells.

Sialuria is present in Salla disease, a new type of storage disorder described by Aula et al. [44, 45] with onset of symptoms at one year of age, characterized by severe mental retardation, and impaired motor function and ataxia.

Some of the diseases in which a disturbance of glycoconjugate metabolism is apparent have been grouped under the inclusive term of "genetic mucolipidoses". This term, coined by Spranger and Wiedeman in 1970 [16] was useful for clinical purposes in a transient period but does not correspond to any accepted term of nomenclature of glycoconjugates.

In 1976 Maroteaux and Humbel [46] proposed calling these diseases "Oligosaccharidoses" deriving from the nature of the urinary excreted material.

However, the term "glycoproteinoses" seems preferable for these diseases, since the urinary oligosaccharide excretion is the consequence of an accumulation of glycoconjugates in the tissues and these are both the result of enzyme defects of glycoprotein catabolism [47]. Admittedly the term is not entirely inclusive since glycolipids also accumulate in several of these disorders such as fucosidosis, G_{M1} gangliosidosis and Sandhoff disease.

Thus the responsible enzyme defect in some of these diseases not only results in the accumulation of oligosaccharides and glycopeptides but may also lead to impaired degradation of other sugar containing compounds.

1.3 INTRODUCTION TO GLYCOPROTEIN METABOLISM

A. General criteria

Glycoproteins are widely distributed macromolecules which have a vast number of functional implications.

Glycoproteins are found in serum and urine, in almost all secretions, in intracellular and plasma membranes, including the membranes that line lysosomes, in the extracellular spaces of connective tissue, in the blood group substances of the erythrocyte membrane. Many enzymes and hormones are glycoproteins.

Of all the biologically occurring macromolecules the glycoproteins, which consist of carbohydrate moieties covalently linked to a polypeptide backbone, represent the most diverse class. Therefore, it is difficult to depict glycoproteins as a series of mutually interrelated compounds.

The protein portions of glycoproteins are as different and complex as carbohydrate-free proteins in general.

The carbohydrate portion of glycoproteins covers from as low as 1% of the total weight to over 85% of the total weight of the molecule.

The sugars most commonly occurring in human glycoproteins are sialic acid, L-fucose, D-galactose, D-mannose, D-glucose, N-acetyl-D-glucosamine and N-acetyl-D-galactosamine [1-5] (Fig. 1).

Other sugars like arabinose and D-xylose are rarely found.

A larger number of glycoproteins contain sialic acid while others are sialic-acid free.

Sialic acid is always at the nonreducing terminus of the carbohydrate chains, unless it is penultimate to another sialic acid which is terminally located [6]. L-fucose is also a terminal sugar; D-glucose is rarely present and occurs only in collagen-type proteins or in glycoproteins which are being processed in the Golgi-endoplasmic reticulum complex.

D-xylose is found only in the carbohydrate-protein linkage region, especially in the large proteoglycan complexes.

The chemical nature of the protein-carbohydrate linkage can be different. In all cases the sugar is glycosidically linked to the aminoacid.

The most common sugar-aminoacid linkage occuring in serum glycoproteins and especially interesting as far as the diseases discussed in this book are concerned involves N-acetylglucosamine, linked to the amide nitrogen of asparagine.

Other linkages involve N-acetylgalactosamine linked to the hydroxyl group of serine of threonine as in soluble blood group substances and mucins, and galactose linked to the hydroxyl group of hydroxylysine in collagen [1-6].

B. Biosynthesis of glycoproteins

The sequence of events in the biosynthesis of glycoproteins has been reasonably well if not completely elucidated.

The protein moiety of a glycoprotein is synthesized at the level of the ribosomes linked to rough endoplasmic reticulum.

The inner part of the sugar portion is biosynthesized, linked to a lipid carrier, dolichol phosphate, by sequential addition of the individual sugar residues. At this stage, occurring in the rough endoplasmic reticulum, a part of the N-acetylglucosamine residues and all of the mannose residues are linked. The

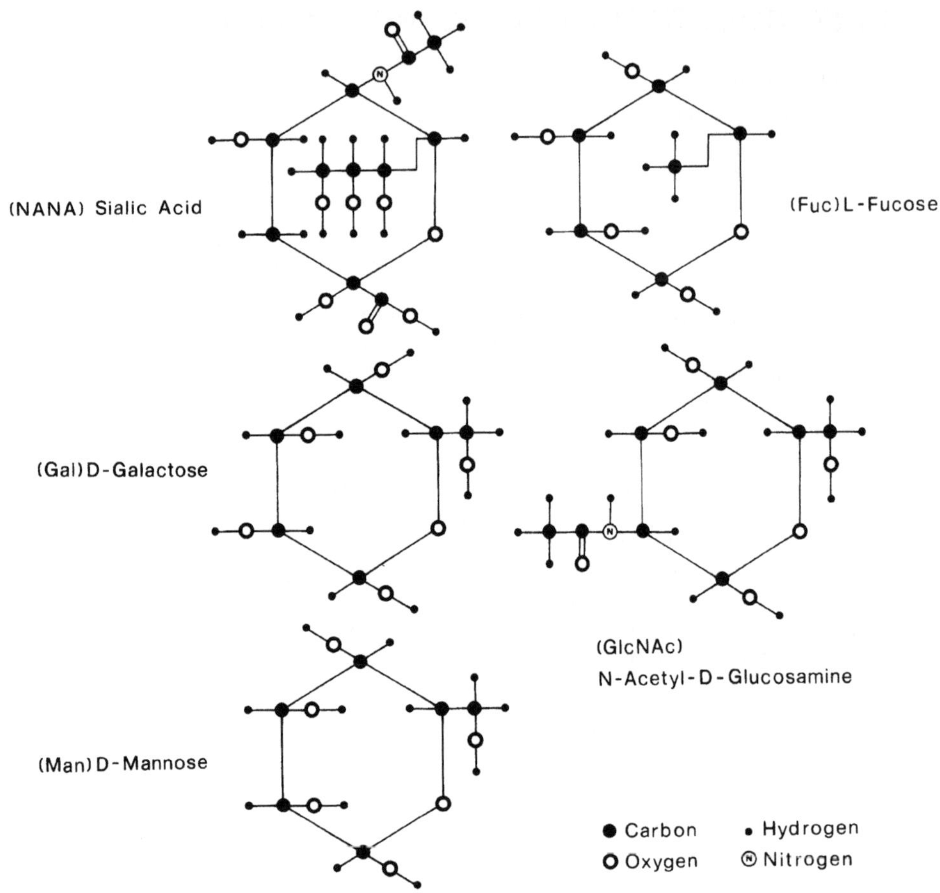

Fig. 1.1 The oligosaccharides attached to proteins are composed of a few simple components.

Sialic acid, fucose, galactose, mannose, N-acetyl-D-glucosamine are particulary involved in the diseases treated in this book.

The most complex sugar, sialic acid, includes both an aminoacetyl and a carboxil group, along with several extra carbons and hydroxil groups. L-fucose is unusual for its L-configuration and has six-carbons and five oxygen atoms. D-galactose and D-mannose are simple sugars of the same composition ($C_6H_{12}O_6$) and differ slightly in structure (stereoisomers). N-acetylglucosamine is a simple 6-carbon sugar with an aminoacetyl (—$NHCOCH_3$) group attached.

oligosaccharide is then transferred to the acceptor aminoacid of the polipeptide and the biosynthesis continues and terminates in the smooth endoplasmic reticulum and in the Golgi apparatus where the remaining residues of N-acetylglucosamine and all of the galactose, fucose and sialic acid residues are introduced.

In general, the addition of sugars to the preformed polypeptide is strictly sequential and is determined by the properties and specificity of the glycosyltransferases which have an intracellular location and work in concert.

This type of organization under genetic control, provides a biochemical explanation for cellular differentiation [3, 7].

C. Catabolism of glycoproteins

The degradation of glycoproteins is thought to be predominantly lysosomal even if glycosidases are found in other cellular sites and particularly in plasma membrane and cytosol. Most, if not all, of the lysosomal enzymes that catalyze the catabolism of glycolipids, glycosaminoglycans and glycoproteins are glycoproteins themselves. They are generally composed mainly of mannose and N-acetylglucosamine and also contain several other sugars in small amounts. These minor sugars appear to include galactose and L-fucose. Sialic acid seems to occur more often in the plasma forms of lysosomal enzymes than in the forms purified from tissues.

Lysosomes contain a full battery of enzymes required for full degradation of the glycoproteins. This battery includes proteases, working on the protein portion, and specific glycosidases, working on the carbohydrate portion of glycoproteins.

The commonly occurring acidic glycosidases are: sialidase, α-L-fucosidase, β-D-galactosidase, α-D-mannosidase, β-D-glucosidase, α-N-acetylglucosaminidase, β-N-acetylglucosaminidase, α-N-acetylgalactosaminidase [1, 4, 5, 7].

A partial degradation of the protein core by a variety of proteolytic enzymes, appears to occur as the first step of glycoprotein degradation. The process leads to formation of glycopeptides, which are attacked by N-aspartyl-β-glucosaminidase (Fig. 2).

This lysosomal enzyme cleaves the aspartylglucosylamine linkage between the polypeptide and the carbohydrate moiety of glycopeptides, liberating aspartic acid, ammonia and the oligosaccharide [8-10].

The appearance in the urine of patients suffering from glycoproteinoses and related diseases of aminoacid-free oligosaccharides containing N-acetylglucosamine at the reducing end strongly suggests the action of an endo-β-N-acetylglucosaminidase which cleaves the N-acetylglucosaminyl-N-acetylglycosaminelinkage [2, 11].

The degradation of the liberated carbohydrate chain is carried on by exoglycosidases which sequentially remove the single monosaccharide residues from the nonreducing end of the carbohydrate chain.

D. Processing and compartmentalization of lysosomal enzymes

While the degradative pathways of glycoproteins have been reasonably well if not completely elucidated as outlined above, there are still a number of uncertain points regarding the processing and the compartmentalization mechanism of lysosomal hydrolases.

Considerable information has been obtained concerning the processes by which lysosomal enzymes are initially synthesized in precursor form, "proenzymes", then phosphorylated and transferred from the site of synthesis to the lysosomal or extracellular destination.

Absorptive pinocytosis of lysosomal enzymes by fibroblasts was first discovered through studies on the uptake of enzymes for enzyme deficient fibroblasts.

The uptake process was found to be specific and it was postulated [12] and successively proved [13] that there were "recognition markers" on the enzymes and specific receptors for these recognition markers on the cell surfaces of fibroblasts.

The common recognition marker on lysosomal enzymes was shown to reside in the oligosaccharide chains of the enzymatic glycoprotein and to consist of

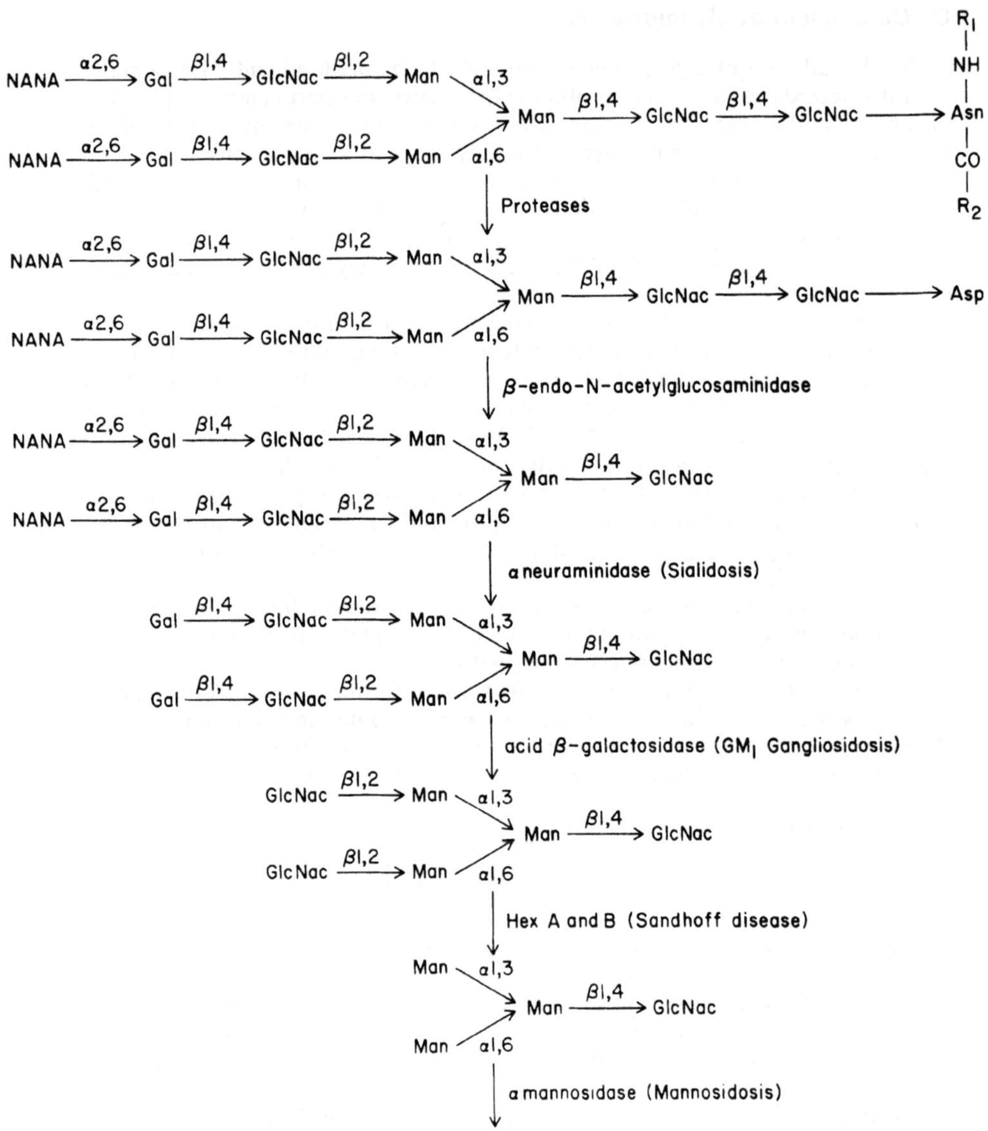

Fig. 1.2 Partial scheme of glycoprotein degradation catabolism showing position of some of known inborn errors. (For the sites of fucosidosis and aspartyl-glycosaminuria, see Chapters 3 and 5).

mannose-6-phosphate.

A number of investigations have established the role of carbohydrate recognition systems for pinocytosis [14-17]. The systems that have been discovered are cell-specific and recognize the following terminal sugar residues: mammalian hepatocytes, mannose, galactose (also N-acetylgalactosamine, glucose) and

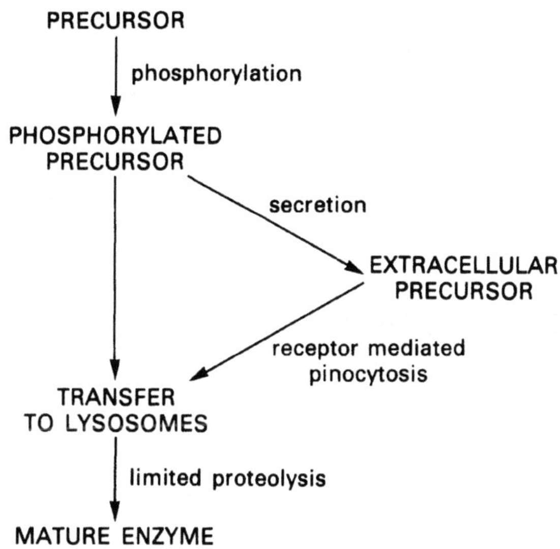

Fig. 1.3 Scheme of biosynthesis of lysosomal enzymes and their secretion-recapture in normal cells.

The enzymes synthesized in «precursor» form of higher molecular weight, are phosphorylated, reach their lysosomal destination and after a limited proteolysis become «mature enzyme». The phosphorylated mannose residue serves also as the recognition marker for lysosomal enzyme uptake and packaging within lysosome.

L-fucose; mammalian reticuloendothelial cells, mannose (also N-acetylglucosamine, glucose); human fibroblasts, mannose-6-phosphate.

Considerable evidence has been recently produced to show that phosphomannosyl is a recognition marker or signal not only for uptake of exogenous enzymes but also for cellular transport of newly synthesized lysosomal enzymes to their lysosomal destination ("traffic signal").

Thus, one can infer that, while small amounts of enzymes are secreted by normal cells and some of these enzymes may be "recaptured", most of the newly synthesized enzymes produced in precursor form reach lysosomes without leaving the cells.

Moreover the biosynthesis of lysosomal enzymes seems to include a series of other modifications to ensure their active form of "mature enzyme" (Fig. 3).

Some lysosomal enzymes (β-hexosaminidase, α-glucosidase, cathepsin D, β-galactosidase and possibly other enzymes) are first produced as phosphorylated precursors with a substantially higher molecular weight than the final active products and may require a limited proteolytic shortening after transport into the lysosomes from an initial non-lysosomal site [19-21].

Details of recent fascinating progress made in the study of the mechanism for segregating acid hydrolases and effecting their transfer to lysosomes have been overviewed by Neufeld [22] and by Sly and Greek [26].

As shown in figure 3 the view is emerging that the newly synthesized acid hydrolase enzymes in a phosphorylated precursor form have a special role in internal and external transfer to lysosomes in this respect. The mechanism of phosphorylation of newly synthesized acid hydrolases involves a phosphotransferase which transfers N-acetylglucosamine-1-phosphate from uridine 5' diphosphate (UPD) N-acetylglucosamine to a mannose residue and a phosphodiesterase which cleaves the phosphodiester bond releasing N-acetylglucosamine and exposing the mannose-6-phosphate recognition marker [18, 23-26].

Therefore the interaction between carbohydrate-recognizing receptors and specific target saccharide residues of the ligand seems to be the code molecular system for recognition (see also Chapter 8, section 10).

1.4 DIFFERENTIAL DIAGNOSIS

A. Clinical assessment

A summary of the main clinical features and biochemical findings of the 4 types of glycoproteinoses is given in table 1.

In table 2 the main clinical features and biochemical findings of the other genetic lysosomal diseases in which an alteration of glycoprotein catabolism is associated are summarized.

An early and precise diagnosis of the glycoproteinoses is a prerequisite for prevention, due to their incurable nature. Early onset of symptoms occurs in severe forms of sialidosis, in I-cell disease, in G_{M1} gangliosidosis, in Sandhoff disease and in Salla disease. Most of these diseases become evident within the first 9 months after birth.

Patients with "moderate" or "mild" forms of sialidosis, fucosidosis, mannosidosis, aspartyglucosaminuria, G_{M1} gangliosidosis and pseudo-Hurler polydystrophy exhibit a more tardive clinical onset and a more prolonged course.

Neurological signs The majority of patients shows lack of behavioral development or profound mental retardation and specific clinical manifestations.

Mental retardation, when present, is slowly progressive and generally moderate in sialidosis type 1, aspartylglucosaminuria, "mild" forms of G_{M1} gangliosidosis and pseudo-Hurler polydystrophy and is usually profound and severe but with appreciable variations in fucosidosis, mannosidosis, I-cell disease, infantile G_{M1} gangliosidosis, Sandhoff disease and Salla disease.

Neuronal signs may be delayed for some time postnatally or even unapparent as in some cases of sialidosis type 1, in aspartylglucosaminuria, in some cases of juvenile or adult gangliosidosis G_{M1} and in pseudo-Hurler-polydystrophy.

The decision as to whether a child is suffering from a static encephalopathy or from a progressive disease depends on the accuracy of the parents' and of the medical experts' observations.

A correct differentiation between mental retardation and neuronal dysfunction which are the consequence of perinatal insults or of a genetic metabolic disorder is important for good medical and psychological care of the patients and their families.

In the first case, parents can be reassured about the risk in a future pregnancy whereas in the latter, the risk is 25% for recurrence since all the disorders reviewed here are inherited as autosomal recessive traits.

Seizures are rarely present in glycoproteinoses but occur in the most severe types of fucosidosis, in infantile and juvenile G_{M1} gangliosidosis and in Sandhoff diseases.

Intention myoclonus and ataxia are classically present in sialidosis but, in the most severe types, seizures and lethargy also occur.

Severe dementia, dysarthria, clumsiness, dysdiadochokinesis and ataxia are present in Salla disease.

Hypotonia or floppiness and signs of upper motor neuron involvement (Babinski sign, clonus etc.) are seen in the first stages of fucosidosis, infantile G_{M1} gangliosidosis and Sandhoff disease and are followed by spasticity and increased deep tendon reflexes. In the terminal stages of diffuse cerebral degeneration, patients tend to lie in a state of decorticate rigidity, unresponsive to their environment [1].

Ocular signs Ocular signs such as red macular spot in sialidosis, Sandhoff disease and infantile G_{M1} gangliosidosis, lens opacities in some cases of mannosidosis and aspartylglucosaminuria, dilatation and tortuosity of arterial retinal vessels in some cases of fucosidosis and some cases of sialidosis, occasional corneal opacities and glaucoma in I-cell disease and occasional corneal opacities in G_{M1} gangliosidosis often provide very useful hints for diagnosis.

Extraneuronal signs They may often help in making the diagnosis. Patients may have coarse features or gargoyle-like facies, dwarfism, kyphosis or scoliosis, joint limitation or joint hyperflexibility and mild to severe dystosis multiplex.

These signs appear early and are severe in I-cell disease, in the infantile sialidosis type 2 and in infantile G_{M1} gangliosidosis.

Patients are generally normo or macrocephalic; microcephaly has been noticed only in some cases of I-cell disease.

Teleangiectasias and angiokeratoma are detectable in early childhood in fucosidosis type 2. Angiokeratoma corporis diffusum therefore suggests the diagnosis of fucosidosis type 2 but has also been found in some cases of combined deficiency of sialidase and β-D-galactosidase and in very occasional cases of aspartylglucosaminuria.

Hyperhydrosis may be present in fucosidosis type 1 and hypohydrosis or anhydrosis in fucosidosis type 2. Thick skin occurs in fucosidosis and it is characteristic and remarkable in I-cell disease. There is also classically present in this disease a hypertrophy of the gingiva and sometimes hirsutism.

Reduced elasticity of the skin is usually detectable in aspartylglucosaminuria.

Oedema, occasionally with ascites, occurs in the most severe types of sialidosis and in infantile G_{M1} gangliosidosis.

Congenital sialidosis should be considered in cases resembling hydrops fetalis.

Hepatosplenomegaly is found in infantile G_{M1} gangliosidosis and may be present, but is generally moderate, in other diseases. Cardiomegaly has been reported in fucosidosis and heart disease has been described in I-cell disease.

In addition, patients affected with the various glycoprotein storage diseases are susceptible to infections throughout life.

Skeletal abnormalities similar to those that occur in the mucopolysaccharidoses are discernible by X-ray in the majority of patients with Sialidosis type 2, fucosidosis, mannosidosis, in I-cell disease and pseudo Hurler polydystrophy, in infantile G_{M1} gangliosidosis. Scoliosis with mild radiological bone abnormalities is present in aspartyglucosaminuria.

Table 1.1 Differential diagnosis of glycoproteinoses.

Disease	Main clinical features	Enzyme defect	Storage products
Sialidosis type 1 (normomorphic)	Visual loss, cherry-red spot, or cherry-red spot myoclonus syndrome; slowly progressive, prolonged course	Sialidase	Sialyloligosaccharides
Sialidosis type 2 (dysmorphic)	Gargoyle facies, growth retardation, dysostosis multiplex, cherry-red spot, myoclonus, neuropathy, moderate dementia, occasionally angiokeratoma. Death before or in adulthood. Severe infantile variant with hydrops fetalis, oedema, ascites, nephrotic syndrome. Death in first years of life.	Sialidase Combined β-galactosidase sialidase	Sialyloligosaccharides
Fucosidosis type 1 (infantile)	Recurrent respiratory infections, muscular hypotonia followed by spasticity and cerebral rigidity, progressive dementia, mild dysostosis multiplex, thick skin, excessive sweating. Death in first years of life.	α-L-fucosidase	Fucoglycoconjugates
Fucosidosis type 2 (adult)	Coarse features, dysostosis multiplex, ataxia, progressive mental-motor retardation, spasticity, teleangiectasies, angiokeratoma corporis diffusum, hypo or anhydrosis, recurrent respiratory infections, retarded growth. Death in adulthood.	α-L-fucosidase	Fucoglycoconjugates
Mannosidosis	Coarse features, dysostosis multiplex, hepatosplenomegaly, recurrent infections, motor and mental retardation, hearing loss, ocular changes. Death in first years of life or after puberty or in adulthood in milder variant.	α-D-mannosidase	Mannosyloligosaccharides
Aspartylglucosaminuria	Slowly coarsening features, mild bone changes, clumsiness, muscular hypotonia, hyperflexible joints, slowly progressive mental retardation, crystal like opacities of the lens. Slowly progressive evolution.	N-aspartyl-β-glucosaminidase	Glycoasparagines

Table 1.2 Differential diagnosis of lysosomal diseases with which an alteration of glycoprotein catabolism is associated.

Disease	Main clinical features	Enzyme Defect	Storage products
G_{M1} gangliosidosis (infantile)	Muscular hypotonia followed by spasticity and decerebrate rigidity, severe dementia, seizures, sometimes gargoyle facial features, hepatosplenomegaly, dysostosis multiplex, cherry-red spot (50% cases); seldom cloudy cornea, death before 2 years.	β-galactosidase	Ganglioside G_{M1}, galacto-oligosaccharides
G_{M1} gangliosidosis (juvenile)	Onset 0,5-2 years. Progressive severe dementia, seizures, loss of motor coordination then spasticity, mild dysostosis, death 3-10 years.	β-galactosidase	Ganglioside G_{M1}, galacto-oligosaccharides
G_{M1} gangliosidosis (adult)	Dysarthria, ataxia, dyskinesia dystonia, spasticity intellectual deterioration.	β-galactosidase	Ganglioside G_{M1}, galacto-oligosaccharides
G_{M2} gangliosidosis (Sandhoff disease)	Enlarged head, spasticity and decerebrate rigidity, profound dementia, seizures, cherry-red spots, blindness, hyperacusia, death before 3-5 years. Sometimes late onset and slower course.	β-N-acetylhexosaminidase A and B	Ganglioside G_{M2}, G_{A2}, globoside, hexosamine-oligosaccharides
I-cell disease (Mucolipidosis type II)	Gargoyle facies, severe dysostosis multiplex, dwarfism, severe dementia, tight skin, gingiva hyperplasia, occasional corneal opacities. Rapidly progressive; death in few years.	N-acetylglucosaminylphosphotransferase Intracellular multiple lysosomal deficiencies, extracellular increased activities of the same enzymes	Glycopeptides, glycolipides, mucopolysaccharides in mesenchymal tissues
Pseudo-Hurler polydystrophy (Mucolipidosis type III)	Mild dysostosis multiplex, short stature, severely stiff joints, no or mild mental retardation. Survival to adulthood.	N-acetylglucosaminylphosphotransferase Intracellular multiple enzyme deficiencies; extra cellular increased activities of the same enzymes	Glycopeptides, glycolipides, mucopolysaccharides in mesenchymal tissues
Salla disease	Onset 6-18 months. Severe dementia, coarse facial features, clumsiness, ataxia, prolonged course.	Unknown	Free sialic acid

B. Clinical and biochemical heterogeneity

In glycoproteinoses, excepting aspartyglucosaminuria there is a considerable degree of clinical and biochemical heterogeneity from patient to patient.

The "same" metabolic defect may be associated, even among siblings, with severe infantile, juvenile and even mild adult forms of the disease. For instance a patient with infantile sialidosis may manifest severe clinical symptoms shortly after birth and die within 5 years whereas patients with the adult form of the same disease have only a red macular spot with visual loss.

However, members of the same family affected with a glycoproteinosis frequently present the same manifestations at nearly the same age.

The differences in clinical symptoms have often been correlated with different levels of residual enzyme activity but sometimes no such relationship can be demonstrated.

Somatic cell hybridization techniques have been used successfully to study the genetic background of clinical or biochemical heterogeneity of glycoproteinoses as in other genetic metabolic diseases [2-4].

The heterogeneity of the clinical features among patients with the "same" disease is in part determined by different mutations affecting the specific enzyme activity in different ways. Many lysosomal enzyme defects appear to follow the "one gene-one enzyme-many substrates" model of O'Brien [5] in which variations of the phenotype may depend on the relative rate at which the mutant enzyme cleaves each of a number of natural substrates for the enzyme (oligosaccharides, glycopeptides, glycolipids or glycosaminoglycans).

Each of these compounds, when stored, may give rise to different clinical effects, i.e. gangliosides-neuronal degenerations, oligosaccharides-hepatosplenomegaly, glycosaminoglycans-bone abnormalities and corneal opacities, etc. Since numerous different mutations are likely in each disorder, and since patients may be genetic compounds (heterozygous for 2 different allelic mutations), phenotypic variation is the rule rather than the exception.

C. Racial and genetic considerations

The glycoproteinoses are rare and are generally panethnic diseases but some occur in a significant number in defined populations. For instance the incidence of aspartylglucosaminuria is low except in Finland where the incidence has been estimated to be more than 1 in 26,000 newborns [6].

All patients with Salla disease are of a geographically restricted area of northern Finland [7].

The majority of patients with fucosidosis have southern Italian ancestors [8].

A significant proportion of patients with type 1 sialidosis are Italian, whereas the majority of patients with type 2 sialidosis are Japanese [9].

An increased frequency of infantile G_{M1} gangliosidosis has been noted in Malta [10].

These ethnic predilections suggest increased heterozygote frequencies for some of these mutations among Finnish and Italians. However, some of these diseases are "new", having been discovered only recently, and no extensive epidemiological studies have yet been carried out.

All of the diseases considered in this book are inherited as an autosomal recessive trait. The pattern of inheritance is usually suggested by classical criteria deriving from pedigree studies: the patient is the offspring of parents who are normal, the disease tends to affect the siblings, both male and female, in a family where there is a high proportion of consanguineous marriages.

The definitive pattern of inheritance is based on the demonstration of intermediate levels of enzyme activity in obligate heterozygotes.

D. Preliminary laboratory tests

Suspicion of a glycoproteinosis on the basis of clinical symptoms will usually be followed by relatively simple, inexpensive tests which require little time like peripheral blood and bone marrow examinations and the study of oligosaccharide excretion in the urine. The presence of cytoplasmatic vacuoles in circulating white cells and of "foam cells" in bone marrow, the absence of excessive urinary excretion of glycosaminoglycans and the demonstration of excessive urinary excretion of oligosaccharides with a specific pattern may yield valuable information for patients under consideration for a glycoprotein storage disease.

The normal urinary content of oligosaccharides is very low and ranges from 0.1 to 4 mg/l for each constituent.

Normal human urine contains small amounts of 20-30 carbohydrates and a large number (nearly one hundred) of different oligosaccharides and glycopeptides, some of which are related to the ABH substances, secretor status and diet, or have a glycogen origin [11-16].

In patients with oligosacchariduria, for instance fucosidosis, the daily excretion of fucooligosaccharides is markedly increased and may be 20-80 times the normal values; in sialidosis the sialyloligosaccharides may be 500-600 times the normal values [17].

As can be seen in table 3 and figure 4 the types of oligosaccharides excreted in the urine are useful parameters in the differential diagnosis of the various glycoproteinoses and the other lysosomal diseases in which there is associated an alteration of glycoprotein catabolism, but the variation among patients with the same type of glycoproteinosis may be considerable.

Using this procedure, the number of diagnostic possibilities and therefore the number of enzyme analyses which the laboratory has to carry out, may be considerably reduced.

Table 1.3 Urinary carbohydrates in glycoproteinoses and related diseases.

Sialidosis	sialyloligosaccharides
Fucosidosis	fucooligosaccharides fucoglycopeptides
Mannosidosis	oligosaccharides rich in mannose and N-acetylglucosamine
Aspartylglucosaminuria	aspartylglucosamine and other glycoasparagines
G_{M1} gangliosidosis	galactooligosaccharides
G_{M2} gangliosidosis (Sandhoff's variant)	oligosaccharides containing N-acetylglucosamine
I-cell disease and pseudo-Hurler polydystrophy	mild sialyloligosacchariduria
Salla disease	free sialic acid

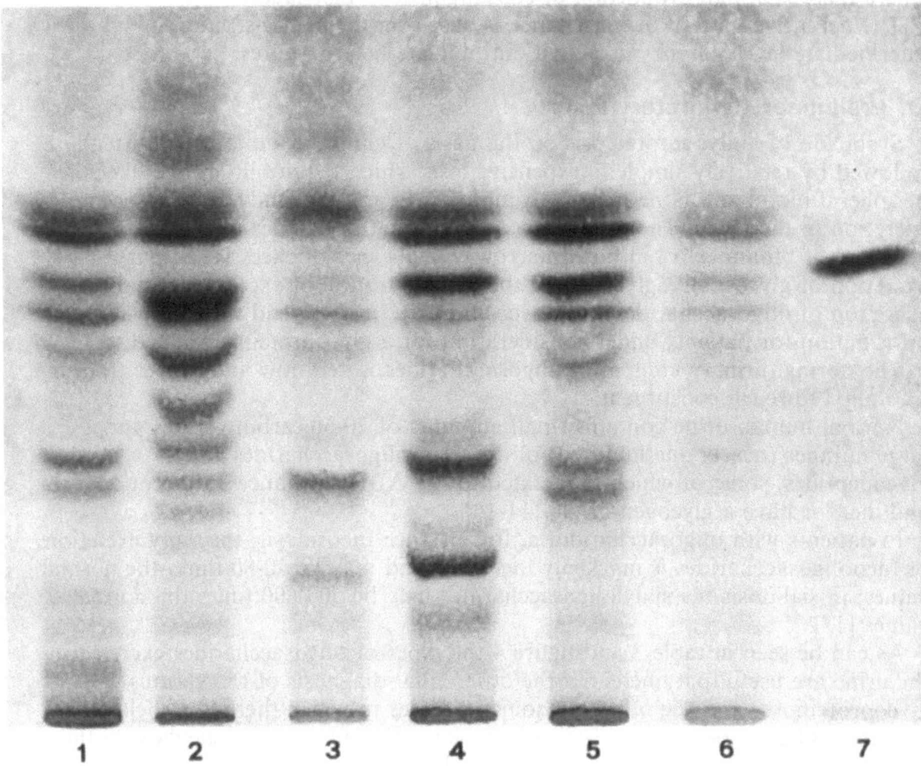

Fig. 1.4 Thin-layer chromatography of urinary oligosaccharides of patients with different forms of glycoproteinosis and of Sandhoff disease with which an alteration of glycoprotein catabolism is associated.

1, fucosidosis; 2, mannosidosis; 3, aspartylglucosaminuria; 4, Sandhoff disease; 5 sialidosis; 6, normal control; 7, lactose.

A number of screening tests such as high voltage electrophoresis, paper and thin layer chromatography, are available for the demonstration of the various types of oligosaccharides excreted in patients with different types of glycoproteinosis.

One dimensional thin-layer chromatography on silica gel was first described by Palo and Savolainen [18] for the screening of oligosaccharides in aspartylglucosaminuria.

This method was modified by Humbel and Collart to include the screening of fucosidosis, mannosidosis and G_{M1} gangliosidosis [19].

For a better separation of the various oligosaccharides a modification of the most frequently used method of Humbel and Collart was proposed by Humbel and Dhondt [20], Sewell [21] and Sewell et al. [22], and O'Brien [23].

Further analyses such as column chromatography, gas-liquid chromatography, mass spectrometry and nuclear magnetic resonance spectroscopy, are necessary for separation, characterization and structural studies of abnormal urinary glycoconjugates.

These analyses are limited to the most specialized laboratories.

E. Strategy for enzyme tests

In practice, diagnostic laboratory tests are usually performed at the level of the primary gene product, the enzyme.

Currently enzyme assay is the most direct and precise test and is an essential part of diagnostic investigation.

The molecular defect is expressed in nearly all organs and cell types and the diagnosis can be established by assaying leukocytes, purified mononuclear cells, cultured skin fibroblasts, tears, saliva, serum, plasma and hair follicles.

Leukocytes separated from peripheral blood provide one of the best sources of intra-cellularly located enzymes and have been frequently used.

Cultured skin fibroblasts are frequently used for enzyme assays. They have the advantage that the skin biopsies from which they are grown are easily transported to other laboratories.

Therefore fibroblasts are useful in a number of cases in which the clinician suspects a diagnosis of glycoproteinosis and sends a skin biopsy to a specialized centre with a request for biochemical analysis. Storage of the fibroblasts in a cell bank can be carried out and investigations are possible over many years at no inconvenience to the patient. Metabolic studies (turnover studies, enzyme uptake, cross correction) can also be performed.

In some instances reinvestigation of fibroblasts of patients with unexplained clinical symptoms has led to the elucidation of the responsible metabolic defect. Moreover the expression of the metabolic defect should be characterized in fibroblasts from a previous affected child or relative from the family under study and the level of residual enzymatic activity established, since genetic heterogeneity exists with apparently identical metabolic diseases.

Fibroblasts are useful for somatic cell hybridization studies in the case of clinical and biochemical heterogeneity. These studies may be important for a correct understanding and classification of variants of the same disease.

Fibroblast cultures from affected children may also be stored in a cell bank to provide control material for a future prenatal diagnosis.

As a general rule a lysosomal enzyme deficiency that can be demonstrated in cultured fibroblasts from skin biopsies of children or adults is likely to be demonstrable also in cultured cells from amniotic fluid.

The usefulness of tears for enzymatic diagnosis of glycoproteinoses and other genetic lysosomal diseases has been recognized by many authors [24-27].

The collection of tears is easy and harmless and some enzymes such as α-fucosidase and α-mannosidase are relatively stable when tears are allowed to dry on the paper strips at room temperature (10-20 minutes).

If the filter papers are mailed in an envelope to temperate countries, over 50% of the activity is still present after 1 week and if the paper strips are kept below -18°C, no detectable loss of activity occurs within 1 year.

Enzyme assays of tears have been used lately for diagnosis and/or confirmation of fucosidosis [26] and mannosidosis [27, 28].

Saliva may be used for enzyme assays [25, 29] and has been used for confirmation of fucosidosis [30].

Serum or plasma may be useful for diagnosing only some glycoproteinoses and related diseases (aspartylglucosaminuria, Sandhoff disease and I-cell disease).

Serum and plasma have proven to be inadequate for diagnosis of fucosidosis, since there exists a group of phenotypically normal individuals who have low serum and plasma levels of α-L-fucosidase but normal activity in leucocytes [31, 32]. Hair follicles obtained by plucking hair have been used as a convenient source of tissue for enzyme assays, particularly in X-linked diseases.

The analytical methods for the demonstration of enzyme deficiences vary greatly in complexity.

The enzymatic defect of homozygotes for glycoproteinoses and related diseases can be easily demonstrated using artificial substrates, such as those of a chromogenic (e.g. the p-nitrophenylglycosides) or a fluorogenic (e.g. the 4-methylumbelliferylglycosides) nature which are commercially available and sufficiently sensitive.

A number of biological and methodological problems may be encountered due to the occurrence of several variants of the disorders.

Any abnormal pattern may be further investigated using more complex methodology or specially synthesized or isolated compounds, or using only natural substrates which are available in few laboratories.

The natural substrates are highly specific but sometimes elaborate procedures are required for their preparation and they are often arduous to obtain and difficult to use routinely. It is desirable that in the future other types of more sensitive artificial substrates be available.

Detailed data of enzymes of the four major types of glycoproteinoses and of the other diseases with which an alteration of glycoprotein degradation is associated are given for each of the diseases in this book.

F. Histology and electron microscopy

For a long period the only basis for the diagnosis and classification of congenital lysosomal disorders was careful clinical description in conjunction with microscopic and ultrastructure morphology of tissue biopsies (brain, liver, peripheral nerves and rectum) or of autopsy material.

Extensive histologic and electron microscopic studies of tissues from the various types of glycoproteinoses showed some characteristic abnormalities. Although it is difficult to determine the exact nature of stored material by these techniques, they are useful for diagnosis. However, brain biopsies and biopsies of liver, peripheral nerve and rectum are not done as commonly as in previous years.

Fine structural examination of the stored material in more easily obtained tissues such as conjunctiva and skin is often pathognomonic of some of these diseases.

As can be seen in table 4, adapted from Libert with his permission, the types of inclusions and their topographic distribution among the different conjunctival cells are often specific enough to make a differential diagnosis of the various glycoproteinoses and of the other diseases reported in this book.

Further data on significant lesions with specific structures and topography of conjunctival and cutaneous biopsies are reported for each of the diseases in this book.

Conjunctival biopsy gives results similar to skin biopsy. Libert, [35] who studied both tissues in parallel, cannot affirm that one is better than the other, even if there are some differences.

Table 1.4 Ultrastructural conjunctival alterations in glycoproteinoses and in some other lysosomal diseases with which an alteration of glycoprotein catabolism is associated (adapted from Libert J., with permission [Introd. differ. diagnosis 35]).

		Epithelium	Inclusion Type	Fibroblasts	Inclusion Type	Blood Capillary Endothelial Cells	Inclusion Type	Lymphatic Capillary Endothelial Cells	Inclusion Type	Sensory Nerves	Inclusion Type
Sialidosis	type 1	—		—		—		—	—	++	(MCB)
	type 2	+	(FG)	+++	(FG)	+	(FG)	++	(FG)	++	(MCB)
Fucosidosis		+++	(C+D)	+++	(C+D)	+++	(C+D)	++	(C+D)	±	(MCB)
Mannosidosis		±	(FG)	+++	(FG)	+	(FG)	+	(FG)	+	(FG)
Aspartylglucosaminuria		++	(FG)	++	(FG)	++	(FG)	++	(FG)	++	(FG)
I-Cell disease (Mucolipidosis type II)		—		+++	(FG)	+	(JR)	—		+	(FG)
Pseudo-Hurler (Mucolipidosis type III)		—		++	(FG)	+	(JR)	—		+	(FG)
G_{M1} Gangliosidosis		+++	(FG)	+++	(FG)	+++	(FG)	++	(FG)	++	(MCB)
Sandhoff disease		++	(FG)	+	(MCB+FG)	+	(MCB)	—		++	(MCB)

— = no lesions; ± = moderate; + = definite; ++ = considerable amounts; +++ = large amounts; (C+D) = Clear and dark inclusions; (MCB) = Membranous cytoplasmatic bodies; (FG) = Clear inclusions containing a fibrillo-granular reticulum; (JR) = Juxtaposed lamellar rings.

Therefore these techniques may be of great assistance in singling out those cases in which a glycoprotein storage disorder is suspected [33-35].

1.5 CARRIER DETECTION

Detection of heterozygotes for different glycoproteinoses and other related diseases as well, plays a relevant role in genetic counselling.

The most frequent use of enzyme testing for detecting heterozygotes is amongst parents and other relatives. Statistically these individuals have a much greater chance of carrying the mutation and the patient's relatives, when married, of bearing affected children themselves.

When a patient dies without enzymatic confirmation of a clinical diagnosis, the autosomal recessive trait of the parents may provide the only biochemical indication of the child's disease.

Identification of carriers by large-scale screening can be considered in those communities or populations where the frequency of the abnormal gene is suspected to be particularly high.

Since the gene frequency in the general population for each glycoproteinosis is, with some exceptions, low and the cost of performing the required test is considerable, carrier detection for glycoproteinoses remains highly individualized and it will be generally offered only to those with affected relatives.

Carriers have levels of enzyme activity approximately half those of normal individuals but the enzyme activity levels vary considerably and may overlap with the range found in normal individuals or even with patients. Therefore, in a proportion of subjects it may be difficult to determine the genotype biochemically. Variations in activity levels depend in part on the choice of the enzyme source, in part on the sensitivity of the assay methods, and in part on the nature of the mutation.

Leukocytes, separated from peripheral blood, have been frequently used for heterozygote detection.

Although mixed leukocytes have proved satisfactory in the detection of the carrier state, purified preparations of mononuclear cells from peripheral blood often permit better discrimination in some situations as, for instance, the detection of carriers for aspartylglucosaminuria and fucosidosis [1, 2]. The use of phytohemoagglutinin-stimulated lymphocytes, calculating the ratio between the test enzyme and other lysosomal enzymes has been reported by a number of authors [3, 4].

Nevertheless this method is time-consuming and should be taken into consideration in those cases in which a simple leukocyte preparation does not give clear discrimination between carrier and non-carrier states.

Cultured fibroblasts have been criticized and used only in limited studies concerning heterozygote detection because it is an expensive and time consuming technique.

Serum or plasma have been used in the mass screening programs for bovine mannosidosis. Variations in plasma α-mannosidase activity were noted at different times of the year and at different ages whereas age is not a variable factor with neutrophyl or lymphocyte assays [5, 6].

Serum and plasma cannot be used for the detection of heterozygotes of glycoproteinoses because the values in the parents are often not distinguishable from those of normals.

Table 1.5 Preferred substrates and enzyme sources for heterozygote detection in glycoproteinoses and related diseases.

Disease	Substrate for enzyme assay	Preferred enzyme source
Aspartylglucosaminuria	2-acetamido-1-β-L-aspartamido--1,2-dideoxy-β-D-glucose	Lymphocytes
Fucosidosis	4-MUB-α-L-fucopyranoside	Lymphocytes
Mannosidosis	4-MUB-α-D-mannopyranoside	Leukocytes, tears
Sialidosis	4-MUB-α-D-N-acetylneuraminate	Leukocytes, fibroblasts
G_{M1} gangliosidosis	4-MUB-β-D-galactopyranoside	Leukocytes
Sandhoff disease	4-MUB-2-acetamido-2-deoxy--β-D-glucopyranoside	Serum, tears, leukocytes
I-cell disease	4-MUB-2-acetamido-2-deoxy--β-D-glucopyranoside	Serum
Pseudo-Hurler polydystrophy	4-MUB-2-acetamido-2-deoxy--β-D-glucopyranoside	Serum

In addition it is known that 8% of normal subjects have a low serum α-fucosidase and normal activity in their leukocytes [7, 8].

Urine is unsuitable for heterozygote diagnosis because results are too variable. The use of tears for detection of heterozygotes for fucosidosis and mannosidosis has been reported by a number of authors [9-12]. However definitive detection of heterozygotes also with this method may be difficult.

Lysosomal enzymes may be assayed using synthetic substrates. Use of natural substrates may be reserved for the study of unusual subjects in establishing a correct mutant genotype that is not satisfactorily resolved with synthetic substrates. In order to correct overlapping between normal and heterozygous, enzymatic activity has also been related to another enzyme whose activity is highly correlated with the test enzyme [13, 14].

However the use of a reference enzyme in heterozygote detection often affords no particular advantage, probably due to independent normal variation of the two enzymes under study.

Considerable information on enzyme assays that positively identify carriers of the trait for fucosidosis, mannosidosis, aspartylglucosaminuria, G_{M1} gangliosidosis and Sandhoff disease has been accumulated.

For the other disorders such as sialidosis, I-cell disease and pseudo-Hurler polydystrophy the number of heterozygotes studied is still small and experience is dispersed among different laboratories.

1.6 PRENATAL DIAGNOSIS

Extensive surveys have shown that prenatal diagnosis, when conducted in experienced units, is a safe and accurate test widely acceptable whenever there is a risk of a severe disorder in the fetus and the parents consider selective abortion to be an acceptable procedure.

The lack of effective treatment for glycoproteinoses and the high risk of having other defective children emphasize the essential role of prenatal diagnosis for these diseases.

Prenatal detection is indicated for families with previous affected offspring and couples at "high risk" for one of the glycoproteinoses and related diseases detected by carrier screening programs in selected communities.

The requisites for prenatal diagnosis are that:

- The biochemical defect for which a family is at risk be known with certainty.
- The specific defect be demonstrable in amniotic fluid cells.
- The affected fetus be clearly and accurately distinguishable from carrier and normal fetuses.

Each of the disorders in this volume except Salla disease, can be detected prenatally by analysis of enzyme activity in cultured amniotic fluid cells.

Prenatal diagnosis of the glycoproteinoses has so far been performed in a limited number of cases because these disorders have been discovered only recently and are rare.

Moreover the clinical onset of certain diseases, such as juvenile and adult sialidosis and the aspartylglucosaminuria, may be delayed so that additional children are born before a diagnosis is made in the first affected sib.

A number of recommendations may be envisaged for the general approach to the prenatal diagnosis.

- Ascertain that the diagnosis for the disorder at risk is correct by enzyme assay of affected patients and parents.
- Perform amniocentesis, after ultrasound examination, using the transabdominal approach at the 15-17th week of gestation.
- Remove 15-20 ml of amniotic fluid and after centrifugation, carry-out culture of amniotic cells, saving the fluid itself for assay.
- Perform biochemical analyses after adequate amniotic cell growth.
- Confirm the diagnosis, if positive, using fetal tissue, or if negative, using blood sample obtained from the infant after delivery.

Additional recommendations in some instances are:

- Compare the deficient enzyme to a lysosomal reference enzyme and, when possible, to the enzyme analyses of skin fibroblasts from the index patient and from the heterozygous parents.
- Perform analysis of amniotic fluid cells from a control sample of the same gestational period.

Cultured cells are the most reliable source for prenatal diagnosis of the glycoproteinoses and related diseases [1-6]. Increased activities of several lysosomal hydrolases in the amniotic fluid supernatant and a multiple lysosomal enzyme deficiency in cultured amniotic fluid cells have been successfully utilized for prenatal diagnosis of "I-cell disease" [7-14]. The most common disorders of those considered in this book for which prenatal diagnosis is requested are the gangliosidoses [1-6, 13, 14] Prenatal diagnoses of several cases of I-cell disease [7-14] and a limited number of sialidosis [15] fucosidosis [13, 14, 16, 17] mannosidosis [13, 14, 18] and aspartylglucosaminuria [19] have been performed. Only rarely have errors occurred [16] but a number of problems have been encountered in establishing a correct prenatal diagnosis especially for mannosidosis and fucosidosis [13, 14, 17, 18, 20]. As it is clear that some of these diseases are infrequent and the total number of pregnancies at risk that have been monitored for each of these diseases is rather small, it is necessary to establish a practice of cooperation between the different centres.

Of course, optimal use of the scarce opportunities to gain experience would require a certain degree of centralization so as to guarantee optimal accuracy and safety of the procedures.

In the following chapters the experience with prenatal diagnosis of each of the diseases considered in this book and extensive evaluation of assay procedures for the single subjects will be given in detail.

1.7 THE CHROMOSOMAL LOCALIZATION OF THE MUTANT GENES

Rapid progress in mapping the human genome has been made in the last 10 years.

More than 350 genes have been assigned to specific chromosomes by studies of clones derived from interspecific somatic cell hybrids (e.g. man-mouse, man-chinese hamster) and by family linkage studies, linkage disequilibrium, nucleic acid hybridization and restriction enzyme mapping [1, 2].

The mutant genes of over 40 autosomally inherited diseases have been specifically localized. Among these are the enzymes involved in the glycoproteinoses such as fucosidosis for which the mutation is mapped to the short arm of chromosome 1 [3] and mannosidosis for which the mutation is on the short arm of chromosome 19 [4].

Gene localization studies revealed the presence of two structural β-galactosidase loci on the human chromosome 3 and 22 [5]. The locus for acid β-galactosidase mutated in infantile G_{M1}-gangliosidosis appears to be on the short arm of chromosome 3 [6, 7].

Sandhoff disease is caused by a mutation in the structural gene coding for the β-subunit of β-hexosaminidase A and B on the long arm of chromosome 5 [8] (see also individual diseases).

The application of recent methods for dissecting the human genome by restriction endonuclease which produces specific DNA fragments and identification of specific sequences by hybridization with nucleic acid probes has opened new horizons in human genetics.

With this fine structure analysis, the mutation in some genetic diseases can be demonstrated at the level of the DNA itself.

It is technically possible to do this if the gene defect is known, because a probe can be identified.

Of course obtaining DNA probes will be a slow and difficult process but it is likely that in the future more gene mutations will be identified at the level of the genome.

Theoretically it will also be possible to use this approach to diagnose mutations for which the basic biochemical defect is not yet known.

The inheritance of polymorphisms of DNA sequence linked to the disease in affected families can be studied. If such a linkage can be determined, it should be possible to move along the genome by cloning techniques and to move towards the coding sequence responsible for the heritable defect.

Since recombinant DNA technology offers the possibility of studying any individual gene from a person with a hereditary defect, the nature of the lesion can be determined [9].

The genetic abnormality for some of these diseases could well be defined in the future even before the primary gene product is identified. The usefulness of identification of heterozygous carriers of recessive disorders and of prenatal diagnosis is obvious. Other uses will be in human population genetics (population, structure and selection, connection between geographically isolated communities). In addition, mapping information and knowledge of the regulatory mechanism will probably be critically important in the effort to replace mutant genes with normal ones.

Since most of the glycoproteinoses are fatal, gene replacement may be the ultimate solution to therapy. Such work will require the use of animal models, several of which exist (mannosidosis-cattle and cat, Sandhoff disease-cat, G_{M1} gangliosidosis-dog, cat, cattle) prior to experimentation in humans.

SELECTED REFERENCES

Historical review

1. Durand P., Borrone C., Della Cella G.: *A new mucopolysaccharide lipid storage disease.* Lancet *2*, 1313, 1966.
2. Durand P., Philippart M., Borrone C., Della Cella G., Bugiani O.: *Una nuova malattia da accumulo di ceramidi tetraesosidi.* Minerva Pediat. *19*, 2187-2196, 1967.
3. Durand P., Borrone C., Della Cella G.: *Fucosidosis.* J. Pediat. *75*, 665-674, 1969.
4. Philippart M.: *Fucosidosis: a novel neurovisceral sphingolipidosis.* Neurology *19*, 304, 1969.
5. Van Hoof F., Hers H.G.: *Mucopolysaccharidosis by absence of α-fucosidase.* Lancet *1*, 1198, 1968.
6. Durand P., Borrone C., Della Cella G., Philippart M.: *Fucosidosis.* Lancet *1*, 1198, 1968
7. Öckerman P.A.: *A generalized storage disorder resembling Hurler's syndrome.* Lancet, *2*, 239-241, 1967.
8. Öckerman P.A.: *Deficiency of beta-galactosidase and alpha-mannosidase: primary enzyme defects in gargoylism and a new generalized disease.* Acta Paediat. Scand. *177*, 35-36, 1967.
9. Öckerman P.A.: *Mannosidosis: isolation of oligosaccharide storage material from brain.* J. Pediat. *75*, 360-365, 1969.
10. Pollit R.J., Jenner F.A., Merskey H.: *Aspartylglucosaminuria, an inborn error of metabolism associated with mental defect.* Lancet *2*, 253-255, 1968.
11. Pollit R.J., Jenner F.A.: *Enzymatic cleavage of 2-acetamido-1 (β-L-aspartamido)-1,2-dideoxy-β-D-glucose by human plasma and seminal fluid. Failure to detect the heterozygous state of aspartylglucosaminuria.* Clin. Chim. Acta *25*, 413-416, 1969.
12. Autio S.: *Aspartylglucosaminuria. Analysis of thirty-four patients.* J. Med. Defic. Res. Monograph Series *1*, 1-93, 1972.
13. Palo J., Riekkinen P., Arstila A., Autio S.: *Biochemical and fine structural studies on brain and liver biopsies in aspartylglucosaminuria.* Neurology (Minneap.) *21*, 1198-1204, 1971.
14. Palo J., Savolainen H.: *Biochemical diagnosis of aspartylglucosaminuria.* Ann. Clin. Res. *5*, 156-162, 1973.
15. Spranger J., Wiedemann H.R., Tolksdorf M., Graucob E., Caesar R.: *Lipomuco-polysaccharidose. Eine neue Speicherkrankeit.* Z. Kinderheilkd. *103*, 285-306, 1968.
16. Spranger J., Wiedemann H.R.: *The genetic mucolipidoses; diagnosis and differential diagnosis.* Humangenetik *9*, 113-139, 1970.
17. Strecker G., Hondi-Assah T., Fournet B., Spik G., Montreuil J., Maroteaux P., Durand P., Farriaux J.P.: *Structure of the three major oligosaccharides excreted in the urine of five patients with three distinct inborn diseases: I-cell disease and two new types of mucolipidosis.* Biochim. Biophys. Acta *444*, 349-358, 1976.

18. Strecker G., Peers M.C., Michalski J.C., Hondi-Assah T., Fournet B., Spik G., Montreuil J.P., Maroteaux P., Durand P.: *Structure of nine sialyloligosaccharides accumulated in the urine of eleven patients with three different types of sialidosis: mucolipidosis II and two new types of mucopolidosis.* Eur. J. Biochem. *75*, 391-403, 1977.
19. Cantz M., Gehler J., Spranger J.: *Mucopolidosis I: increased sialic acid content and deficiency of α-N-acetylneuraminidase in cultured fibroblasts.* Biochem. Biophys. Res. Commun. *74*, 732-738, 1977.
20. Durand P., Gatti R., Cavalieri S., Borrone C., Tondeur M., Michalski J.C., Strecker G.: *Sialidosis (Mucolipidosis I).* Helv. Paediat. Acta *32,* 391-400, 1977.
21. O'Brien J.S.: *Neuraminidase deficiency in the cherry-red spot-myoclonus syndrome.* Biochem. Biophys. Res. Commun. *79*, 1136-1141, 1977.
22. Lowden J.A., O'Brien J.S.: *Sialidosis: a review of human neuraminidase deficiency.* Am. J. Hum. Genet. *31*, 1-18, 1979.
23. Kelly T.E., Graetz G.: *Isolated acid neuraminidase deficiency: a distinct lysosomal storage disease.* Am. J. Med. Genet. *1*, 31-46, 1977.
24. Maroteux P., Humbel R., Strecker G., Michalski J.C., Mande R.: *Un nouveau type de sialidose avec atteinte rénale: la néphrosialidose.* Arch. Franç. Pédiat. *35*, 819-829, 1978.
25. Alysworth A.S., Thomas G.H., Hood J.C., Malouf N., Libert J.: *A severe infantile sialidosis: clinical, biochemical and microscopic features.* J. Pediat. *96*, 662-668, 1980.
26. Johnson W.G., Thomas G.H., Miranda A.F., Driscoll J.M.: *Congenital sialidosis, a new form of* α*-L-neuraminidase deficiency. Its possible relation to hydrops fetalis.* Neurology *30*, 377, 1980.
27. Wenger D.A., Tarby T.J., Wharton C.: *Macular cherry-red spots and myoclonus with dementia: coexistent neuraminidase and* β*-galactosidase deficiencies.* Biochem. Biophys. Res. Commum. *82*, 589-595, 1978.
28. Wolfe L.S., Senior R.G., Ng Ying Kin N.M.K.: *The structure of oligosaccharides accumulating in the liver of* G_{M1}*gangliosidosis type 1.* J. Biol. Chem. *249*, 1829-1838, 1974.
29. Tsay G.C., Dawson G.: *Oligosaccharide storage in brains from patients with fucosidosis,* G_{M1} *gangliosidosis and* G_{M2} *gangliosidosis (Sandhoff's disease).* J. Neurochem., *27*, 733-740, 1976.
30. Lundblad A., Sjöblad S., Swensson S.: *Characterization of a penta and an octasaccharide from urine of a patient with juvenile* G_{M1} *gangliosidosis.* Arch. Biochem. Biophys. *188*, 130-136, 1978.
31. Cantz M., Kresse H.: *Sandhoff's disease: defective glycosaminoglycan catabolism in cultured fibroblasts and its correction by* β*-N-acetylhexosaminidase.* Eur. J. Biochem. *47*, 581-590, 1974.
32. Brunngraber E.G., Brown D.D., Aro A.: *Glycoproteins in brain tissue of O-variant of* G_{M2} *gangliosidosis.* J. Neurochem. *22*, 125-128, 1974.
33. Berra B., Brunngraber E.G.: *Neuropathological changes in brain glycoproteins.* Biochem. Exp. Biol. *13*, 31-40, 1977.
34. Strecker G., Montreuil J.: *Déscription d'une oligosaccharidurie accompagnant une gangliosidose* G_{M2} *à déficient total en N-acetyl-hexosaminidase.* Clin. Chim. Acta *33*, 395-401, 1971.
35. Ng Ying Kin N.M.K., Wolfe L.S.: *Oligosaccharide accumulating in the liver from a patient with* G_{M2} *gangliosidosis variant O.* Biochem. Biophys. Res. Commun. *66*, 123-130, 1975.
36. Strecker G., Herlant-Peers M.C., Fournet B., Montreuil J.: *Structure of seven oligosaccharides excreted in the urine of a patient with Sandhoff's disease (*G_{M2} *gangliosidosis variant O).* Eur. J. Biochem. *81*, 165-171, 1977.
37. Sandhoff K., Andreae U., Jatzkewitz H.: *Deficient hexosaminidase activity in an exceptional case of Tay-Sachs disease with additional storage of kidney globoside in visceral organs.* Pathol. Eur. *3*, 278-285, 1968.
38. Thomas G.H., Tiller G.E., Reynolds L.W., Miller C.S., Bace J.W.: *Increased levels of sialic acid associated with a sialidase deficiency in I-cell disease (Mucolipidosis II) fibroblasts.* Biochem. Biophys. Res. Comm. *71*, 188-195, 1976.

39. Okada S., Kato T., Yabunchi H., Nishigaki M., Kobata A., Chiyo H., Furuyama J.: *Hypersialyloligosacchariduria in mucolipidosis: a method for diagnosis.* Clin. Chim. Acta *86*, 159-167, 1978.
40. Leroy J.G., De Mars R.I.: *Mutant enzymatic and cytological phenotypes in cultured human fibroblasts.* Science *157*, 804-806, 1967.
41. Maroteaux P., Lamy M.: *La pseudopolydystrophie de Hurler.* Press Méd. *74*, 2889-2892, 1966.
42. Wiesmann U., Vassella F., Herschkowitz N.N.: *I-cell disease: leakage of lysosomal enzymes into extracellular fluids.* New Engl. J. Med. *285*, 1090-1091, 1971.
43. Leroy J.G., Ho M.W., Mc Brinn M.C., Zielke K., Jacob I., O'Brien J.S.: *I-cell disease: biochemical studies.* Pediat. Res. *6*, 752-757, 1972.
44. Aula P., Autio S., Raivio K., Rapola J., Thoden C.J., Koskela S.L., Yamashina J.: *Salla disease, a new lysosomal disorder.* Arch. Neurol. *36*, 88-94, 1979.
45. Renlund M., Chester M.A., Lundblad A., Aula P., Raivio K.O., Autio S., Koskela S.L.: *Increased urinary excretion of free N-acetylneuraminic acid in thirteen patients with Salla disease.* Eur. J. Biochem. *101*, 245-250, 1979.
46. Maroteaux P., Humbel R.: *Les oligosaccharidoses: un concept nouveau.* Arch. Franç. Pédiat. *33*, 641-643, 1976.
47. Strecker G.: *Glycoprotéins et glycoprotéinoses.* In "Les Oligosaccharidoses" (J. P. Farriaux Ed.) 13-30, Crouant et Roques, Lille, 1977.

Introduction to the Glycoprotein metabolism

1. Ginsburg V., Neufeld E.H.: *Complex heterosaccharides of animals.* Ann. Rev. Biochem. *38*, 371-386, 1969.
2. Montreuil J.: *Recent data on the structure of the carbohydrate moiety of glycoproteins. Metabolic and biological implications.* Pure Appl. Chem. *42*, 431-477, 1975.
3. Kornfeld R., Kornfeld S.: *Comparative aspects of glycoprotein structure.* Ann. Rev. Biochem. *45*, 217-237, 1976.
4. Roseman S.: *Sugars of the cell membrane.* Hosp. Pract. *10*, 61-70, 1975.
5. Spiro R.G.: *Glycoproteins.* Adv. Protein. Chem. *27*, 349-467, 1973.
6. Suzuki K.: *Catabolism of sialylcompounds in nature.* In "Biological Roles of Sialic Acid". A Rosemberg and C.L. Schengrund (Eds), 159-181, Plenum Press, New York, 1976.
7. Schachter H., Roden L.: *The biosynthesis of animal glycoproteins.* In "Metabolic conjugation and Metabolic Hydrolysis. Fishman W.H. (Ed.) vol. 3, 2-149 Academic Press, New York, 1973.
8. Spik G.: *Etude des propriétés des glycosidases impliqués dans le catabolisme des glycoprotéines.* In "Les oligosaccharidoses". J.P. Farriaux (Ed.) 31-41, Crouant et Roques, Lille 1977.
9. Ogushi T., Yamashina I.: *Distribution of a glycopeptide-degrading enzyme in tissue and cells.* Biochim. Biophys. Acta *156*, 417-419, 1968.
10. Makino M., Koijma T., Yamashina I.: *Studies on enzymes acting on glycopeptides.* J. Biochem. (Tokyo) *63*, 186-192, 1968.
11. Strecker G., Hondi-Assah T., Fournet B., Spik G., Montreuil J., Farriaux J.P., Maroteaux P., Durand P.: *Structure of 9 sialyl-oligosaccharides excreted in urine of 11 patients with 3 different types of sialidosis: mucopolidosis II and two new types of mucopolidosis.* Eur. J. Biochem. *75*, 391-403, 1977.
12. Hickman S., Neufeld E.F.: *A hypothesis for I-cell disease: defective hydrolases that do not enter lysosomes.* Biochem. Biophys. Res. Commun. *49*, 992-999, 1972.
13. Neufeld E.F., Sando G.N., Garvin A.J., Rome L.H.: *The transport of lysosomal enzymes.* J. Supramol. Struct *6*, 95-101, 1977.
14. Ashwell G., Morrell A.: *The role of surface carbohydrates in the hepatic recognition and transport of circulating glycoproteins.* Adv. Enzymol. *41*, 99-127, 1974.
15. Kaplan A., Achord D.T., Sly W.S.: *Phosphohexosyl components of a lysosomal enzyme*

are recognized by pinocytosis receptors on human fibroblasts. Proc. Natl. Acad. Sci. U.S.A. *74*, 2026-2030, 1977.
16. Natowicz M.R., Chi M.M.Y., Lowry O.H., Sly W.S.: *Enzymatic identification of mannose 6-phosphate on the recognition marker for receptor-mediated pinocytosis of β-glucuronidase by human fibroblasts.* Proc. Natl. Acad. Sci. U.S.A. *76*, 4322-4326, 1979.
17. Schlesinger P.H., Doebber T.W., Mandell B.F., White R., Schryver C.de, Rodman J.S., Miller M.J., Stahl P.: *Plasma clearance of glycoproteins with terminal mannose and N-acetylglucosamine by liver non-parenchimal cells.* Biochem. J. *176*, 103-109, 1978.
18. Hasilik A., Klein U., Waheed A., Strecker G., Figura Von K.: *Phosphorylated oligosaccharides in lysosomal enzymes: identification of α-N-acetylglucosamine (1) phospho (6) mannose diester groups.* Proc. Natl. Acad. Sci. U.S.A. *77*, 7074-7078, 1980.
19. Skudlarek M., Swank R.T.: *Biosynthesis of two lysosomal enzymes in macrophages. Evidence of a precursor of β-galactosidase.* J. Biol. Chem. *254*, 9939-9942, 1979.
20. Hasilik A., Neufeld E.F.: *Biosynthesis of lysosomal enzymes in fibroblasts. I. Synthesis as precursors of higher molecular weight.* J. Biol. Chem. *255*, 4937-4945, 1980.
21. Hasilik A., Neufeld E.F.: *Biosynthesis of lysosomal enzymes. II Phosphorylation of mannose residues.* J. Biol. Chem. *255*, 4946-4950, 1980.
22. Neufeld E.F.: *The uptake of enzymes into lysosomes: An overview. In: Enzyme Therapy in Genetic Diseases.* R.J. Desnick (Ed.) Birth Defects: Orig. Art. Ser., XVI, 77-84, Alan R. Liss Inc. The National Foundation, 1980.
23. Reitman M.L., Kornfeld S.: *UDP-N-acetylglucosamine: glycoprotein N-acetylglucosamine-1-phosphotransferase+ proposed enzyme for the phosphorylation of the high mannose oligosaccharide units of lysosomal enzymes.* J. Biol. Chem. *256*, 4275-4281, 1981.
24. Hasilik A., Waheed A., Von Figura K.: *Enzymatic phosphorylation of lysosomal enzymes in the presence of UDP-N-acetylglucosamine. Absence of the activity in I-cell disease.* Biochem. Biophys. Res. Commun. *98*, 761-767, 1981.
25. Varki A., Kornfeld S.: *Identification of a rat liver α-N-acetylglucosaminyl phosphodiesterase capable of removing "blocking" α-N-acetylglucosamine residues from phosphorylated high mannose oligosaccharides of lysosomal enzymes.* J. Biol. Chem. *255*, 8389-8401, 1980.
26. Sly W.S., Creek K.: *The role of mannose-6-phosphate recognition marker and its receptor in enzyme transport and in processing of oligosaccharides on acid hydrolases.* In "Sialidases and Sialidoses" G. Tettamanti, P. Durand, S. Di Donato (Eds), Perspec. Inher. Metab. Dis., Edi. Ermes s.r.l. Milano *4*, 453-462, 1981.

Differential diagnosis

1. Rapin J.: *Progressive genetic-metabolic diseases of the central nervous system in children.* Pediat. Ann. *5*, 55-116, 1976.
2. Hoeksema J., Van Diggelen O.P., Galjaard H.: *Intergenetic complementation after fusion of fibroblasts from different patients with β-galactosidase deficiency.* Biochem. Biophys. Acta *566*, 72-79, 1979.
3. Galjaard H.: *Genetic Metabolic Disease. Early Diagnosis and Prenatal analysis.* Elsevier/North Holland, Amsterdam-New York 1980.
4. Hoogeveen A.T., Verheijen F.W., D'Azzo A., Galjaard H.: *Genetic heterogeneity in human neuraminidase deficiency.* Nature *285*, 500-502, 1980.
5. O'Brien J.S.: *Molecular genetics of G_{M1} β-galactosidase.* Clin. Genet. *8*, 303-313, 1975.
6. Aula P., Autio S., Raivio K.O., Rapola J.: *Aspartylglucosaminuria.* Chapter 5, 123-152 this book.
7. Renlund M., Chester M.A., Lundblad A., Aula P., Raivio K.O., Autio S.: *Increased urinary excretion of free N-acetylneuraminic acid in thirteen patients with Salla disease.* Eur. J. Biochem. *101*, 245-250, 1979.
8. Durand P., Borrone C., Costantino G., Cavalieri S., Filocamo M., Romeo G.:

Detection of carriers and prenatal diagnosis for fucosidosis in Calabria. Hum. Genet. *51*, 195-201, 1979.
9. Lowden J.A., O'Brien J.S.: *Sialidosis: a review of human neuraminidase deficiency.* Am. J. Hum. Genet. *31*, 1-18, 1979.
10. O'Brien J.S.: *The gangliosidoses.* In "The Metabolic Basis of Inherited Disease". J.B. Stanbury, J.B. Wyngaarden, D.S. Frederickson (Eds), 4th ed., 841-866, Mc Graw-Hill, New York, 1978.
11. Lundblad A.:*Isolation and characterization ofa urinary oligosaccharide characteristic of blood-group 0 (H)-secretors.* Biochim. Biophys. Acta *165*, 202-206, 1968.
12. Lundblad A.: *Blood-group specific oligosaccharides in urine.* In "Blood and Tissue Antigens". D. Aminoff (Ed.), 427-435, Academic Press, New York, 1970.
13. Lundblad A., Hallgren P., Rudmark A., Svensson S.: *Structures and serological activities of three oligosaccharides isolated from urines of nonstarved secretors and from secretors on lactose diet.* Biochemistry *12*, 3341-3345, 1973.
14. Strecker G., Montreuil J.: *Isolement et étude de la structure de 16 oligosaccharides isolés de l'urine humaine.* C.R. Acad. Sc. Paris *277*, Série D., 1393-1396, 1973.
15. Strecker G., Trentesaux-Chauvet C., Riazi-Fazard T., Fournet B., Bouquelet S., Montreuil J.: *Mise en évidence d'une oligosaccharidurie associée à diverses mélituries et détermination de la structure des oligosaccharides excrétés. Hypothése concernant l'origine des oligosaccharides des liquides biologiques.* C.R. Acad. Sc. Paris *277*, Série D, 1569-1572, 1973.
16. Lennartson G., Lundblad A., Sjöblad S., Svensson S., Ockerman P.A.: *Quantitation of a urinary tetrasaccharide by gas chromatography and mass spectrometry.* Biomed. Mass Spectrom. *3*, 51-54, 1976.
17. Strecker G., Peers M.C., Michalski J.C., Hondi-Assah T., Fournet B., Spick G., Montreuil J., Farriaux J.P., Maroteaux P., Durand P.: *Structure of nine sialyloligosaccharides accumulated in the urine of eleven patients with three different types of sialidoses. Mucopolidosis II and two new types of mucolipidosis.* Eur. J. Biochem. *75*, 394-403, 1977.
18. Palo J., Savolainen H.: *Studies on serum and urinary glycopeptides and glycosaminoglycans in aspartylglucosaminuria.* Clin. Chim. Acta *36*, 431-437, 1972.
19. Humbel R., Collart M.: *Oligosacharides in the urine of patients with glycoprotein storage diseases.* Clin. Chim. Acta *60*, 143-145, 1975.
20. Humbel R., Dhondt J.L.: *Dépistage des oligosaccharidurіes.* In "Les oligosaccharidoses". J.P. Farriaux (Ed.), 119-122, Crouant et Roques, Lille, 1977.
21. Sewell A.C.: *An improved thin-layer chromatographic method for urinary oligosaccharide screening.* Clin. Chim. Acta *92*, 411-414, 1979.
22. Sewell A.C., Gehler J., Spranger J.: *Comprehensive urinary screening for inborn errors of complex carbohydrate metabolism.* Klin. Wochenschr. *57*, 581-584, 1979.
23. O'Brien J.S.: *The cherry-red spot-myoclonus syndrome: a newly recognized inherited lysosomal storage disease due to acid neuraminidase deficiency.* Clin. Genet. *14*, 55-60, 1978.
24. Carmody P.J., Rattazzi M.C., Davidson R.G.: *Tay-Sachs disease. The use of tears for the detection of heterozygotes.* N. Engl. J. Med. *289*, 1072-1074, 1973.
25. Singer J.D., Cotlier E., Krimmer R.: *Hexosaminidase A in tears and saliva for rapid identification of Tay-sachs disease and its carriers.* Lancet *2*, 1116-1119, 1973.
26. Libert J., Van Hoof F., Tondeur M.: *Fucosidosis: ultrastructural study of conjunctiva and skin and enzyme analysis of tears.* Invest. Ophthalmol. *15*, 626-639, 1976.
27. Van Hoof F., Libert J., Aubert-Tulkens G., Serra M.V.: *The assay of lacrimal enzymes and the ultrastructural analysis of conjunctival biopsies: new techniques for the diagnosis of inborn lysosomal diseases.* Metab. Ophthalmol. *1*, 165-171, 1977.
28. Cotlier E., Kivlin J., Del Monte M.A.: *Mail screening for inborn errors of metabolism with tears.* Discussion by F. Van Hoof In "The Eye and Inborn Errors of Metabolism". D. Bergma, A.J. Bron, E. Cotlier (Eds). Birth Defects: Orig. Art. Ser. Inc., National Foundation XII, 105-106, Alan R. Liss, New York 1976.
29. Menguy R., Masters Y.F., Desbaillets L.: *Human salivary glycosidases.* Proc. Soc.

Exp. Biol. Med. *134*, 1020-1026, 1970.
30. Den Tandt W.R., Jaeken J.: *Confirmation of metachromatic leukodystrophy and fucosidosis by enzyme analysis of saliva.* Neuropädiatrie *11*, 189-190, 1980.
31. Ng W.G., Donnell G.N., Koch R., Bergren W.R.: *Biochemical and genetic studies of plasma and leucocytes α-L-fucosidase.* Am. J. Hum. Genet. *28*, 42-50, 1976.
32. Gatti R., Cavalieri S., Romeo G.: *Relationship between α-fucosidase deficiency in plasma and α-fucosidase in leukocytes.* Hum. Genet. *48*, 23-30, 1979.
33. Martin J.J., Ceuterick C.: *Morphological study of skin biopsy specimens: a contribution to the diagnosis of metabolic disorders with involvement of the nervous system.* J. Neurol. Neurosurg. Psychiat. *41*, 232-248, 1978.
34. Martin J.J., Ceuterick C., Libert J.: *Malattie neurometaboliche ereditarie. Morfologia e diagnosi.* Prospet. Pediat. *9*, 449-467, 1979.
35. Libert J.: *Diagnosis of lysosomal storage diseases by the ultrastructural study of conjunctival biopsies.* Pathol. Annual *15*, 37-66, 1980.

Carrier detection

1. Aula P., Raivio K., Autio S.: *Enzymatic diagnosis and carrier detection of aspartylglucosaminuria in cultured fibroblasts.* Pediat. Res., *10*, 625-629, 1976.
2. Durand P., Gatti R., Borrone C., Costantino G., Cavalieri S., Filocamo M., Romeo G.: *Detection of carriers and prenatal diagnosis for fucosidosis in Calabria.* Hum. Genet. *51*, 195-201, 1979.
3. Wolfe L.S., Callahan J., Fawcet J.S., Andermann F., Scriver C.R.: *G_{M1} gangliosidosis without chondrodystrophy and visceromegaly.* Neurology *20*, 33-44, 1970.
4. Singer H.S., Schafer I.A.: *White cell β-galactosidase activity.* N. Eng. J. Med. *282*, 571, 1970.
5. Jolly R.D., Thompson K.G., Tse C.A., Munford R.E., Merral M.: *Identification of mannosidosis heterozygotes: factors affecting normal plasma α-mannosidase levels.* N.Z. Vet. J. *22*, 155-162, 1974.
6. Jolly R.D., Tse C.A., Greenway R.M.: *Plasma α-mannosidase activity as a means of detecting mannosidosis heterozygotes.* N.Z. Vet. J. *21*, 64-69, 1973.
7. Ng W.G., Donnell N.G., Koch R., Bergren W.R.: *Biochemical and genetic studies of plasma and leukocyte α-L-fucosidase.* Am. J. Hum. Genet. *28*, 42-50, 1976.
8. Gatti R., Cavalieri S., Romeo G.: *Relationship between α-fucosidase deficiency in plasma and α-fucosidase activity in leukocytes.* Hum. Genet. *48*, 23-30, 1979.
9. Libert J., Van Hoof F., Tondeur M.: *Fucosidosis: ultrastructural study of conjunctiva and skin and enzyme analysis of tears.* Invest. Ophthalmol. *15*, 626-639, 1976.
10. Van Hoof F.: *Diagnostic enzymatique des oligosaccharidoses.* In "Les Oligosaccaridoses. J.P. Farriaux (Ed.), 129-132, Crouant et Roques, Lille, 1977.
11. Goldberg M.F.: *The use of tears for heterozygotes detection and genetic counseling.* Invest. Ophthalmol. *13*, 159-168, 1974.
12. Van Hoof F., Libert J., Aubert-Tulkens G., Serra M.V.: *The assay of lacrimal enzymes and the structural analysis of conjunctival biopsies: new techniques for the study of inborn lysosomal diseases.* Metab. Ophthalmol. *1*, 165-171, 1977.
13. Matsuda J., Arashima S., Anakura M., Oka Y.: *α-L-fucosidase and α-mannosidase activity in the white blood cells in the disease and carrier state of fucosidosis.* Clin. Chim. Acta *48*, 9-13, 1973.
14. Thompson K.G., Jolly R.D., Winchester B.G.: *Mannosidosis. Use of reference enzymes in heterozygote detection.* Biochem. Med. *15*, 233-240, 1976.

Prenatal diagnosis

1. O'Brien J.S., Okada S., Fillerup D.L., Veath M.L., Adornato B., Brenner P.H., Leroy J.G.:

Tay-Sachs disease: prenatal diagnosis. Science *172*, 61-64, 1971.
2. Schneck L., Friedland J., Valenti C., Adachi M., Amsterdam D., Volk B.M.: *Prenatal diagnosis of Tay-Sachs disease*. Lancet *1*, 582, 1970.
3. Lowden J.A., Cutz E., Conen P.E., Rudd N., Doran T.: *Prenatal diagnosis of* G_{M1} *gangliosidosis*. New Engl. J. Med. *288*, 225-228, 1973.
4. Desnick R.J., Kriviat W., Sharp H.L.: *In utero diagnosis of Sandoff's disease*. Biochem. Biophys. Res. Commun. *51*, 20-26, 1973.
5. Galjaard H.: *Genetic Metabolic Diseases. Early Diagnosis and Prenatal Analysis*. Elsevier/North Holland. Amsterdam-New York 1980.
6. Milunsky A.: *Prenatal diagnosis of hereditary biochemical disorders of metabolism*. In "The Genetic Disorders and the Fetus. Diagnosis, Prevention and Treatment". A. Milunski (Ed.), 209-367, Plenum Press New York-London, 1979.
7. Huijing F., Warren R.J., Mc Leod A.G.: *Elevated activity of lysosomal enzymes in amniotic fluid of a fetus with mucolipidosis II (I-cell disease)*. Clin. Chim. Acta *44*, 453-455, 1973.
8. Warren R.J., Condron C.J., Hollister D., Huijing F., Neufeld E.F., Hall C.W., Mc Leod A.G.W., Lorincz A.E.: *Antenatal diagnosis of mucolipidosis II (I-cell disease)*. Pediat. Res. *7*, 343-349, 1973.
9. Aula P., Rapola J., Autio S., Raivio K., Karijalainen O.: *Prenatal diagnosis and fetal pathology of I-cell disease*. J. Pediat. *87*, 221-226, 1975.
10. Matsuda I., Arashima S., Mitsuyama T., Oka Y., Ikeuchi T., Kaneko Y., Ishikawa M.: *Prenatal diagnosis of I-cell disease*. Humangenetik *30*, 69-73, 1975.
11. Gehler J., Cantz M., Stoeckenius M., Spranger J.: *Prenatal diagnosis of mucolipidosis II (I-cell disease)*. Eur. J. Pediat. *122*, 201-206, 1976.
12. Kleijer W.J., Niermeijer M.F., Galjaard H.: *Experience with 136 prenatal diagnoses of metabolic disease*. In "Prenatal Diagnosis". J.D. Murken, S. Stengel-Rutkowski, E. Schwinger (Eds.), 85, Enke Publ. Stuttgart, 1979.
13. Durand P., Borrone C.: *Le diagnostic prenatal des maladie métaboliques*. Méd. et Hyg. *38*, 3533-3541, 1980.
14. Durand P., Borrone C., Gatti R., Filocamo M.: *Results of prenatal diagnosis for genetic metabolic disease in Italy*. In "Recent Advances in Prenatal Diagnosis". C. Orlandi, L. Bovicelli, P. Polani (Eds.), 249-257, John Wiley & Sons Pub., 1981.
15. Kleijer W.J., Hoogeveen A., Verheijen F.W., Niermeijer M.F., Galjaard H., O'Brien J.S., Warner T.G.: *Prenatal diagnosis of sialidosis with combined neuraminidase and β-galactosidase deficiency*. Clin. Genet. *16*, 60-61, 1979.
16. Matsuda I., Arashima S., Oka Y., Mitsuyama T., Ariga S., Ikeuchi T., Ichida T.: *Prenatal diagnosis of fucosidosis*. Clin. Chim. Acta *63*, 55-60, 1975.
17. Poenaru L., Dreyfus J.C., Boué J., Nicolesco H., Ravise N., Bamberger J.: *Prenatal diagnosis of fucosidosis*. Clin. Genet. *10*, 260-264, 1976.
18. Poenaru L., Girard S., Thepot F., Madelenat P., Hurauxrendu C., Vinet M.C., Dreyfus J.C.: *Antenatal diagnosis in three pregnancies at risk for mannosidosis*. Clin. Genet. *16*, 428-432, 1979.
19. Aula P.: personal communication.
20. Butterworth J., Guy G.J.: *α-L-fucosidase of human skin fibroblasts and amniotic fluid cells in tissue culture*. Clin. Genet. *12*, 297-302, 1977.

The chromosomal localization of the mutant genes

1. Mc Kusick V.: *The anatomy of human genome*. Am. J. Med. *69*, 267-276, 1980.
2. Mc Kusick V.: *Mendelian Inheritance in Man. Catalogs of Autosomal Dominant, Autosomal Recessive and X-linked Phenotypes*. 5.th Ed. Johns Hopkins University Press. Baltimore, 1978.
3. Turner V.S., Turner B.M., Kucherlapati R.P., Ruddle F.H., Hirschorn K.: *Assignement of the human α-L-fucosidase gene locus to chromosome 1 by use of a "Clone panel"*. Baltimore Conference Third International Worskshop on Human Gene Mapping. Birth

Defects: Orig. Art. Ser. XII, National Foundation, New York *7*, 238-240, 1976.

4. Champion M.J., Brown J.A., Shows T.B.: *Assignement of lysosomal alpha-mannosidase (MAN-B) and confirmation of the peptidase D (PEPD) genes to chromosome 19 in man.* Winnipeg Gene Mapping Cong., 1977.
5. Wit J. de, Hoeksema H.L., Bootsma D., Westerveld A.: *Assignement of structural β-galactosidase loci to human chromosome 3 and 22.* Hum. Genet. *51*, 259-267, 1979.
6. Hoeksema H.L., Wit J. de, Westerveld A.: *The genetic defect in the various types of human β-galactosidase deficiency.* Human Genet. *53*, 241-247, 1980.
7. Naylor S.L., Lalley P.A., Elliott R.W., Brown J.A., Shows T.B.: *Evidence for homologous regions of human chromosome 3 and mouse chromosome 9 predicts location of human genes.* Proceedings Am. Soc. Hum. Genet. 31st Annual Meeting *159 A*, 480, New York, 1980.
8. Gilbert F., Kucherlapati R.P., Creagan R.P., Murnane M.J., Darlington G.J., Ruddle F.H.: *Tay-Sachs' and Sandhoff's disease. The assignement of genes for hexosaminidase A and B to individual human chromosomes.* Proc. Natl. Acad. Sci. U.S.A. *72*, 262-267, 1975.
9. Botstein D., White R.L., Skolmick M., Davis R.W.: *Construction of a genetic linkage map in man using restriction fragment length polymorphism.* Am. J. Hum. Genet. *32*, 314-331, 1980.

2 Sialidosis

John S. O'Brien

NAME OF DISEASE	Sialidosis
PREVALENCE	Unknown
MODE OF INHERITANCE	Autosomal recessive
MUTANT GENE PRODUCT	Glycoprotein sialidase
CHROMOSOMAL LOCATION	?
EXPRESSION	
1. Altered DNA structure	?
2. Disturbed protein function	Enzyme activity defect
3. Disrupted cell and organ function	Accumulation of sialylated oligosaccharides and glycoconjugates leads to brain dysfunction and somatic changes

2.1 INTRODUCTION

Defects in the activity of neuraminidase (N-acetylneuraminic acid hydrolase EC 3.2.1.18) have been described in several different clinical disorders. In 1977 Cantz et al. [1] and Spranger et al. [2] first demonstrated a profound deficiency of acid neuraminidase in one of their patients who had been previously described under the classification of "mucopolidosis I" [3, 4]. The patient had a neurodegenerative disorder with Hurleroid features, skeletal dysplasia and cherry-red spots in the macula. Signs of the illness developed in the first year of life. In addition to the neuraminidase defect these patients accumulated sialic acid-containing material in fibroblasts. Parents had activities of neuraminidase which were intermediate between patients and controls implicating the deficiency as a primary defect. Durand et al. [5], O'Brien [6], Thomas et al. [7] and Rapin et al. [8] described a different group of patients with neuraminidase deficiency. These young adults presented with decreasing visual acuity, cherry-red spots in the macula and myoclonus and excreted sialyloligosaccharides in their urine. Their phenotype differed from that of mucolipidosis I patients in that intelligence was normal, somatic bony abnormalities were absent, and their survival was longer. Family studies indicated that patients with the "cherry-red spot-myoclonus" syndrome were homozygous for a recessively transmitted autosomal mutation affecting acid neuraminidase [6, 7]. Another patient, an infant with Hurleroid features but with normal mental development and acid neuraminidase deficiency, was described by Kelly and Graetz [9].

In addition to these two groups of patients a dysmorphic group of young adults, who were previously though to have a primary defect in β-galactosidase activity, have now been included in the clinical spectrum of patients with neuraminidase deficiency [11]. These patients were normal at birth and throughout early childhood. Dysmorphic features developed towards the end of the first decade and were accompanied by psychomotor degenerative changes. Myoclonus and frequently a cherry-red spot was described.

Lowden and O'Brien [11] reviewed the clinical phenotypes of primary human neuraminidase deficiency and, like Durand et al. [5] used the term "sialidosis" to describe patients with this disorder. In the present review I will discuss the clinical genetics of sialidosis, the chemistry of the compounds stored, the molecular genetics of acid neuraminidase, and laboratory methods for antenatal diagnosis and diagnosis of homozygotes and heterozygotes.

2.2 CLINICAL PHENOTYPES

Thirty-seven patients with proven or probable primary acid neuraminidase deficiency were reviewed by Lowden and O'Brien [11] to arrive at their classification. In the interim several additional patients have been described, most of whom fall into the subtypes already given in their scheme. Several patients

Table 2.1 Clinical features of Sialidosis

	TYPE 1		TYPE 2	
	Juvenile	*Infantile*	*Infantile*	*Congenital*
Age Onset	12.4±4.7 yrs	10.4±5.3 yrs	0-10 mon.	birth
Sex	8 male/ 10 female	8 male/ 6 female	2 male/ 3 female	3 male/ 2 female
Race/ethnic origin	7/10 Italian	10/14 Japanese	5/5 Caucasian	
Seizures	6/13	5/13	4/5	+
Myoclonus	16/18	10/13	4/5	
Ataxia	3/12	9/12	3/5	
Pyramidal tract signs	0/18	4/13	3/5	+
Mental retardation	2/18	8/14		+
Cherry-Red Spots	17/17	12/12	4/5	
Corneal clouding/lens opacity	3/17	10/12	4/5	
Loss visual acuity	10/10	11/11	1/1	
Hearing loss	0/14	6/10	3/4	
Coarse facies	0/14	14/14	5/5	+
Vertebral changes	0/14	13/13	5/5	+
Growth disturbance	0/14	12/12	4/5	
Visceromegaly	1/14	0/12	5/5	+
Foam cells	4/8	10/11	5/5	+
EEG abnormal	6/11	4/7	3/4	+

* All details were not available for all cases, thus the denominator in each figure above indicates the number of instances in which this clinical feature was mentioned in the case reports [11].

however differ from those described. Four siblings, 2 males, 2 females described by Johnson et al. [12] presented with congenital sialidosis, with polyhydramnios, premature birth, early death, congenital ascites, skin telangiectasis, vascular disease, congenital hepatosplenomegaly and severe encephalopathy with siezures. This is the most severely affected phenotype of sialidosis described thusfar and alerts to the presence of sialidosis in babies with hydrops fetalis or congenital lipidosis. A single additional patient has been studied by Lowden [13]. Nephrosialidosis is the term described by Maroteaux et al. [14] for sialidosis presenting with severe renal disease, especially nephrosis.

The Lowden-O'Brien classification (Table 1) divides sialidosis into two categories; normomorphic without somatic changes (Type 1) and dysmorphic with somatic changes (Type 2):

—Sialidosis Type 1, a group with normal physical appearance and normal body proportions, cherry-red spots, myoclonus and usually normal levels of β-galactosidase in body tissues and fluids. Some patients have severe renal disease.

—Sialidosis Type 2, a dysmorphic group with short stature, bony abnormalities and often deficient β-galactosidase in tissues or body fluids. Sialidosis Type 2 can be subclassified into congenital, infantile and juvenile subgroups.

A. Sialidosis Type 1 - Normosomatic group

The clinical features have been well described by Rapin et al. [8]. Affected patients present between 8-15 yrs with decreasing visual acuity, myoclonus or both. On examination they have normal features, normal growth and usually normal intellect. Cherry-red spots are seen in the macular region of the fundus. The cornea and lens are usually clear but occasionally punctuate lens opacities are noted [5]. Visual acuity gradually decreases during the course of the disorder. In one patient [8], the cherry-red spots faded in her 20th year.

The myoclonus begins in the limbs and is gradually progressive. It is an action-type myoclonus induced by stimulation, movement or emotion. It usually decreases or abates during sleep. As the disease progresses, myoclonus may prevent the patient from walking, sitting and eventually from feeding. The massive spasms are poorly controlled with anticonvulsants, diazepam, haloperidol, levodopa, conjugated estrogens, ketogenic diet or magnesium [8, 15]. One patient benefitted from L-tryptophan therapy [16] but others have not [17]. In association with the progression of the myoclonus, tendon stretch reflexes increase, but, as in Fabry's disease, several patients complained of burning pain in their extremities, exacerbated by hot weather [8]. Sensorimotor peripheral polyneuropathy may be present; it is accompained by segmental demyelination and distinctive inclusions in the cytoplasm of Schwann cells [18]. Because of the severity of the myoclonus, the assessment of intellectual capacity is sometimes difficult. Nevertheless it appears that intelligence is not impaired or is minimally impaired. Seizure disorders of the grand mal type followed by loss of consciousness have been present in about half the patients. They were often initiated by a massive myoclonic spasm.

Electrophysiologic abnormalities were examined extensively in two patients by Engel et al. [19]. They found low voltage fast activity in the EEG. Similar observations were recorded in other cases [8, 15, 20].

One patient had an enlarged liver [5] but all other physical findings were normal in this group. Of 10 Type 1 patients whose ethnic origins were reported, 7 are either Italian or of Italian ancestry.

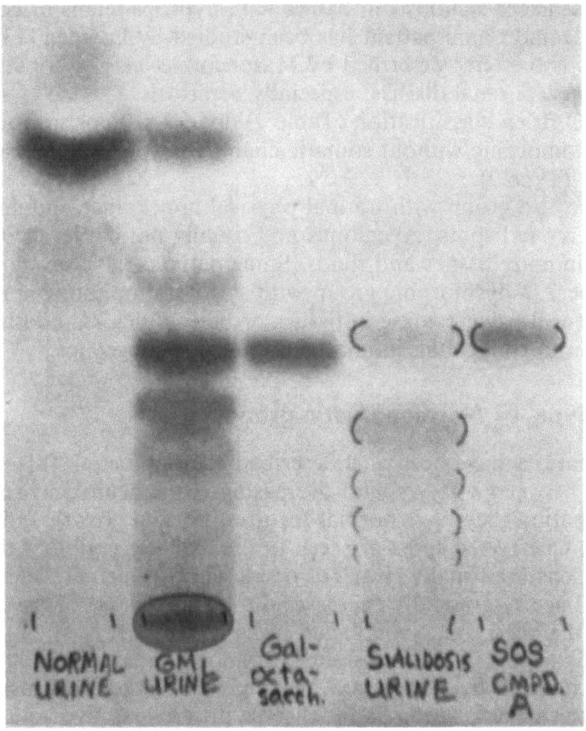

Fig. 2.1 Thin layer chromatography of oligosaccharides. Urine (10 μl) or extracts of purified oligosaccharides were spotted on silica gel plates (0.25 mm, Brinckman, activated by heating at 120°C prior to use) and development was in n-propanol-acetic acid-water 3:2:2 for 2 hours. Plates were sprayed with orcinol (1% in 50% H_2SO_4) heated at 120° for 10 min and purple staining oligosaccharides were visualized. Lanes are: Normal Urine - urine from a normal child: G_{M1} Urine - urine from a 2 years old infant with G_{M1} gangliosidosis, Type 1; Gal-octasacch. - galactosyl-octasaccharide isolated from liver of patient with G_{M1} gangliosidosis; Sialidosis Urine - urine from a patient with sialidosis, Type 1 and SOS, Compd. A - sialyloligosaccharide containing six sugars (same as Strecker's Cmpd. II, Table 2) isolated from urine of patient with sialidosis. Monosaccharides migrate to the top of the chromatogram and glycoproteins do not migrate in this system. Resorcinol positive sialic acid-containing compounds are outlined in pencil.
Note different migration of galactosyl - and sialyl - oligosaccharides in these disorders. Method is described in Holmes and O'Brien [50].

Two of the 8 patients have died; one at 21 years and the other at 28 years. The oldest living patients are however over 30 years [7, 8].

In nine patients, blood smears and/or bone marrow preparations were examined; vacuolated lymphocytes or foam cells were noted in two [5, 20]. Vacuolation was described in the liver of one patient who died with diabetes [21] and cytoplasmic inclusions and membrane-bound vacuoles were reported in liver biopsies from one patient who also had foamy histiocytes in the bone marrow [5] and from one with normal bone marrow [8]. A cerebral biopsy of the latter patient revealed cortical thinning and "neuronal lipidosis" with membrano-vesicular and lysosome-like bodies in neurons [22]. Although many of the cerebral vacuoles appeared empty or filled with scanty amounts of amorphous granular material, lipofuscin bodies were also prominent. In frozen sections of liver, the vacuoles stained with Alcian-blue and the periodic-acid Schiff (PAS) reagent indicating they contained a negatively-charged carbohydrate-containing compound [5, 8]. Many lysosomal hydrolases, including β-galactosidase, β-N-acetylhexosaminidase, β-glucosidase, α-mannosidase, α-fucosidase, β-glucuronidase and sphingomyelinase were assayed in serum, leucocytes, fibroblasts and occasionally tissue samples in eight of the patients. In one case β-galactosidase activities were less than 10% of normal values in WBC and fibroblasts [23] but in all other cases examined these lysosomal hydrolases were within normal limits.

In eight Type 1 patients neuraminidase has been assayed and the enzyme defect has been demonstrated. Furthermore, urines from nine patients have been examined for oligosaccharides by thin layer chromatography and the typical increase in resorcinol-positive spots were found (Fig. 1). In two, oligosaccharides were isolated in quantity from urine, quantified and structurally characterized [5, 24].

B. Sialidosis Type 2 - Dysmorphic group

The clinical features in these patients are more variable than those of the Type 1 patients. All present with coarse facial features which vary in severity from thick lips, a flattened nasal bridge and mild hypertelorism to a typical Hurleroid appearance. The facial changes are often accompanied by increased head circumference and short stature. Radiographic changes are similar to those seen in mucopolysaccharidoses but in the juvenile group may be less profound than classic Hurleroid changes. They are usually not seen until other dysmorphic features of the disorder become pronounced. They include beaking of the lumbar vertebrae, usually L_1 or L_2, broadening of the ribs, thickening of the skull and loss of the fine texture of the bones with increasing coarse trabeculation.

Cherry-red spots have been noted in all but one of the cases reviewed. That child was only eight months old when reported [9], and the lesion may not have developed. Myoclonus was present in 75% of the patients. Ataxia was common in Type 2 patients and rare in Type 1.

Many other Hurleroid features were observed in the dysmorphic group. They include lens opacities, corneal clouding, visceromegaly (particularly liver), joint stiffness, inguinal and umbilical hernias and sensorineural hearing loss. Most Type 2 patients have low normal to moderately retarded intelligence.

The dysmorphic patients can be divided into those with congenital onset, infantile onset and juvenile onset (Table 1). As described above, five patients with congenital onset sialidosis have been studied in two families; consanguinity was not evident [12, 13]. Of the 14 juvenile onset cases, 10 occurred in Japanese patients. The earliest symptoms, joint stiffness and coarse features in some, or decreased visual acuity in others, developed between 8-15 yrs and became progressively more pronounced over the next 10 yrs. All but three of the juvenile onset patients are still alive. The oldest is 25 years. Hepatosplenomegaly is found only in the congenital and the infantile subgroups. In the juvenile onset patients it usually occurs in the absence of clinical splenomegaly and does not result in abnormal liver function tests. In the congenital onset patients it is accompained by splenomegaly and liver function may be impaired since ascites is often present.

Many features of the dysmorphic type of sialidosis are, however, variable and inconsistent. For example, of the two infants who had dysmorphic signs at birth [9, 22], one grew at the 98 percentile for age and was of apparently normal mentality while the other had marked growth retardation, failed to thrive throughout her life and dysplayed profound psychomotor retardation.

Mental retardation appears to be more severe in Type 2 patients. While only two Type 1 patients are mildly to moderately retarded five of the juvenile, four of the infantile, and all 5 of the congenital onset Type 2 patients are severely retarded.

Various lysosomal hydrolases have been measured in serum, fibroblasts, leucocytes and liver in six juvenile and three infantile cases. In all juvenile cases studied, β-galactosidase was low in tissues but usually normal in plasma. β-Galactosidase activity was normal in 4 congenital onset patients [12] and in two infantile patients [3, 9] and deficient in a third [13]. Neuraminidase activities were less than 10% of normal in cultured fibroblasts from all patients and excretion of sialyloligosaccharides (SOS) was increased in all urines examined.

Included in the juvenile onset subgroup is one patient who has normal or nearly normal intellect, slightly coarse features, cherry-red spots, myoclonus and mild bony changes of the osteochondrodystrophy type. Her samples were referred to our laboratory for lysosomal hydrolase assays and neuraminidase deficiency was documented [18]. Assignation to Type 2 was made on the basis of bone changes and short stature (147 cm) at 18 years. The heterogenous nature of the Type 2 group will almost certainly result in further arbitrary classification of other new patients.

2.3 GENETICS

The 18 Type 1 patients reviewed by Lowden and O'Brien [11] derive from 13 families and in two, matings were consanguineous. Among the 19 Type 2 patients, there are three sets of sibs and two further patients who had affected sibs recorded but on whom no details were available. In six of the families the parents were consanguineous. Of the 37 patients reviewed almost half, or 17, were male.

To establish the disorder as an autosomal recessive mutation it is customary to demonstrate a gene-dose response in obligate heterozygotes. In five families [1, 7, 9, 18, 25] intermediate activity has been demonstrated in parents of probands with sialidosis Type 1 (three patients) and Type 2 (two patients). Thomas et al. have demonstrated intermediate deficiencies of neuraminidase in relatives (niece, nephew, siblings and children) of probands with Type 1 and Type 2 congenital sialidosis [7, 12]. These results provide strong evidence that sialidosis is transmitted as an autosomal recessive trait.

The ethnic origins of patients with each type of sialidosis add further support for autosomal recessive inheritance. Seven of ten patients with Type 1 sialidosis are Italian whereas 10 of the 14 juvenile onset Type 2 patients are Japanese. These ethnic predilections suggest increased heterozygote frequencies for this mutation among Italians and Japanese. They also reinforce the validity of the listed groupings. The heterozygote frequencies for Type 1 or 2 sialidosis among Italians of Japanese have yet to be determined. In the infantile onset Type 2 disorder, three patients are French, one Portuguese and in the fifth the background is not listed. Clinically this latter group is the most heterogeneous.

2.4 NATURE OF THE STORAGE COMPOUNDS

The gamut of compounds stored in sialidosis has not been completely defined. A complete lipid analysis of a brain biopsy of one Type 1 patient [22] revealed an increase in cholesterol esters, a slight increase in cerebrosides and a 3-5-fold increase in the free amino acids in grey matter. The ganglioside content and pattern were normal. A similar result was obtained upon analysis of postmortem brain tissue in a Type 1 patient [26]. In a patient with infantile Type 2 sialidosis, lipid analysis of a brain biopsy was normal [27]. In a third patient with juvenile onset Type 2 sialidosis [28] lipid analysis of grey matter, including thin layer chromatography of the gangliosides, was normal.

As yet, the only biochemical study [26] reported on extraneural tissues is analysis of sialyloligosaccharides in viscera in a Type 1 patient; a remarkable increase was found, similar to that in urine (Fig. 1).

Using urine from the patients of Spranger et al. [3], Durand et al. [5], Strecker et al. [24] have conducted a spectacular isolation and characterization of 12 oligosaccharides some of which are increased more than 100 times normal. The structures (Table 2) are closely related and possibly many derive from a common precursor which could resemble compound X. Significantly, in molar terms, more than 80% of the sialyl linkages are of the α 2→6 configuration. It is believed that these SOS originate as the oligosaccharide side chains of glycoproteins, although only compounds II, III, V and IX (which contain α2→6 sialyl linkages) have been demonstrated in glycoproteins [29]. Their appearance in urine as amino acid-free oligosaccharides containing N-acetyl-glycosamine at the reducing end strongly implies the action of an endo-N-acetylglucosaminidase which cleaves the side chain from the glycoprotein at a chitobiose (N-acetyl-glucosaminyl-N-acetylglycosamine) linkage. Although such endo-β-N-acetyl-glucosaminidases have been shown to exist in mammalian tissues [30, 31] none have been found as yet which act upon glycoproteins containing sialyloligosaccharide side chains as complex as those excreted in sialidosis [30] (Fig. 2).

The metabolic block appears to involve the first step in sialyloligosaccharide degradation, catalyzed by lisosomal neuraminidase (Fig. 2).

The relationship between lipofuscin accumulation in brain, described in patients with Type 1 [22] and Type 2 [27] sialidosis, and the neuraminidase deficiency remains unclear. Although it bears a histochemical resemblance, it may differ from the lipofuscin which is wear and tear pigment that increases nonspecifically in many cerebral degenerative disorders. The magnitude of its accumulation in sialidosis may imply a more specific abnormality. Biochemical analysis of SOS in different brain regions has been carried out in a Type 1 patient [26]; the pattern was similar to that in urine except for a higher proportion of less polar SOS (Fig. 3). The relationship of the accumulation of SOS in brain and myoclonus is not clear.

Table 2.2 Urinary sialyloligosaccharides in Sialidosis

Structure	Relative Excretion (mg/L) Normal	Sialidosis
I. αNANA-(2→3)-β-Gal-(1→4)-β-GlcNAc-(1→2)-α-Man-(1→3)-β-Man-(1→4)-GlcNAc	0.1-0.3	4, 16
II. αNANA-(2→6)-β-Gal-(1→4)-β-GlcNAc-(1→2)-α-Man-(1→3)-β-Man-(1→4)-GlcNAc	0.2-0.5	140, 145
III. αNANA-(2→6)-β-Gal-(1→4)-β-GlcNAc-(1→2)-α-Man-(1→3) α-Man-(1→6) >β-Man-(1→4)-GlcNAc	0.1-0.3	14, 22
IV. αNANA-(2→3)-β-Gal-(1→4)-β-GlcNAc-(1→2)-α-Man-(1→3) β-Gal-(1→4)-β-GlcNAc-(1→2)-α-Man-(1→6) >β-Man-(1→4)-GlcNAc	0.1	5.8, 8
V. αNANA-(2→6)-β-Gal-(1→4)-β-GlcNAc-(1→2)-α-Man-(1→3) β-Gal-(1→4)-β-GlcNAc-(1→2)-α-Man-(1→6) >β-Man-(1→4)-GlcNAc	0.1-0.5	20, 18
VI. αNANA-(2→6)-β-Gal-(1→4)-β-GlcNAc-(1→2) αNANA-(2→3)-β-Gal-(1→4)-β-GlcNAc-(1→4) >α-Man-(1→3)-β-Man-(1→4)-GlcNAc	0.1	8, 16
VII. αNANA-(2→3)-β-Gal-(1→4)-β-GlcNAc-(1→2)-α-Man-(1→3) αNANA-(2→3)-β-Gal-(1→4)-β-GlcNAc-(1→2)-α-Man-(1→6) >β-Man-(1→4)-GlcNAc	0.1-0.5	2.5, 5
VIII. αNANA-(2→3)-β-Gal-(1→4)-β-GlcNAc-(1→2)-α-Man-(1→3) αNANA-(2→6)-β-Gal-(1→4)-β-GlcNAc-(1→2)-α-Man-(1→6) >β-Man-(1→4)-GlcNAc	0.1-0.5	5, 5
IX. αNANA-(2→6)-β-Gal-(1→4)-β-GlcNAc-(1→2)-α-Man-(1→3) αNANA-(2→6)-β-Gal-(2→4)-β-GlcNAc-(1→2)-α-Man-(1→6) >β-Man-(1→4)-GlcNAc	0.2-0.4	110, 130
X. αNANA-(2→3)-β-Gal-(1→4)-β-GlcNAc-(1→4) αNANA-(2→6)-β-Gal-(1→4)-β-GlcNAc-(1→2) >α-Man-(1→3) αNANA-(2→6)-β-Gal-(1→4)-β-GlcNAc-(1→2)-α-Man-(1→6) >β-Man-(1→4)-GlcNAc	0.1-0.5	10, 12

Note: Data from Strecker et al. [24] and Durand et al. [5].

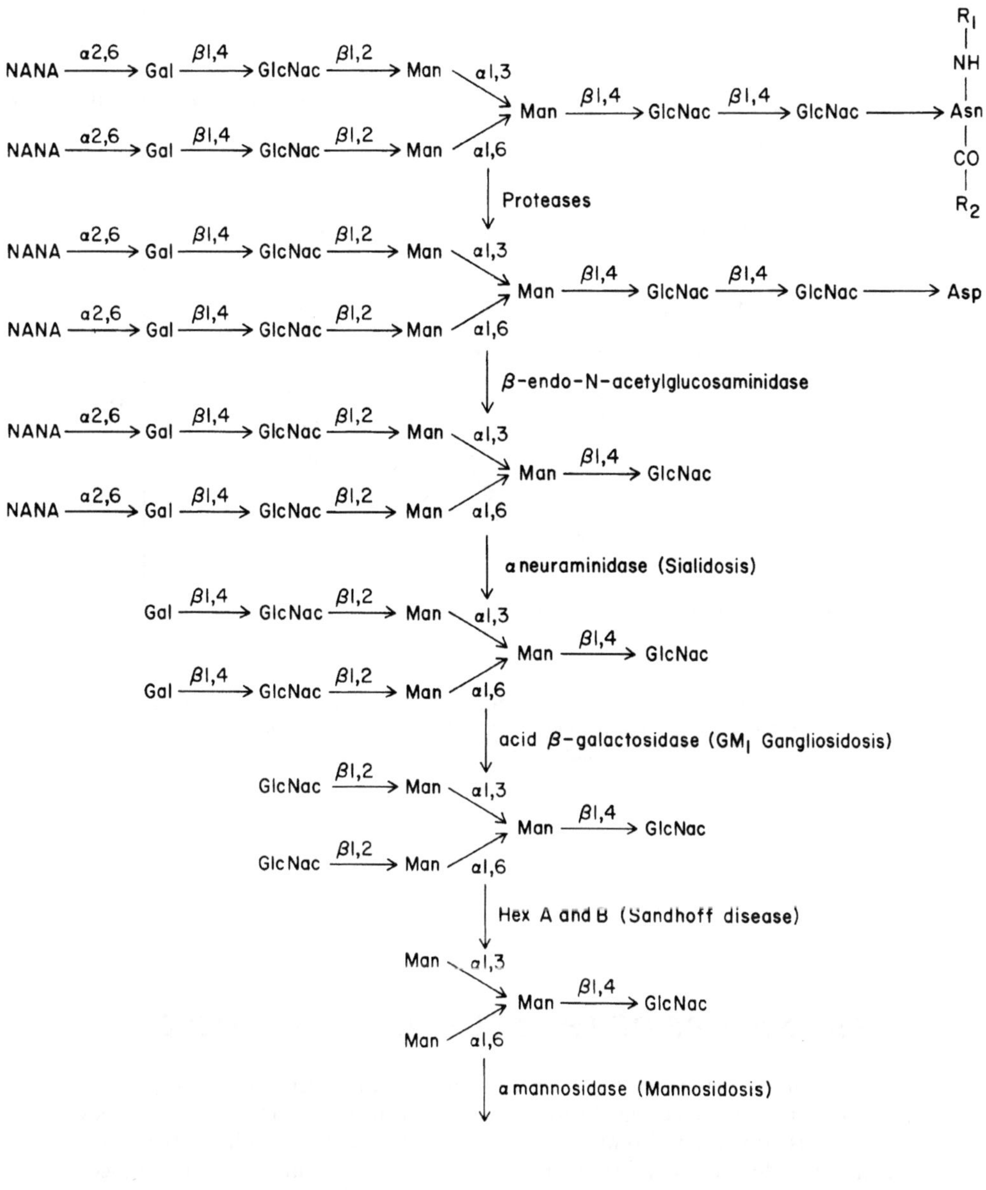

Fig. 2.2 Scheme for degradation of asparaginyl linked oligosaccharide in glycoproteins. Diseases resulting from block at each step are given.

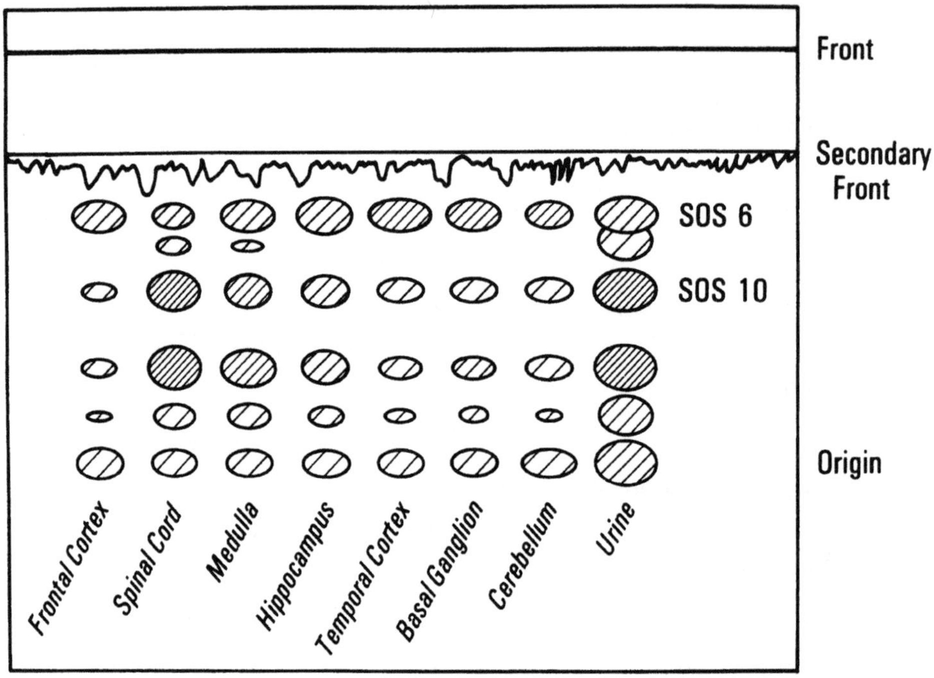

Fig. 2.3 Schematic diagram of sialyloligosaccharide (SOS) patterns in sialidosis Type 1 run on silica gel plates and stained with resorcinol as described [50]. SOS-6-hexasaccharide; SOS-10-decasaccharide.

2.5 THE NATURE OF THE NEURAMINIDASE DEFECT

Neuraminidase deficiency has been demonstrated in over two dozen patients with sialidosis. The defect has been demonstrated in fibroblasts and in leucocytes using various substrates. Comparisons of one assay method with another are difficult since different conditions have been used. In the majority of instances assays are performed on whole homogenates prepared by brief sonication of the cells; prolonged sonication causes marked losses of activity [32], indicating a particulate or multimeric enzyme.

Affected patients have low levels of activity using sialyloligosaccharides (0-10% of control) and normal activity when gangliosides are used as substrates. Residual activity is higher in the patients' extracts when sialic acid (NeuNAc), linked via an α2→3 glycosidic bond to the adjacent hexoside is used as substrate, compared to the α2→6 linked derivatives. Control fibroblasts also cleave the α2→3 linked derivatives at a higher rate than the α2→6 linked derivatives. O'Brien [25] demonstrated a profound deficiency using the disaccharide α-L-N-acetylneuraminosyl (2→6) N-acetylgalactosaminitol as substrate, indicating that the neuraminidase in question cleaves this linkage which is a constituent of mucous glycoproteins, as well as the more common linkage to galactose, which is frequent in serum glycoproteins. Both O'Brien [25] and Strecker [24, 33] demonstrated that the mutant cells fail to cleave compound II, a naturally excreted sialylhexasaccharide with α2→6 linked NeuNAc.

Strecker et al. [24] demonstrated that over 80% of the NeuNAc linkages in the urinary SOS from affected patients are of the α2→6 type, and suggested that at least two neuraminidases are present in normal fibroblasts, one hydrolyzing α2→6 linkages and the other hydrolyzing α2→6 linkages, with sialidosis patients having a normal activity for the latter. This explanation does not appear to be tenable for the following reasons. A striking reduction of activity for α2→3 linked substrates has been demonstrated in sialidosis patients indicating that both α2→3 and α2→6 linkages are cleaved by the same enzyme. Further, parents of sialidosis patients have half-normal activity when both pure α2→3 and α2→6 neuraminlactose are used as substrates [25, 26]. The most tenable explanation at present is the one gene: one enzyme: many substrate model proposed for β-galactosidase mutants [34]. The neuraminidase which is affected by the mutation normally cleaves both α2→3 and α2→6 linkages but the mutant enzyme retains higher residual activity for the former, giving rise to preferential storage and excretion of α2→6 linked oligosaccharides.

It appears most likely that the mutations which give rise to the different types of sialidosis are allelic. The different phenotypes may be explained by a more extreme deficiency of neuraminidase in the dysmorphic group [25, 26]. Predictably the dysmorphic group will excrete and store larger quantities of oligosaccharides (not yet demonstrated).

Gangliosides G_{M3} and G_{D1a} which contain only α2→3 linked NeuNAc are apparently hydrolyzed by a different enzyme from that which cleaves oligosaccharides.

Thus, it is clear that there are at least two neuraminidase isoenzymes, one active on sialyloligosaccharides and the other active on gangliosides.

The subcellular distribution and substrate specificity of neuraminidases has been studied extensively. In liver four separate subcellular species of neuraminidase have been described. The enzyme is found in lysosomes [35, 36], plasma membrane [37], Golgi membranes [38] and in cytosol [39] and all forms hydrolyze α2→3 neuraminlactose. Only the lysosomal and plasma membrane enzymes hydrolyze gangliosides. Although the pH optima for most of these neuraminidase is 4.1 - 4.4, the soluble enzyme has a pH optimum of 5.5 [39] and appears to be the only form in human serum [40].

Few attempts have been made to separate human neuraminidases into different molecular forms. One contributing factor is the difficulty in solubilizing the enzyme [41]. Tuppy and Palese [42] however solubilized pig kidney particulate neuraminidase and found an active band of enzyme migrating on polyacrylamide gels. Venerando et al. [43] have separated two active species of neuraminidase from the cytosol of pig brain. These enzymes differed slightly in their pH optima and Km's. Further studies of the different molecular forms of neuraminidase are necessary in order to clarify the relationship of each of them to the defect(s) in sialidosis.

2.6 EFFECT OF THE NEURAMINIDASE DEFECT ON OTHER LYSOSOMAL HYDROLASES

Patients with sialidosis have increased concentrations of bound sialic acid [4, 9]. Some of the bound neuraminic acid occurs in sialyloligosaccharides as has been found in fibroblasts, from a Type 2 patient [11] but some is also bound to glycoproteins, including lysosomal hydrolases. Using fibroblasts from a patient with Type 1 sialidosis, Swallow et al. [44] have demonstrated that several lysosomal hydrolases and adenosine deaminase have altered elctrophoretic mobilities (more electronegative). The charge abnormality is normalized by treatment with neuraminidase. This observation is similar to that of Vladutiu and Rattazzi [45] and Champion and Shows [46] for lysosomal hydrolases in I-cell disease. A major difference however between sialidosis and I-cell disease is that lysosomal hydrolases are not elevated in serum [8] or media [7] in sialidosis as they are in I-cell disease or in mucolipidosis III. Further, treatment of I-cell disease β-hexosaminidase with neuraminidase does not correct the uptake defect even though the abnormality of charge is corrected [47] indicating that neuraminidase deficiency is not the primary abnormality in I-cell disease but is a secondary defect. In keeping with this proposal is the finding that parents of I-cell patients have normal neuraminidase activities [48].

A. β-galactosidase deficiency

In several sialidosis patients (most with Type 2) a deficiency of acid β-galactosidase has been described using G_{M1} ganglioside, 4-methylumbelliferyl-β-D-galactopyranoside and p-nitrophenyl-β-D-galactopyranoside as substrates. This deficiency led to an initial diagnosis of G_{M1} gangliosidosis in at least three sialidosis patients and has been used to identify many probable cases of sialidosis tabulated in table 1 [11]. The β-galactosidase deficiency may be profound in serum, leucocytes and fibroblasts but is often not present in all tissues studied.

The mechanism responsible for the β-galactosidase deficiency is unclear and it is not certain that patients with combined neuraminidase and β-galactosidase deficiency have the same genetic defect as those with neuraminidase deficiency alone. It is clear that β-galactosidase is normal in these patients' parents indicating that the locus for β-galactosidase is unaffected [23]. Further research on this subject is required to clarify the molecular defect in such patients. Temporarily, they are included in the classification given in table 1.

2.7 NEURAMINIDASE DEFICIENCY IN OTHER DISORDERS

I-cell disease (Mucopolidosis II) and pseudo-Hurler polydystrophy (mucopolidosis III) are closely related disorders in which multiple lysosomal hydrolase deficiencies have been described in fibroblasts and elevated activities are found in media and serum. These changes are not seen in sialidosis.

Thomas et al. [48] found no measurable neuraminidase activity in fibroblasts from three patients with I-cell disease; however, parents of these children had normal fibroblast neuraminidase indicating that the neuraminidase deficiency is not the primary defect in I-cell disease. Strecker et al. [33] confirmed the neuraminidase deficiency in leucocytes from two I-cell disease patients but did not study the obligate heterozygotes. Leucocytes from I-cell disease patients usually have low levels (30-50% of control) of multiple lysosomal hydrolases but enzyme activity is rarely absent. The enzymes from I-cells appear to lack a recognition factor (6-phosphoryl-mannose) which is necessary for packaging within lysosomes (see discussion, Chapters 1 and 8). It is important to include several enzymes when assaying neuraminidase to rule out I-cell disease and pseudo-Hurler polydystrophy.

2.8 METHODS FOR DIAGNOSIS

A useful screening method for sialidosis is thin layer chromatography of urinary SOS. The method and its application are described in figure 1. Plates are stained for total oligosaccharides with orcinol and specifically for SOS with resorcinol. Differentiation of fucosidosis, mannosidosis and G_{M1} gangliosidosis from sialidosis in readly achieved using the appropriate standards. Thus far, I have screened 10 suspected patients with sialidosis and made a provisional diagnosis in each which was verified by enzyme assay. Patients with I-cell disease excrete similar SOS but in quantities which are about 10 times less than those in sialidosis [24].

Table 2.3 4MU-Neuraminidase fibroblast assay.

Substrate	4-methylumbelliferyl α(2–7)-N-acetylneuraminic acid (4MUN) (synthetic: Warner T.G. and O'Brien J.S.) ammonium salt mol. wt. 468.4 — Harvest fibroblasts by scraping after rinsing monolayer with 1 rinse of isotonic saline. — Centrifuge cells at 1000×g in isotonic saline and decant saline. — Suspend pellet in approximately 2 volumes distilled water and sonicate 1 second. — Proteins should be 2–6 mg/ml.
Assay	10 µl enzyme: 10 µl 4.0 mM 4MUN in water; 10 µl 0.5 M Na acetate buffer, pH 4.3 — Incubate 15-30 min at 37°C and — stop reaction with 1 ml glycine-carbonate buffer, pH 10 0.085 M and — read tubes on fluorimeter. For fibroblast V max is 90 n moles/mg/hr with range of 40-200. Caution: Enzyme loses activity on freezing, after sonication for more than 3 seconds or upon heating at 37°C. Substrate blank should be 0.5% in glycine-carbonate buffer. If higher, reduce substrate by extracting substrate in water (4.0 mM solution) with 1 volume of freshly distilled ethyl acetate for 3 extractions. Discard ethyl acetate phase. Note - if ethyl acetate contains even traces of acetic acid and water it will hydrolyze 4MUN.

Warner and O'Brien [32].

Neuraminidase assays have thus far been carried out using fibroblasts and leucocytes. Activity in the latter is approximately 10 times lower than in fibroblasts, necessiting the use of a more sensitive assay. Recently a fluorescent derivative, 4-methylumbelliferyl-β-N-acetylneuraminic acid (4MU-NANA) has become available for diagnosis and heterozygote detection. Using this substrate, O'Brien and Warner [32] have demonstrated higher residual activity in Type 1 patients than in Type 2. Furthermore, an altered Km (5 times elevated) was found in Italian patients with Type 1 sialidosis, but not in non-Italians [26]. Although preliminary, the data indicate at least 2 different mutations in Type 1 sialidosis. Finally, prenatal diagnosis has been successfully accomplished in several families using 4MU-NANA as substrate, a feat not possible with less sensitive substrates [18, 49].

At present I recommend urinary screening of suspected cases for SOS using a random sample (containing preservative) analyzing SOS by thin-layer chromatography (Fig. 1). Patients with I-cell disease and pseudo-Hurler polydystrophy also excrete SOS in urine giving a pattern similar to sialidosis but with about 10 times lower quantities of SOS [32]. Patients with aspartylglucosaminuria also excrete SOS but the compounds excreted have an Rf different from those in sialidosis [51, 52]; the major sialylated AGU compounds migrate just behind sialyllactose [52]. Fibroblast or leucocyte neuraminidase assay of probands and parents can then be carried using the 4MU-NANA assay to achieve diagnosis. Additional lysosomal enzymes such as β-galactosidase should be assayed to rule out I-cell disease and pseudo-Hurler polydystrophy using fibroblasts and serum. The conditions for assay in my laboratory are described in table 3. Important points include short sonication and assays of fresh sonicates, since the enzyme is unstable to prolonged sonication and to freezing [32]. The enzyme activity in fibroblasts has a single pH optimum with a peak at pH 4.3 [32]. Patients with sialidosis have a deficiency of activity at all pH values.

REFERENCES

1. Cantz M., Gehler J., Spranger J.: *Mucolipidosis I: increased sialic acid content and deficiency of a α-N-acetylneuraminidase in cultured fibroblasts.* Biochem. Biophys. Res. Commun. *74*, 732-738, 1977.
2. Spranger J., Wiedemann H.R., Tolksdorf M., Graucob E., Caesar R.: *Lipomucopolysaccharidose.* Z. Kinderheilkd. *103*, 285-290, 1968.
3. Spranger J., Gehler J., Cantz M.: *Mucolipidosis I: a sialidosis.* Amer. J. Med. Genet. *1*, 21-29, 1977.
4. Spranger J., Wiedemann H.: *The genetic mucolipidoses.* Hum. Genet. *9*, 113-139, 1970.
5. Durand P., Gatti R., Cavalieri S., Borrone C., Tondeur M., Michalski J.-C., Strecker G.: *Sialidosis (Mucolipidosis I).* Helv. Paediat. Acta *32*, 391-400, 1977.
6. O'Brien J.S.: *Neuraminidase deficiency in the cherry-red spot-myoclonus syndrome.* Biochem. Bhiophys. Res. Comun. *79*, 1136-1141, 1977.
7. Thomas G.H., Tipton R.E., Chien L.T., Reynolds L.W., Miller C.S.: *Sialidase (α-N-acetylneuraminidase) deficiency: the enzyme defect in an adult with macular cherry-red spots and myoclonus without dementia.* Clin. Genet. *13*, 369-379, 1978.
8. Rapin I., Goldfischer S., Katzman R., Engel J., O'Brien J.S.: *The cherry-red spot-myoclonus syndrome.* Ann. Neurol. *3*, 234-242, 1978.
9. Kelly T.E., Graetz G.: *Isolated acid neuraminidase deficiency: a distinct lysosomal storage disease.* Amer. J. Med. Genet. *1*,31-46, 1977.
10. Wynburn-Mason R.: *On some anomalous forms of amaurotic idiocy and their bearing on the relationship of the various types.* Br. J. Ophthalmol. *27*, 193-207, 1943.

11. Lowden J.A., O'Brien J.S.: *Sialidosis: A review of human neuraminidase deficiency.* Am. J. Hum. Genet. *31*, 1-18, 1979.
12. Johnson W.G., Thomas G.H., Miranda A.F., Driscoll J.M., Wigger J.H., Yeh M.N., Schwartz R.C., Cohen C.S., Berdon W.E., Koenigsberger M.R.: *Congenital Sialidosis: Biochemical studies, clinical spectrum in four sibs, two successful prenatal diagnoses.* Am. J. Hum. Genet. *32*, 43A, 1980.
13. Lowden J.A.: *Personal communication*, 1980.
14. Maroteaux P., Humbel R., Strecker G., Michalski J.C., Mande R.: *Un nouveau type de sialidose avec atteinte renale: la nephrosialidose.* Arch. Franç. Pediatr. *35*, 819-844, 1978.
15. Goldstein M.L., Kolodny E.H., Gascon G.C., Gilles F.H.: *Macular cherry-red spot, myoclonic epilepsy and neuro visceral storage in a 17-year-old girl.* Trans. Am. Neurol. Assoc. *99*, 110-112, 1974.
16. Di Donato S.: *Personal communication*, 1980.
17. Rapin I.: *Personal communication*, 1980.
18. Steinman L., Tharp B.R., Dorfman L.J., Forno L.S., Sogg R.L., Kelts K.A., O'Brien J.S.: *Peripheral neuropathy in the cherry-red spot-myoclonus syndrome (Sialidosis Type I).* Ann. Neurol. *7 (5)*, 450-456, 1980.
19. Engl J.Jr, Rapin I., Giblin D.R.: *Electrophysiological studies in two patients with cherry-red spot-myoclonus syndrome.* Epilepsia *18*, 73-87, 1977.
20. Goldberg M.F., Cotlier E., Fichensher L.G., Kenyon K., Enat R., Borowsky S.A.: *Macular cherry-red spot, corneal clouding and β-galactosidase deficiency.* Arch. Int. Med. *128*, 387-398, 1971.
21. Justice R.M., Wenger D.A., Naidu S., Rosenthal I.M.: *Enzymatic studies in a new variant of G_{M1} gangliosidosis in an older child.* Pediatr. Res. *11*, 407, 1977.
22. Gonatas N.K., Terry R.D., Winkler R.: *A case of juvenile lipidosis: the significance of electron microscopic and biochemical observations of a cerebral biopsy.* J. Neuropath. Exp. Neurol. *22*, 557-580, 1963.
23. Wenger D.A., Tarby J.F., Wharton C.: *Macular cherry-red spots and a myclonus with dementia: co-existent neuraminidase and β-galactosidase deficiencies.* Biochem. Biophys. Res. Comm. *82*, 589-595, 1978.
24. Strecker G., Peers M.-C., Michalski J.-C.: *Structure of nine sialyloligosaccharides accumulated in urine of eleven patients with three different types of sialidosis: mucolipidosis II and two new types of mucolipidosis.* Eur. J. Biochem. *75*, 391-403, 1977.
25. O'Brien J.S.: *The cherry-red spot-myoclonus syndrome: a newly recognized inherited lysosomal storage disease due to acid neuraminidase deficiency.* Clin. Genet. *14*, 55-60, 1978.
26. O'Brien J.S.: *Enzymology of the Sialidoses.* In "Lysosomes and Lysosomal Storage Diseases" (J.W. Callahan and J.A. Lowden Eds.) Raven Press, New York, pp. 263-269, 1981.
27. Berard M., Toga M., Bernard R.: *Pathologic findings in one case of neuronal and mesenchymal storage disease.* Path. Sur. *3*, 172-183, 1968.
28. Suzuki Y., Nakamura N., Shimada Y.: *Macular cherry-red spots and β-galactosidase deficiency in an adult.* Arch. Neurol. *34*, 157-161, 1977.
29. Montreuil J.: *Recent data on the structure of the carbohydrate moiety of glycoproteins. Metabolic and biological implications.* Pure and Appl. Chem. *42*, 431-477, 1975.
30. Tarentino A.L., Maley F.: *Purification and properties of an endo-β-N-acetylglucosaminidase from hen oviduct.* J. Biol. Chem. *251*, 6537-6543, 1976.
31. Nishigaka M., Muramatsu T., Kobata A.: *Endoglycosidases acting on carbohydrate moieties of glycoproteins: demonstration in mammalian tissues.* Biochem. Biophys. Res. Comm. *59*, 638-645, 1974.
32. Warner T.G., O'Brien J.S.: *Synthesis of 2'(4-methylumbelliferyl)-α-D-N-acetylneuraminic acid and detection of skin fibroblast neuraminidase in normal humans and in sialidosis.* Biochemistry *13*, 2783-2787, 1979.
33. Strecker G., Michalski J.C., Montreuil J., Farriaux J.P.: *Defect in neuraminidase associated with mucolipidosis II (I-cell disease).* Biomed. Exp. *25*, 238-240, 1976.
34. O'Brien J.S.: *Molecular genetics of G_{M1} β-galactosidase.* Clin. Genet. *3*, 303-313, 1975.

35. Aronson N.N. Jr., Duve C. de: *Digestive activity of lysosomes. II. The digestion of macromolecular carbohydrates by extracts of rat liver lysosomes.* J. Biol. Chem. *243*, 4564-4573, 1968.
36. Horvat A., Touster O.: *On the lysosomal occurence and properties of the neuraminidase of rat liver and Ehrlich ascites tumor cells.* J. Biol. Chem. *243*, 4380-4390, 1968.
37. Schengrund C.-L., Jensen D.S., Rosenberg A.: *Localization of sialidase in the plasma membrane of cat liver cells.* J. Biol. Chem. *247*, 2742-2746, 1972.
38. Kishore G.S., Tulsiani D.R.P., Bhavanandan V.P, Carubelli R.: *Membrane bound neuraminidases of rat liver. Neuraminidase activity in Golgi apparatus.* J. Biol. Chem. *250*, 2655-2659, 1975.
39. Tulsiani D.R.P., Carubelli R.: *Studies on soluble and lysosomal neuraminidases of rat liver.* J. Biol. Chem. *245*, 1821-1827, 1970.
40. Shauer R., Veh R.W., Wember M.: *Demonstration of neuraminidase activity in human blood serum and human milk using a modified radioactive labelled* α_1*-glycoprotein as substrate.* Hoppe-Seylers Z. Physiol. Chem. *357*, 559-566, 1976.
41. Ohman R., Rosenberg A., Svennerholm L.: *Human brain sialidase.* Biochemistry *9*, 3774-3782, 1970.
42. Tuppy H., Palese P.: *A chromogenic substrate for the investigation of neuraminidase.* FEBS Lett. *3*, 72-75, 1969.
43. Venerando B., Tettamanti G., Cestaro B., Zambotti V.: *Studies on brain cytosol neuraminidase. Isolation and partial characterization of two forms of the enzyme from pig brain.* Biochim. Biophys. Acta *403*, 461-472, 1975.
44. Swallow D.M., O'Brien J.S., Hoogeveen A.T., Bucks D.W.: *Electrophoretic analysis of glycoprotein enzymes in the sialidoses and mucolipidoses.* Ann. Hum. Genet. *45*, 29-37, 1981.
45. Vladutiu G.D., Rattazzi M.C.: *Abnormal lysosomal hydrolases excreted by cultured fibroblasts in I-cell disease (mucolipidosis II).* Biochem. Biophys. Res. Commun. *67*, 956-964, 1975.
46. Champion M.J., Shows T.B.: *Electrophoretic abnormalities of lysosomal enzymes in mucolipidosis fibroblast lines.* Am. J. Hum. Genet. *29*, 149-163, 1977.
47. Vladutiu G.D., Rattazzi M.C.: *I-cell disease: impaired fibroblast uptake of β-hexosaminidase is not due to excess sialic acid.* Am. J. Hum. Genet. *29*, 110A, 1977.
48. Thomas G.H., Tiller G.E. Jr., Reynolds L.W.: *Increased levels of sialic acid associated with a sialidase deficiency in I-cell disease (mucolipidosis II) fibroblasts.* Biochem. Biophys. Res. Commun. *71*, 188-195, 1977.
49. Kleijer W.J., Hoogeveen A., Verheijen F.W., Niermeaijer M.F., Galijard A., O'Brien J.S., Warner T.G.: *Prenatal diagnosis of sialidosis with combined neuraminidase and β-galactosidase deficiency.* Clin. Genet. *16*, 60-61, 1979.
50. Holmes E.W., O'Brien J.S.: *Separation of glycoprotein-derived oligosaccharides by thin-layer chromatography.* Analyt. Biochem. *93*, 167-170, 1979.
51. Sugahara K., Akasaki M., Funakoshi I., Aula P., Yamashina I.: *Structural studies of glycoaspargines from urine of a patient with aspartylglycosylaminuria (AGU).* FEBS Lett. *78*, 284-286, 1977.
52. Warner T.G., O'Brien J.S.: *Unpublished data.*

Paolo Durand
Rosanna Gatti
Carla Borrone

3 Fucosidosis

NAME OF DISEASE	Fucosidosis
PREVALENCE	Italian people
MODE OF INHERITANCE	Autosomal recessive
MUTANT GENE PRODUCT	α-L-fucosidase
CHROMOSOMAL LOCATION	1p32-p34
EXPRESSION	
1. Altered DNA structure	?
2. Disturbed protein function	Small portion of α-L-fucosidase with altered kinetic properties
3. Disrupted cell and organ function	Accumulation of fucoglycoconjugates in lysosomes of most cells, including brain, leads to brain dysfunction and somatic changes

3.1 INTRODUCTION

Fucosidosis is transmitted as an autosomal recessive trait and is characterized by the accumulation of fucoglycoconjugates (fucoglycopeptides, fucooligosaccharides, fucosphingolipids and keratan sulfate) within lysosomes because of a deficiency in the lysosomal α-L-fucosidase.

Fucosidosis might also be classified as an inherited lipidosis because one of the storage products, H-isoantigen, is a pentaglycosylceramide but since there is evidence that the catabolism of chiefly tissue glycoproteins is also deranged it will be considered here as one of the main glycoproteinoses.

The first description of two brothers with the classical form of fucosidosis with a progressive and fatal course was in 1966 [1].

Studies in 1967 and 1969 established the pathological and ultrastructural description of this disease [2-4]. Contemporary studies on the tissues resulted in the identification of fucoglycoconjugates stored and in the recognition of the enzyme defect as absence of activity of the lysosomal hydrolase α-L-fucosidase [3-8].

While these results have been confirmed, distinctive clinical forms [9-11] and additional storage products [12-13] have been reported.

Very few reviews on fucosidosis have been published over the past years [14-17].

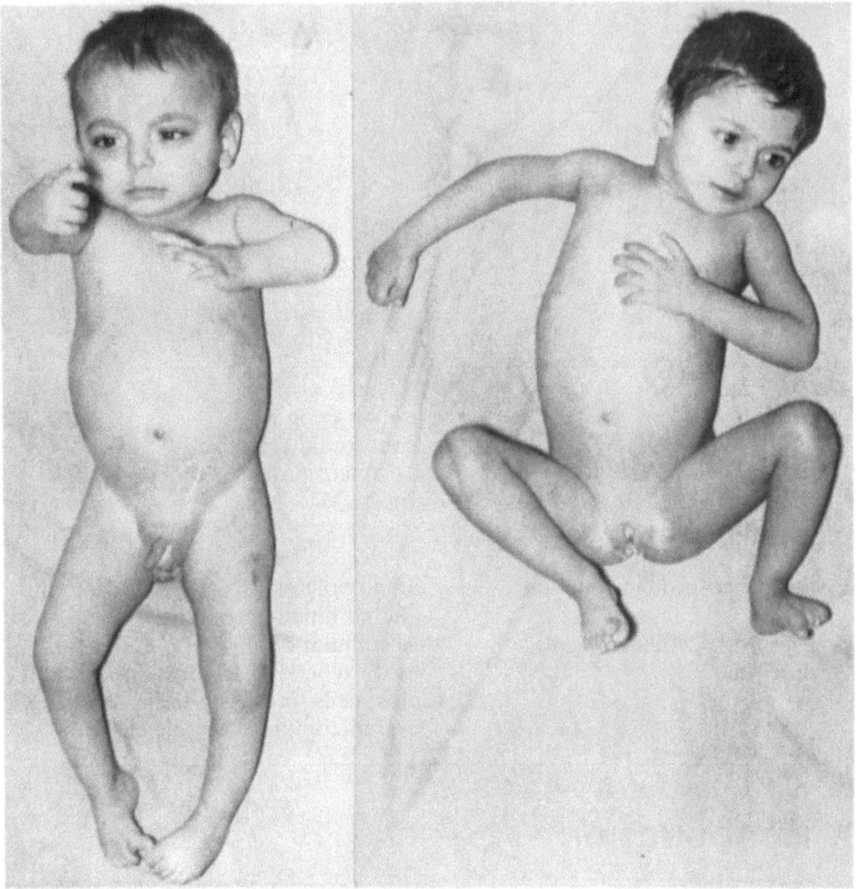

Fig. 3.1 a Fucosidosis type 1. The original two affected siblings at 2 (left) and 3 (right) years. Mental regression was already present [2, 3].

3.2 CLINICAL DESCRIPTION

There is a considerable degree of heterogeneity amongst patients with fucosidosis out of which it has so far been possible to delineate two clinical types - type 1 and type 2 - of the disease.

The differentiation is used today, although it will become clear that it is not always adequate.

A. Fucosidosis type 1 (most severe)

Patients with the type 1 of the disease [3, 4, 18-24] are normal at birth and during the first 5 to 8 months, then general muscular hypotonia appears along with repeated respiratory infections, feeding difficulties and failure to thrive.

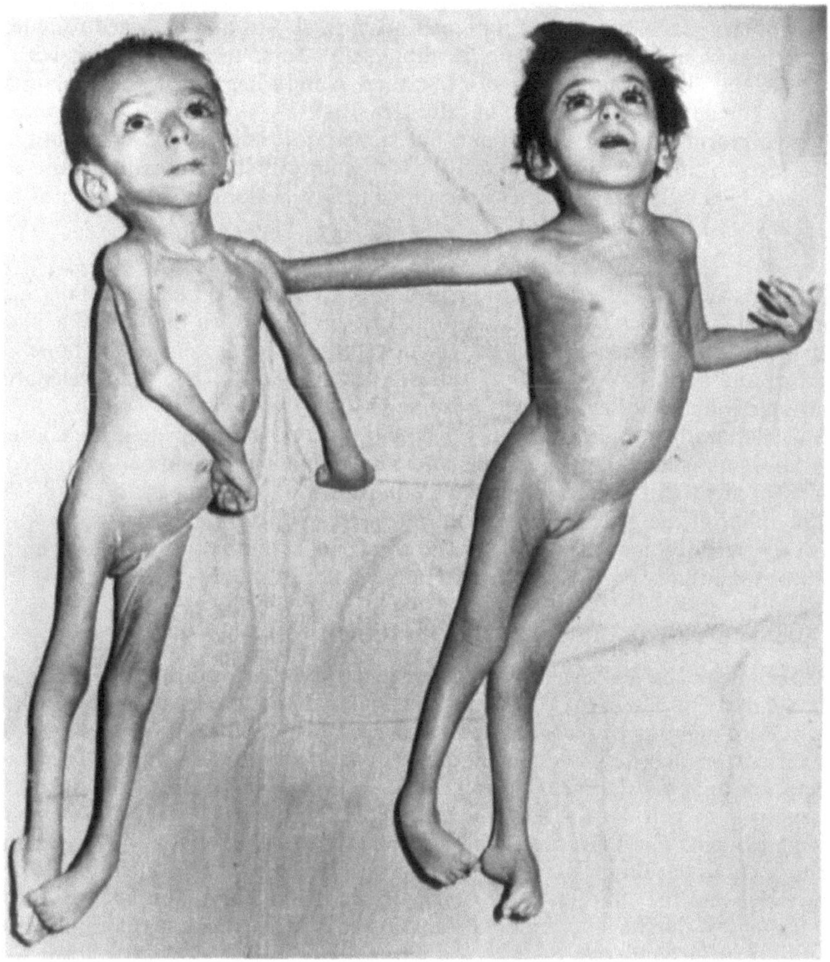

Fig. 3.1 b Fucosidosis type 1. The same patients at 3 1/2 and 4 1/2 years were both emaciated and showed a decerebrate rigidity [2, 3].

Subsequently the patient exhibits a slowing of motor and mental progress, followed, from the age of 1 or 2 years, by general deterioration of neurological function and health (Fig. 1a).

Neurological involvement, the hallmark of the disease, manifests itself as progressive muscle hypertonicity or spasticity. Seizures may occur early or during the later stages of the disease. Gradual mental deterioration usually progresses to dementia.

Deterioration continues to a vegetative state, followed by decerebrate rigidity and progressive emaciation (Fig. 1b). Death from pneumonia usually occurs before 6 years of age.

In addition patients may present slightly coarse facial features with prominent forehead, large face, broad and flattened nose, thick lips and enlarged tongue, thick skin with abundant sweating, cardiomegaly, dorsolumbar kyphosis with other skeletal abnormalities and short stature. A mild hepatosplenomegaly may be present. The cornea and lens are usually clear.

Bone abnormalities may be absent, but sometimes radiological examination shows signs of mild dysostosis multiplex with enlarged skull, dorsolumbar kyphosis, biconvexity, sometimes anterior/superior beaking of the vertebral bodies and delayed skeletal maturation [25, 26].

Laboratory data Vacuolated lymphocytes are found in blood smears. Lymphocytic vacuoles may be found only in a few cells with the toluidine stain after methyl-alcohol fixation. Foamy histiocytes may be present in the bone marrow, and are usually found in liver and other organs. These foam cells are slightly periodic acid Schiff (PAS) positive [3, 4].

Other laboratory findings are: normal acid mucopolysaccharide urine content, high Lewis antigens in both red cells and fluids, in some patients an increased salinity of sweat with return to normal values in the terminal period [3, 18, 19, 24] and inability of the gallbladder to concentrate dyes [3, 23].

The electroencephalogram usually shows irregular low activity waves. The cerebrospinal fluid proteins are generally normal but may be increased [23].

B. Fucosidosis type 2 (less severe, chronic)

Patients with the less severe type 2 classification of fucosidosis [9, 15, 20, 28-33] have a variety of clinical symptoms and usually survive longer than those with type 1. Although the onset of symptoms occurs early in life, the progression of this variant of fucosidosis is slow and allows survival into adulthood. The patients undergo normal development during their first 1 or 2 years after which unsteadiness of gait, ataxia, learning difficulties, emotional lability and dementia become evident. The first most common symptom is difficulty in walking usually due to a combination of leg weakness and unsteadiness. This difficulty is progressive and the child loses the ability to walk and stand, and later even to sit. Impairment of language and mental function is usually noted at the same time as the motor disturbances.

On physical examination, the main abnormalities, besides the ataxia, are muscle hypotonia followed by hypertonicity or spasticity, sometimes strabismus, dysarthria and swallowing disturbances. Coarsening of facial features occurs and becomes more evident during the course of the disease (Fig. 2). The patients suffer from recurrent respiratory infections and at the age of 3 or 4 years develop first telangectasic lesions and then purple pinhead sized raised skin lesions first on the scrotum and lower extremities, later over the abdomen, upper extremities and back; these have the characteristic appearance of angiokeratoma corporis diffusum (Fig. 3).

Hypohydrosis or anhydrosis develop synchronously with the skin lesions and sometimes the patients are unable to control body temperature and may suffer from seizures.

There is either minimal or no enlargement of the liver and spleen. The corneas and lens are clear but vascular thinness, tortuosities and aneurysms in the conjunctiva and in the fundus are found in some patients [18, 27, 34].

Cherry-red macular spot has only been described in one patient [15].

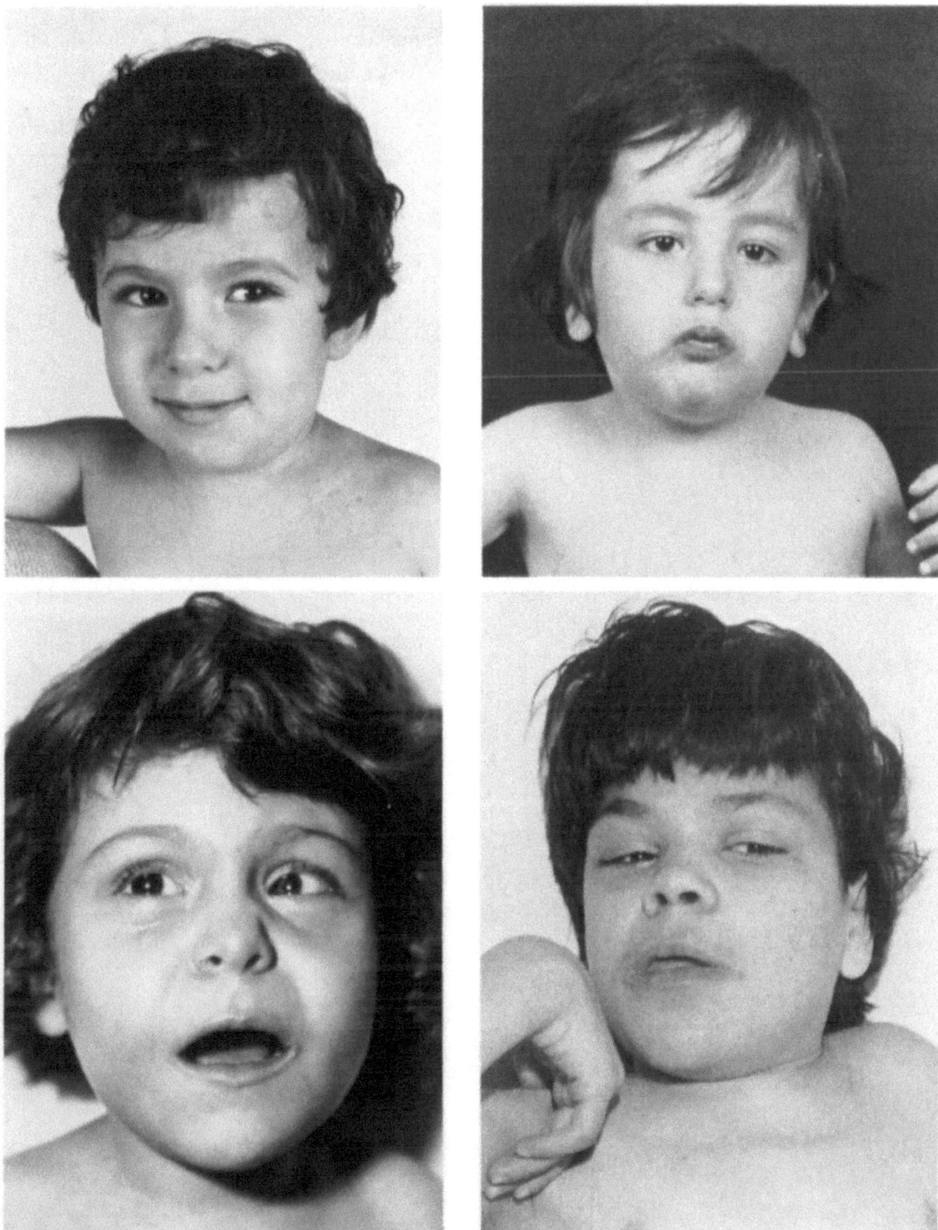

a b

c d

Fig. 3.2 Four cases with fucosidosis type 2 photographed at various ages: **a**, 3 years **b**, 4 years; **c**, 6 years and **d**, 17 years. Only in the two brothers of **c** and **d** coarse features and strabismus are evident.

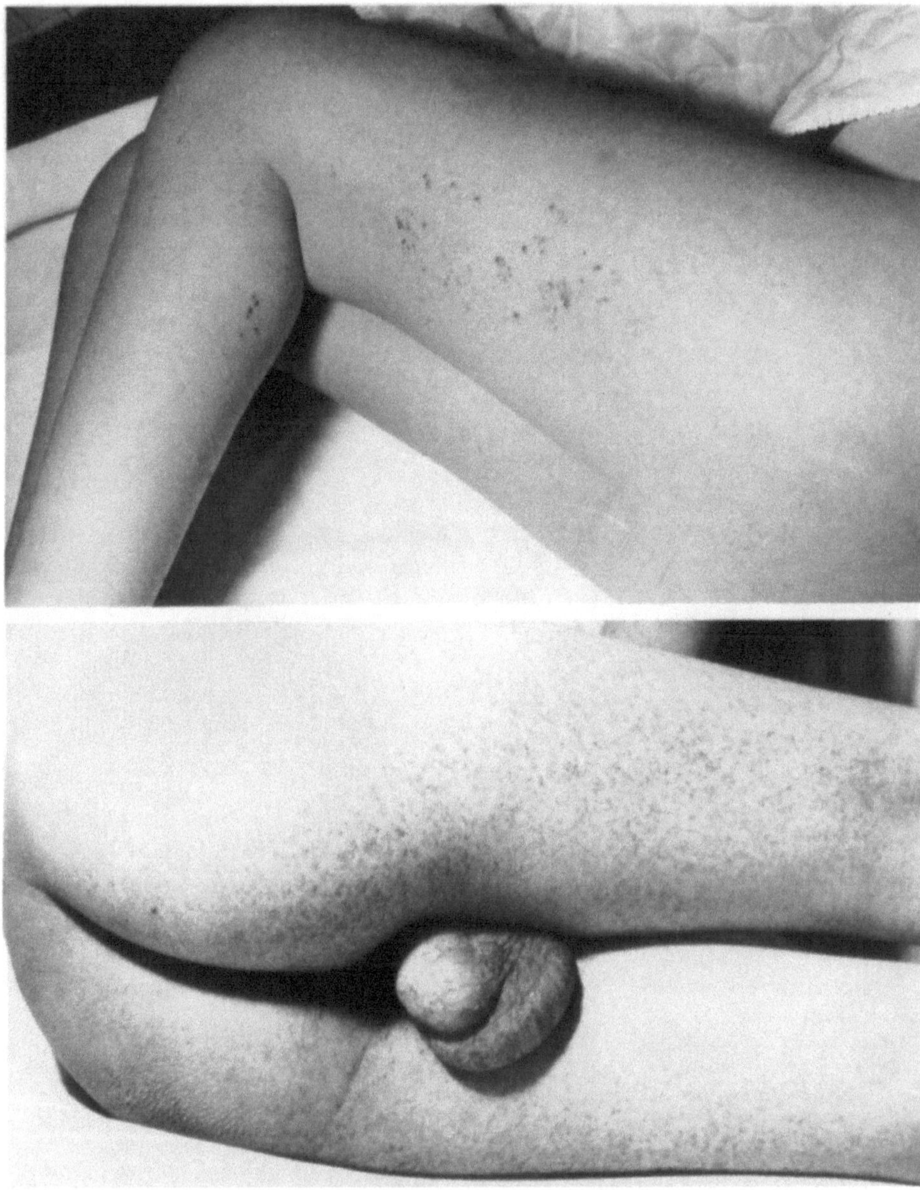

a

b

Fig. 3.3 **a**, clusters of individual punctate angiectasies in the superficial layers of the skin of the legs; **b**, angiokeratoma on the buttocks, scrotum and thighs [27].

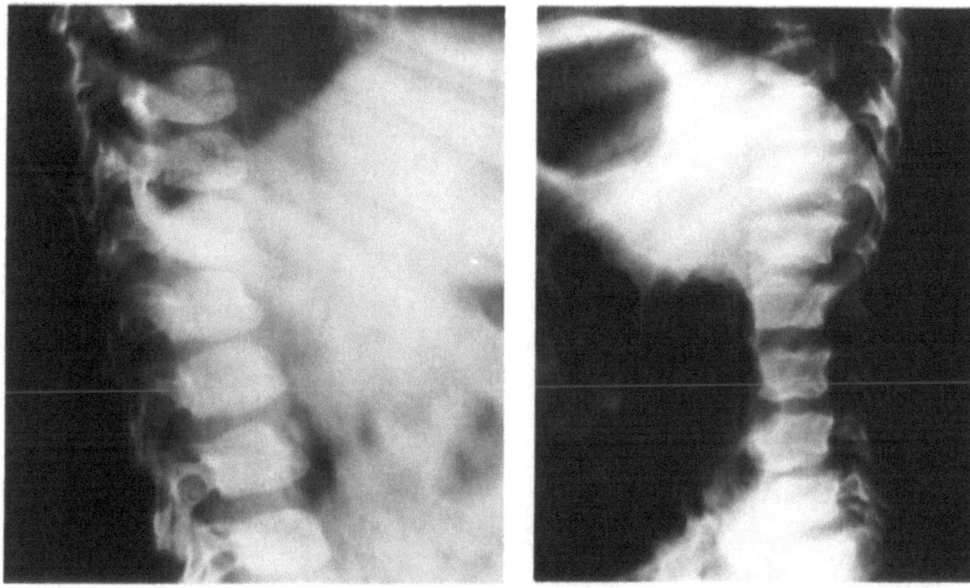

a b

Fig. 3.4 **a**, anterior inferior beaking of the dorso-lumbar vertebrae in a patient of 3 years with fucosidosis type 2; **b**, generalized flattening and irregularity of the margins of the vertebrae. Calcification of vertebral disc between D_{11} and D_{12} in a patient of 17 years with fucosidosis type 2 [27].

Linear growth is retarded and severe kyphoscoliosis with other skeletal abnormalities is usually present.

Radiological examination shows diploic thickening especially over the frontal and supra-orbital region of the skull, and sometimes premature synostosis of sutures. The frontal and sphenoidal sinuses may be undeveloped or hypoplastic.

The rib cage may be normal in contour but occasionally a pigeon breast deformity develops. Medial widening of clavicles, slight broadening and spatulation of the ribs and flattening of glenoids is present. The vertebral column shows flattening and anterior-inferior beaking of lumbar vertebrae and kyphosis at the thoracic-lumbar junction (Fig. 4). The pelvis and hips are initially normal but, with age, are mildly deformed with mild flaring of the iliac wings, deformity and increased obliquity of the acetabular cavities. Flattening of the femoral epiphyses and coxa valga occur with age (Fig. 5). The long bones of the upper extremities show only mild abnormalities. In the lower extremities, demineralization of all bones progresses with age and atrophy of the cortex of the lower end of the femur with expansion of the medullary cavity may be present.

Retardation of bone maturation is present.

Laboratory data Laboratory findings are similar to those found in fucosidosis type 1 but the chloride and sodium content of sweat is normal, except in a few cases [30, 31]. While the total mucopolysaccharide content of urine is

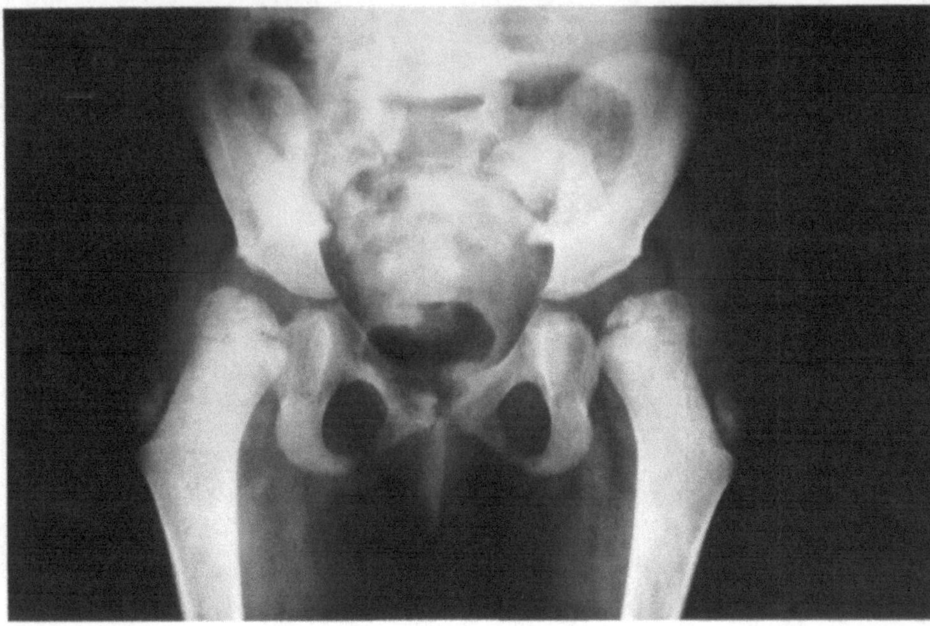

a

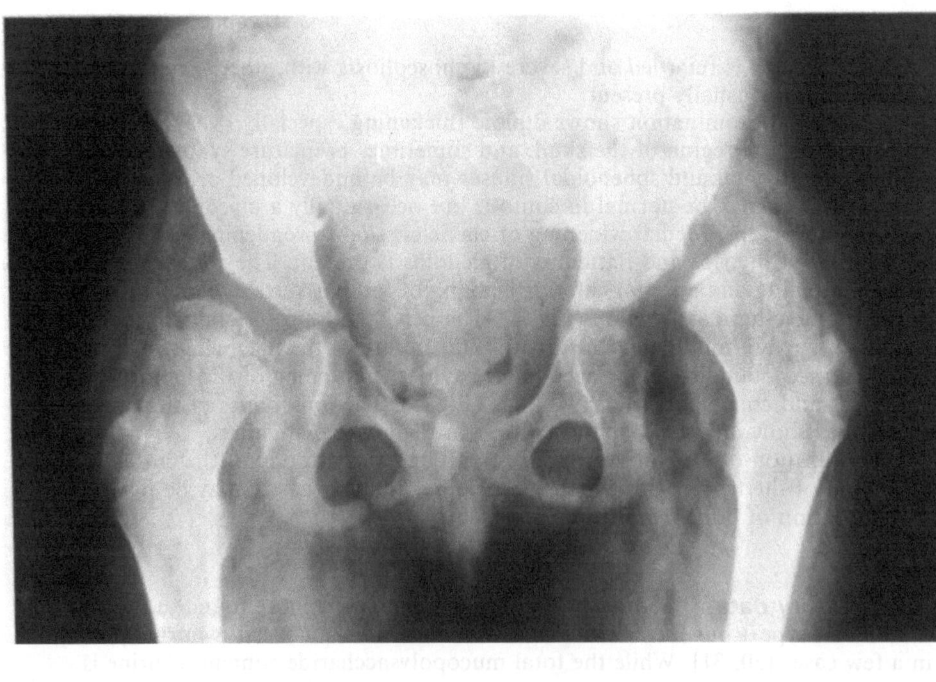

b

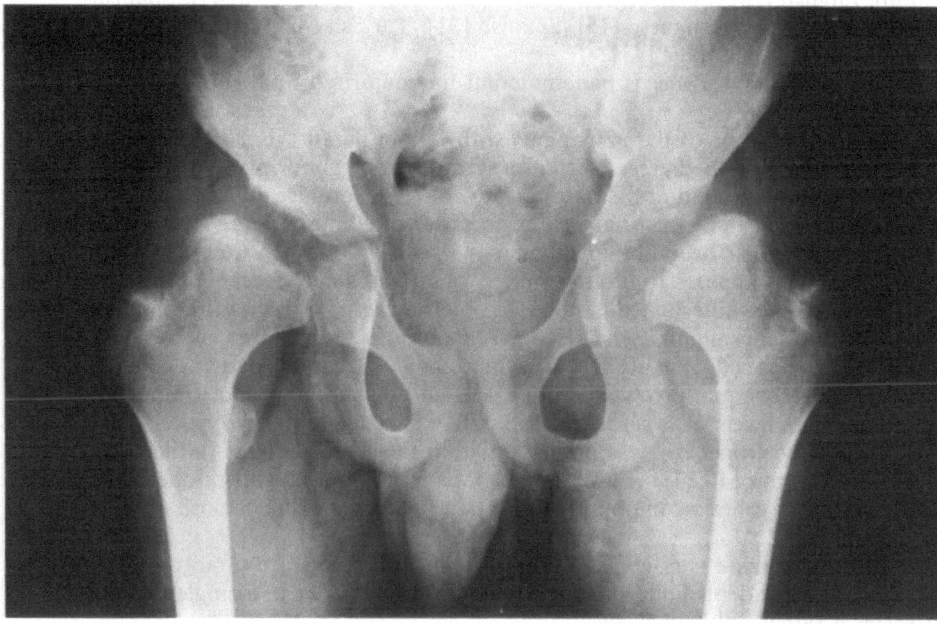

c

Fig. 3.5 **a**, **b**, pelvic and hip changes in the same patient at 3 and 5 years with fucosidosis type 2. Irregularly mineralized acetabular margins with increased slope of acetabular angles, flattened irregularly mineralized femoral heads and coxa valga; **c**, pelvic and hips changes in a patient of 17 years with fucosidosis type 2. Note shallow acetabular cavities with sclerotic roofs, iliac bodies constricted, femoral capital epiphyses sclerotic and destroyed, femoral necks enlarged and shortened.

usually normal, the separation of glycosaminoglycans on thin-layer silicagel reveals, in some cases, a certain quantity of a compound similar to keratansulfate [33, 35, 36].

C. Atypical cases

Single patients with atypical symptoms have been described: some with focal neurological symptoms [22], some with clinical abnormalities limited to the skeleton [37] or to angiokeratomatous rash [20]. Other unusual cases from a clinical and biochemical point of view have been reported [38, 39].

3.3 INCIDENCE AND GENETICS

Since the original description of fucosidosis in 1966 more than 60 cases have been reported in the world literature. Whereas 22 (some of them unpublished) have been of Italian background [3, 10, 21, 30, 33], other patients have been reported with German [18], German-Italian [29], Japanese [19], Mexico-American

[30], English [28], Belgian [21], U.S.A. [20], Dutch [22, 38, 40, 41], Spanish-American [32], Portuguese [15], Spanish [31], Turkish [42], Canadian-French [24] and Algerian parentage [43].

Therefore the disorder is panethnic but the majority of patients have had Italian ancestry.

In particular out of the 22 patients with Italian parentage, 8 have been traced back to two large pedigrees originating from two neighboring villages in the Southern Italian region of Calabria (Figs. 6, 7). A founder effect common to these two pedigrees can be postulated, but no proof of common ancestry between the two pedigrees has been obtained so far [45].

Fucosidosis appears to be transmitted as an autosomal recessive trait.

This mode of inheritance is suggested by the classic criteria of the high incidence of consanguinity between parents of affected children (nearly half the families concerned), the absence of symptoms in parents of patients, the approximately equal involvement of males and females and, most importantly, the intermediate levels of α-L-fucosidase activity in obligate heterozygotes [44, 45].

Clinical heterogeneity was observed within the same pedigree in the two examined by us [45,141] (Figs. 6, 7). Beratis et al. [46] have mentioned the possibility of genetic compounds to explain two types of fucosidosis in the same sibship.

At present there is no explanation for clinical heterogeneity, but the occurence of two independent mutations for the same rare disorder in a single pedigree seems a rather unlikely event.

Complementation studies after cell hybridization of fibroblasts from patients with type 1 and type 2 indicated that the two defects were not complementary; no restoration of enzyme activity was detected [47].

The different forms may be the result of quantitative differences in residual α-L-fucosidase activity or of the interaction of several independent variables as for instance the genetic polymorphism in blood group antigens, glycoproteins and glycolipids available as substrate for the grossly deficient enzyme.

Comparative biochemical studies of the different clinical types of fucosidosis to support this hypothesis must be performed.

The α-L-fucosidase (α-FUC) locus has been mapped to the short arm of chromosome 1 and is closely linked to the Rhesus blood group locus [48-54]. The mapping of genes on chromosome 1 is extensive and the small segment of the short arm of chromosome 1 has been mapped in detail. The α-FUC regional location is on band 32-band 34 of the short arm of chromosome 1 [52]. The Goss-Harris method of mapping which combines features of recombination study in families and synteny tests in hybrid cells when applied to chromosome 1 shows that the order of the loci is: Phosphoglucomutase-1 (PGM-1): Uridine monophosphate kinase (UMPK): Adenylate kinase-2 (AK-2), α-L-fucosidase (α-FUC): Enolase-1 (ENO-1) [53]. On the basis of a family segregating for elliptocytosis (E1) and 6-phosphogluconate dehydrogenase (PGD) as well as the common polymorphism Rh, PGM-1 and α-FUC, Cook et al. [54] concluded that the map of 1p is in the male 1pter-PGD-18%-E1-2%-Rh-2%-α-FUC-25%-PGM-1-centromere. In the female the above intervals are estimated to be 22, 4,2 and 37% respectively.

The precise nature of the molecular gene defect in fucosidosis is unknown. No cross reacting material to antibodies prepared against normal α-L-fucosidase was detected in patients [55, 56].

Recently, partial purification and characterization of the α-L-fucosidase from

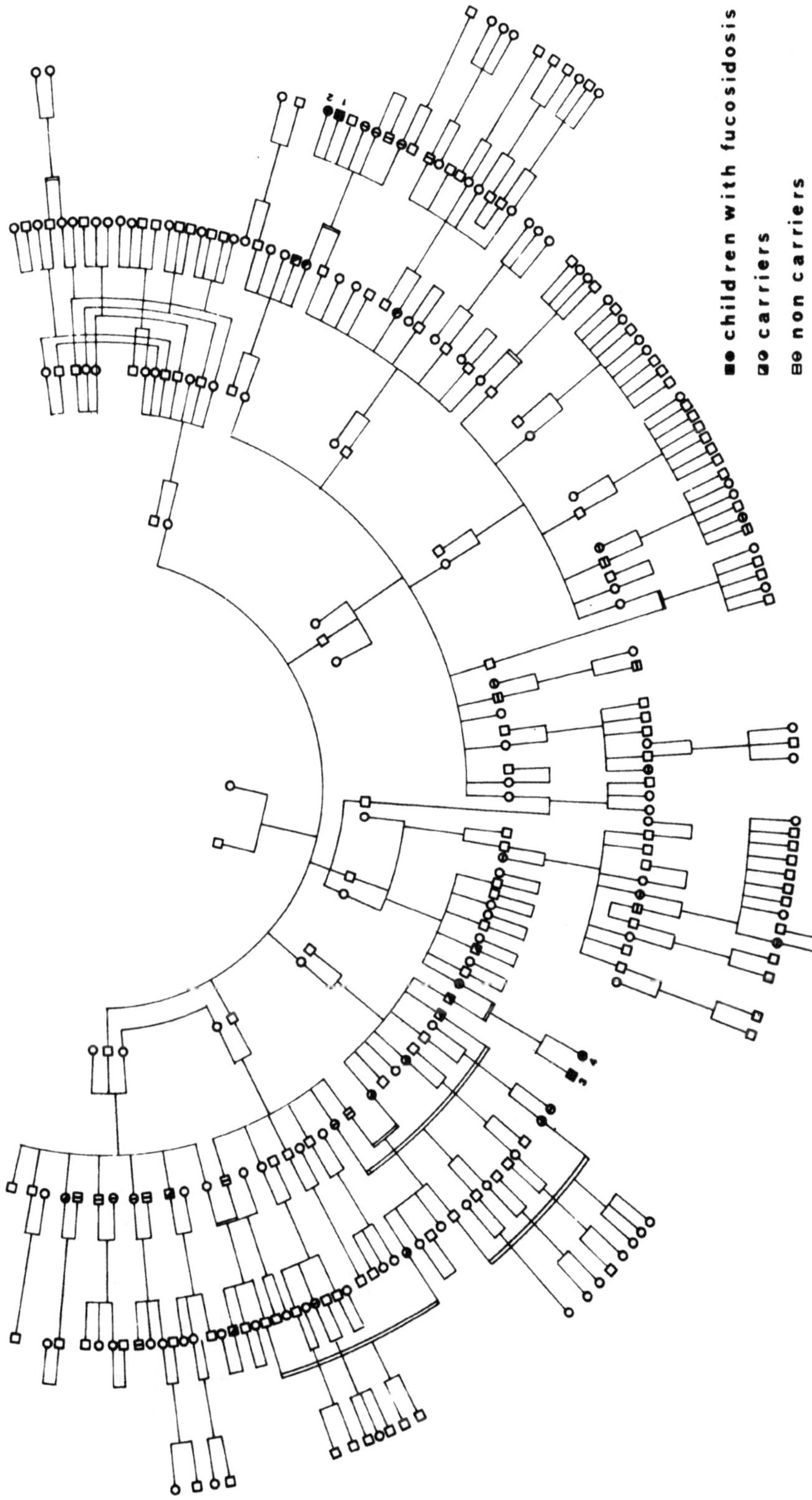

Fig. 3.6 Pedigree no. 1. Patients 1 and 2 are the two original cases of Fig. 1, affected with fucosidosis type 1. Patients 3 and 4 are the cases **c** and **d** of Fig. 2, affected with fucosidosis type 2.

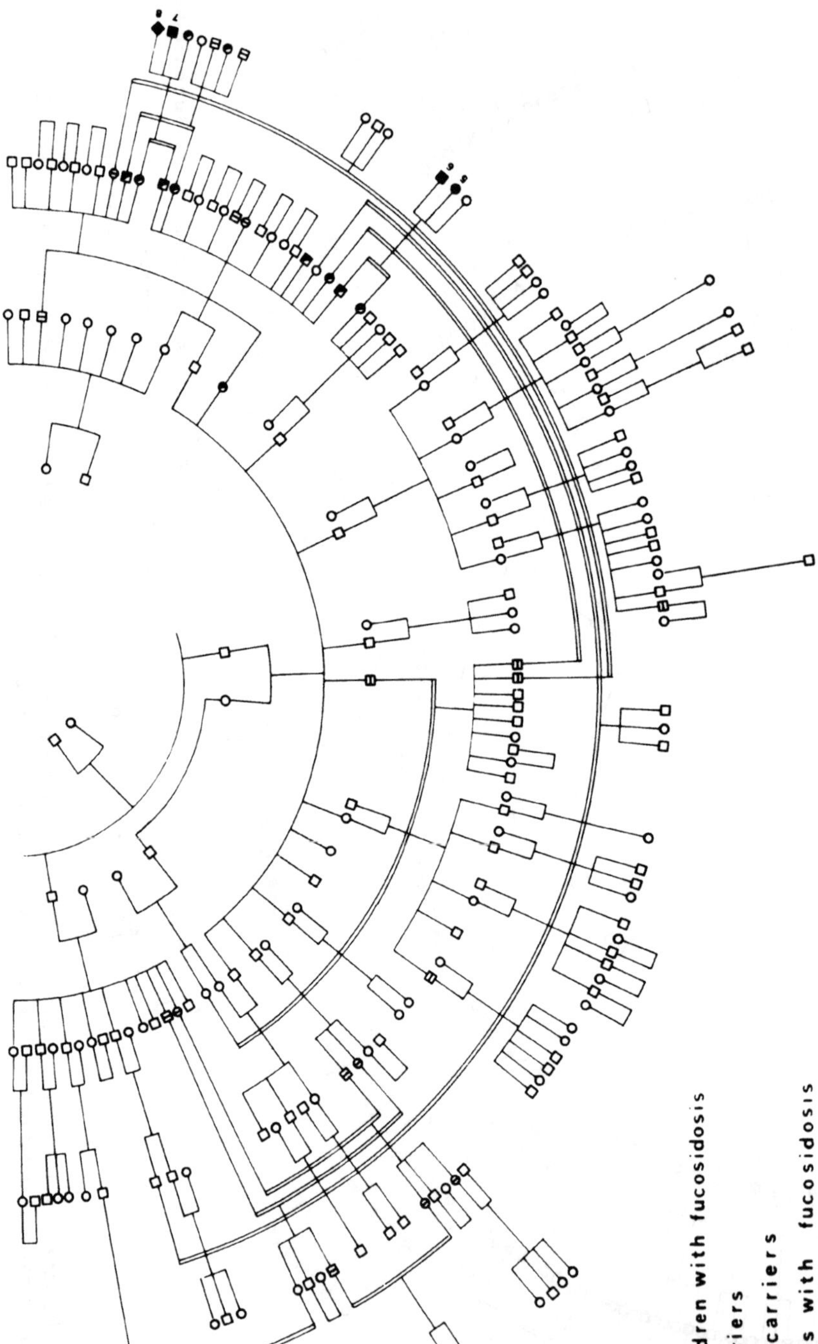

Fig. 3.7 Pedigree no. 2. The patient no. 7 is the patient **b** of Fig. 2, affected with fucosidosis type 2. The members no. 5 and 6 of this pedigree had the type 1 of the disease [4].

the liver of a fucosidosis patient revealed the retention of a selective, small portion of antigenically normal α-L-fucosidase with altered kinetic properties [16]. More detailed data about the present "state of the art" will be illustrated in the sections 3.7 and 3.9.

3.4 PATHOLOGY

The post mortem examinations, even if in a limited number, have furnished a reasonably clear picture [3, 57-60].

Biopsies of conjunctiva, skin, skeletal muscle, liver, brain, tonsils, nasal polips and rectum were performed and examined in many patients [2, 4, 9, 18, 19, 21-23, 28-33, 58, 60-63].

A. Autopsy findings

Gross examination shows mild abnormalities through almost every organ and tissue. The most involved are the brain, heart, lungs, liver, kidneys [3, 57-60].

The leptomeninges are thickened. The weight of the brain may be increased; no abnormality is apparent on external examination except for a yellow colour of the globus pallidus in some patients [57, 60]. The heart presents a cloudy degeneration of myocardium. The liver is enlarged and has a grayish colour. The gall-bladder was strawberry-like in three of the four cases examined [3, 60].

The lungs show diffuse emphysema with areas of atelectasis or edema. The kidneys present a cloudy degeneration.

B. Microscopic and ultrastructural changes

Abnormalities include the presence of large cells, distended with storage material. The parenchimal cells of all organs including exocrine and endocrine glands have this abnormality [3, 58-60]. Extensive electron microscopic studies have shown that the cells of all tissues examined are filled with single membrane limited inclusions of three types: (1) clear inclusions which are prominent and contain a weakly osmiophilic fibrillogranular reticulum; (2) dark inclusions which are less numerous and contain a dense granular material (3) multilaminated bodies comprised of stacked pairs of lamellae within a unit membrane [3, 4, 21, 29, 32, 33, 58, 60, 62].

Specific stainings in ultrastructure (silver proteinate and phosphotungstic acid affinity) suggest that the two former round or oval inclusions of variable size probably represent fucose-rich oligosaccharide chains, either free or bound to peptides. The osmiophilic multilayered bodies and the lamellar structures are characteristic of complex lipids and probably are fucose-glycosphingolipids [21].

Skin lesions Light microscopic examination of the skin reveals dilated capillaries in the papillary plexus. The endothelial cells are slightly swollen and the cytoplasm contains a number of fine PAS-negative vacuoles. Similar changes are present in the endothelial lining of small blood vessels in the upper and deep dermis. Swollen, finely vacuolated cells are uniformly present in the secretory coils of sweat glands and epidermis [18, 20, 21, 28-30, 32, 61-63].

The electron microscopic examination of the skin reveals that clear and dark inclusions are located in the capillary endothelial cells, pericytes, fibroblasts and in dark cells of sweat glands and that lamellar structures are found in Schwann

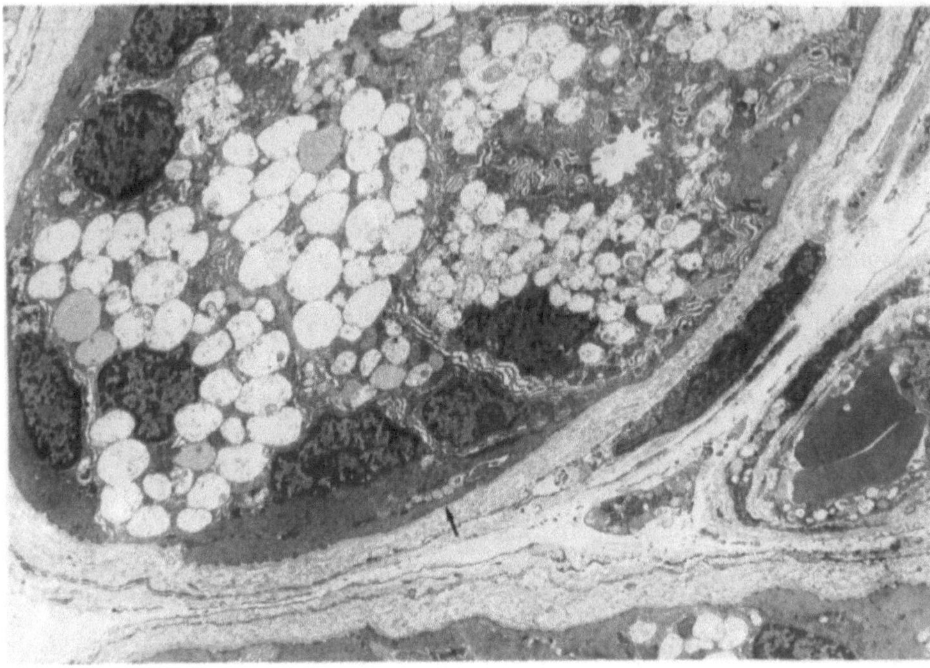

Fig. 3.8 Skin biopsy - Uranyl acetate, lead citrate. The myoepithelial cells of the sweat glands contain small clear inclusions (arrow). The cells in the secretory coil are distended by numerous clear and dense inclusions. Little normal cytoplasm remains between the inclusions. (x 4,000) [Courtesy of Dr. Libert et al., 21].

cells of intradermal nerves and myoepithelial cells of sweat glands. Both types of inclusions occur in some other cells as for instance in the epidermal cells of the stratum Malpighi (Fig. 8) [18, 20, 21, 29, 30, 32, 33, 62, 63].

The histologic findings in skin biopsies may already be present in the first months of life before the appearance of the clinical signs [21]. The severity of changes in the endothelium of blood vessels and in the secretory coil of sweat glands does not seem to correlate with the patient's age [32].

The differences between the types 1 and 2 of fucosidosis are that in type 2 more severe changes occur with considerable variations in the proportions of the lamellar lipid structures.

The characteristic aspect of inclusions in various kinds of cells is helpful in differentiating the angiokeratoma of fucosidosis from that of Fabry's disease; in the latter all the inclusions are in the form of lamellar, dense bodies. Other features differentiating fucosidosis from Fabry's disease are mental and physical retardation and normal renal function in fucosidosis; in fucosidosis vascular abnormalities develop at an early age.

Storage in the vascular walls is considered responsible for circulatory abnormalities and vasomotor disturbances and the development of angioectasia and angiokeratoma.

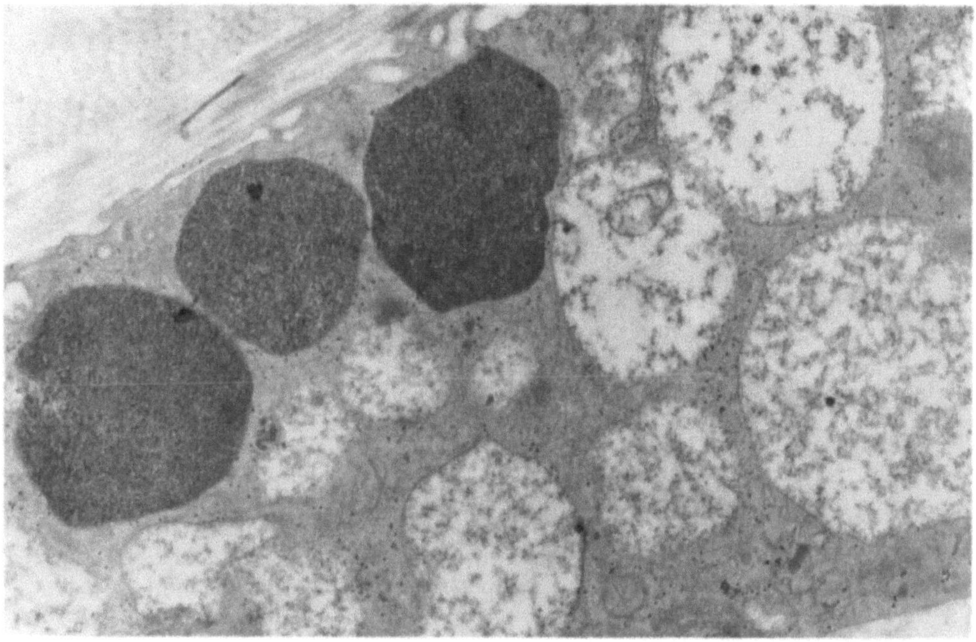

Fig. 3.9 Conjunctival stroma - Silver proteinate. The fibroblasts display two inclusion types. Dark inclusions contain a dense granular material which is strongly contrasted with silver. Clear inclusions show a faintly stained reticulum (x 21,000). [Courtesy of Dr. Libert et al., 21].

On the other hand extensive vacuolation of the secretory cells of sweat duct glands, interfering with their absorptive mechanism, could account for hyperhydrosis with the increased concentration of sweat electrolytes documented in some patients with the type 1 of fucosidosis. Hypohydrosis and anhydrosis found in type 2 might be explained by severe abnormalities in the blood vessels, autonomic nerves and sweat glands [32].

Ocular lesions Light microscopy of conjunctiva reveals that the epithelial cells in the apical layers and the endothelial cells of the capillaries are filled with numerous clear vacuoles of various sizes.

Dilatation of the thin-walled venules with angulation characterizes the retinal findings. The fragility of the capillary walls is probably responsible for the vascular tortuosities and the aneurism-like dilatations which become evident after several years.

The ultrastructure study of conjunctival biopsy shows that all cell types are affected. In the epithelium, small vacuoles appear in the basal layer, and greater inclusions are found in the adjacent layer (Fig. 9).

In the stroma numerous clear and dark inclusions distend the fibroblasts, the endothelial cells of the blood vessels and the lymphatic capillaries, the perivascular cells and the perineural cells. Moreover some isolated lamellar lipid structures or dense aggregates and dense spherules are also observed in the inclusions [21].

Brain lesions On light microscopic examination, the majority of the nerve cells in the cerebral cortex contain large vacuoles which are either empty or contain a homogeneous or granular material which generally does not show affinity for the usual dyes but occasionally is PAS positive. A vacuolation is present in the glial cells of white matter of the cerebral hemisphere and a diffuse astrocytosis with a dense astroglyosis is observed. The oligoglia is probably unable to form normal myelin or to maintain myelin sheaths and hypomyelination is apparent. Numerous small vacuoles containing granular material are also present in the cells of neostriatum, thalamus and hypothalamic nuclei. The pallidum shows numerous sudanophilic foam cells, especially in perivascular spaces.

The brain stem is sometimes irregularly involved and the inferior olives and the medial reticular gray matter are severely degenerated. A fibrous gliosis is evident. Also the cerebellar white matter is involved and diffuse astrocytosis with a mild astrogliosis is found [3, 4, 31, 57-60].

Electron microscopic examination of the brain reveals the presence of two types of inclusions, clear and dark, with a diameter of 0.4-1 μ and delimited by a single membrane in astrocytes, oligodendrocytes, macrophages and endothelial cells. Neurons contain only clear inclusions and only a few lamellar structures [4, 31, 58]. Electrolucent inclusions have been found in the cells of the brain of a 22 week old fetus with fucosidosis [65].

Liver lesions Hepatic cells contain abundant storage material giving them a clear, swollen appearance; this material is PAS negative or only weakly PAS positive. Storage inclusions are also found in Kupffer's cells, histiocytes, epithelial cells of the bile canaliculal system, and in vascular endothelial cells of portal tracts [3, 4, 18, 19, 31, 58, 60].

Electron microscopic examination shows that the hepatocytes contain some clear inclusions with a diameter of 0.5-5 μ, delimited by a single membrane, and in contrast to other cells, particularly neurons, contain a large number of osmiophilic lamellar inclusions characteristic of complex lipids. They are usually organized as rounded structures with a periodicity of 5.5 and 11.6 nm (Fig. 10).

Some of the lamellar material may also be free in the cytoplasm of Kupffer's cells, liver fibroblasts, and epithelial cells of the bile canalicular system; endothelial cells of the blood vessels contain mainly non lamellar inclusions [3, 4, 18, 19, 58, 60].

Other tissues Many other parenchymal organs, including the lungs, heart, spleen and kidneys are involved.

Vacuolated inclusions are found in epithelial cells of the seromucous and mucous glands of the oesophagus and of trachea. Vacuoles are found in the smooth muscle of the bronchus and in macrophages of the alveoli. The involvement of respiratory tract epithelium may explain the frequent occurrence of bronchitis and pneumonitis. Macrophages with vacuolo-granular cytoplasm are found in bone marrow, tonsils and lymphoid tissue of the spleen. Their cytoplasm is often Sudan-positive.

Involvement of all exocrine and some endocrine glands is noted [3, 60].

Abnormal and often enlarged inclusion bodies which apparently accumulate material of a glycoprotein and lipid nature are found in the small-intestinal cells.

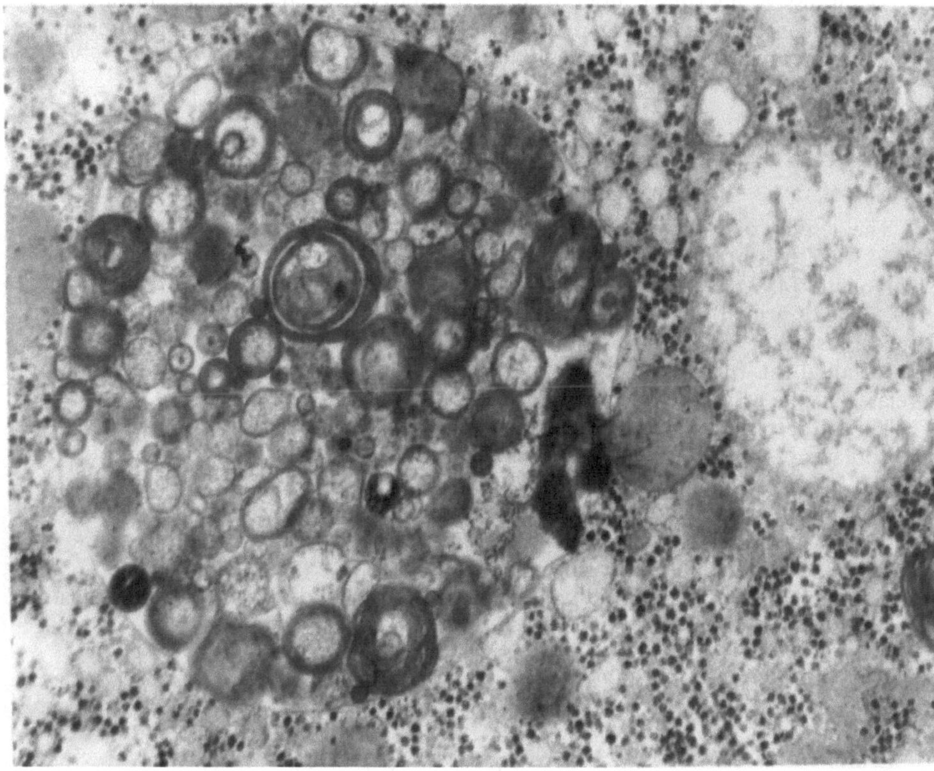

Fig. 3.10 Liver biopsy - Electron micrography of cytoplasmic inclusions in liver. In this hepatocyte, one inclusion filled with numerous rounded lamellar structures and beside it a clear inclusion, showing a faintly stained reticulum. (x 29,000) [1, 3, 14].

The significantly enlarged lysosome-like bodies in the absorptive cells show the same silver-stain affinity as the Golgi apparatus, apical vesicles and tubules and the cell coat [64]. This might indicate a crinophagic function of the lysosome-like bodies in the transport or secretion of cell coat material. PAS-positive material is found in the cytoplasm of macrophages of the lamina propria mucosae and in cells of the submucosa of the rectum.

Involvement of peripheral nerves is shown by abnormalities in sural nerve, intradermal nerve fibres and in the myenteric plexus nerves [21, 30, 32, 58, 62, 63]. The abnormalities consist of demyelination of the sural nerve and storage of isolated or concentric lamellar profiles in the cytoplasm of the Schwann cells. Moderate involvement of skeletal muscle fibres is also found [30].

In conclusion almost all cell types are affected in fucosidosis. The marked and generalized involvement of endothelial cells especially of the blood capillaries is much less prominent in Mucopolysaccharidosis I, II and III.

The presence of both clear and dark inclusions in endothelial cells and in fibroblasts, does not occur in other storage diseases and is pathognomonic of fucosidosis.

As discussed, neurons contain nearly exclusively clear and dark inclusions, whereas hepatocytes contain principally lamellar inclusions, indicating different types of fucosylated compounds in each cell type.

3.5 THE ENZYMATIC DEFECT

The activity of α-L-fucosidase is absent or grossly deficient in liver [3, 4, 18, 19, 21, 23, 66, 67]; brain [3, 4]; lung, pancreas and kidneys [3]; fibroblasts and leukocytes [4, 9, 18-23, 27, 29-33, 44, 47, 66, 68-71]; serum [20, 21, 27, 31, 44, 66, 69, 71, 72]; urine [4, 20, 21, 31, 66]; tears [21, 31] and saliva [73] of patients with fucosidosis.

Residual α-L-fucosidase activity is variable in different patients and is probably related to the assay method and the storage of the material.

Very little residual activity has been reported in liver [66, 67], cultured fibroblasts [27, 68, 70, 71], leukocytes [9, 20, 27, 29, 32, 44, 66, 71], serum [20, 21, 27, 31, 44, 66, 69, 71, 72], urine [9, 20, 21, 31, 66] of some patients, whereas complete lack of activity was found in others [3, 4, 18, 19, 21, 23, 31, 33, 69].

The low residual activity may represent either non specific hydrolysis of the substrate or genuine α-L-fucosidase activity [47].

A correlation, between residual enzymatic activity and severity of the clinical course has not been established [10, 11].

3.6 PROPERTIES OF NORMAL HUMAN α-L-FUCOSIDASE

The characterization of the normal human α-L-fucosidase is the necessary prerequisite for understanding the nature of α-L-fucosidase deficiency in fucosidosis.

α-L-fucosidase is involved in the catabolism of a wide range of fucose-containing glycoconjugates. The presence of α-L-fucosidase has been demonstrated in bacteria [74], protozoa [75], marine organisms [76] and mammals [77, 78].

The enzyme is ubiquitous in human tissues [79, 81], fluids [72, 82, 83, 84], plasma membranes of erytrocytes [85] and most cell types [68, 70, 82, 86].

A. Multiple forms

Human α-L-fucosidase exists in multiple isoelectric molecular forms called isoenzymes. Multiple molecular forms of α-L-fucosidase can be separated by starch gel electrophoresis, polyacrylamide slab gel electrophoresis or by column isoelectric focusing. Tissues studied include liver [16, 67, 87, 88], brain [81, 88, 89], heart [81], kidney [81, 87], placenta [16, 81, 88], spleen [88], many other types of cells [16, 71, 88, 90], urine [87] and serum [16, 71, 87, 90].

The isoelectric profiles for α-L-fucosidase of most types of cells are similar. There are 6-9 forms with isoelectric points (pI's) ranging from pH 5.0 to 7.0.

On isoelectric focusing the serum enzyme is different from the enzyme from tissues; there is a higher proportion of activity at acidic pI values (5.0 - 6.0) [16, 55, 71, 83, 91]. The multiple forms of α-L-fucosidase seem to be related to their different content of sialic acid residues. In fact neuraminidase treatment results in

Table 3.1 Summary of properties of two forms of α-L-fucosidase separated by gel filtration [79] and ion exchange chromatography [93].

	α-L-fucosidase I	α-L-fucosidase II
pH optimum	4.5-5	5.5
pH stability	stable on incubation in the acidic range	labile on incubation in the acidic range
thermolability	heat-stable	heat labile
molecular weight	high	low (50,000)

the disappearance of the acidic isoenzymes and increase of the more neutral forms [16, 55, 71, 83, 91].

The pattern of liver α-L-fucosidase isoenzymes changes during fetal development [92]. Whereas the most neutral form of α-L-fucosidase is prominent in early development, it is greatly diminished later in fetal development and postnatal life.

Polymorphism of α-L-fucosidase has been reported. Family studies performed in leukocytes with the technique of isoelectric focusing in thin-layer acrilamide gel have shown that the three common phenotypes, designated Fu^1, Fu^2, Fu^{2-1}, represent homozygosity or heterozygosity for two autosomal codominant alleles, Fu^1 and Fu^2 [88].

B. Kinetic and physical properties

Two forms of α-L-fucosidase, a higher and a lower molecular-weight form designated Fuc I and Fuc II, respectively, were separated in liver using gel filtration and ion-exchange chromatography [79, 93]. Table 1 summarizes the properties of these two forms.

It was suggested that these two forms, differing in pH optimum, pH stability, thermolability and substrate specificities, result from different states of aggregation of the same enzyme [93].

Purification of human α-L-fucosidase has increased the knowledge of the properties and of the interrelationship between the two forms [16, 94, 95]. The Km and V_{max} of purified liver α-L-fucosidase are 0.22 and 14.1 μM/mg of protein/minute, respectively, using the 4 methylumbelliferyl α-L-fucoside substrate [16, 56]. This substrate is hydrolyzed more readily than p-nitrophenyl-α-L-fucoside, for which the Km and V_{max} values are higher [16, 56, 96].

The pH optimum curve, using the p-nitrophenyl and 4MUB substrates, indicates a major pH optimum in the region of 5.0 and a second optimum at a more neutral pH 6.5 [16].

The purified liver enzyme is relatively thermostable and retains 100% of its activity after preincubation at 45° for 60 minutes [97].

Small but reproducible differences in Km, pH optimum curves and thermostability are also demonstrated with regard to seven isoenzymes separated by preparative isoelectric focusing [98]. The most neutral form has a higher apparent Km and decreased activity at acidic pH values (3.0 - 5.0) when compared to the other isoelectric forms. In addition a trend of increasing thermostability with increasing acidity is apparent.

The purified, apparently homogeneous liver α-L-fucosidase has a molecular weight, determined by gel filtration, of 175.000 ± 18.000 daltons. Under disaggregating conditions the enzyme was found to generate a single subunit of approx. 50,000 mol. wt. [16].

These studies suggest that human liver α-L-fucosidase is a tetramer made of four identical subunits and that the macromolecular form (α-L-fucosidase I) and the low weight form (α-L-fucosidase II) probably represent the tetrameric and monomeric forms of α-L-fucosidase [16, 55]. The interconversion of α-L-fucosidase I and II in experimental conditions [55, 81] and the identical isoelectric patterns of these two physical states [87] would support this hypothesis.

Recent observations indicate that the structure of α-L-fucosidase is more complex than the monomer-tetramer molecular model. Cester et al. [91] suggest the existence of at least two structurally different α-L-fucosidase units with additional diversity due to aggregation and sialic acid content. Alhadeff and Andrews-Smith [99] have found two non identical subunits with very similar molecular weight. Chien and Dawson [100] separated two forms of the enzyme by differential affinity chromatography: α-L-fucosidase I, with molecular weight of 100.000 which consists of two 50.000 dalton units and α-L-fucosidase II, with molecular weight of 50.000. Since these two forms differ with respect to aminoacid and oligosaccharide composition, pH optimum, electrophoretic mobility, Km, thermal stability, natural substrate specificities and uptake by cells, it seems clear that α-L-fucosidase I is not a simple dimer of α-L-fucosidase II. The 100.000-molecular weight α-L-fucosidase form may consist of an active 50.000-molecular weight subunit or A chain (identical to the α-L-fucosidase II form) and a non-identical, less active, 50.000-molecular weight subunit or B chain.

The properties of α-L-fucosidase in other human cells are quite similar to those of the liver enzyme [68, 81, 86, 89]. Only serum α-L-fucosidase is different from the enzyme found in tissues with respect not only to its isoelectric profile unit but also to its subunit, size, molecular weight and sialic acid content [55, 71, 83, 91, 97].

Since neuraminidase treatment leads to an altered isolectric profile and a decreased molecular weight (comparable to the molecular weight of the liver enzyme) it is possible that the presence of sialic acid affects the properties of serum α-L-fucosidase. In fact the amount of sialic acid in the purified serum α-L-fucosidase, was found to be twice the amount of sialic acid found in the liver purified enzyme [83].

C. Substrate specificity

Most of the studies about human α-L-fucosidase have been carried out with synthetic substrates. Only a few reports are available on the substrate specificity of mammalian α-L-fucosidases for natural fucose-containing glycoconjugates [78].

Wiederschain et al. have found that human α-L-fucosidases are highly specific for L-fucosyl residue and α-anomeric configuration [79]. Recently Chien and Dawson [100] have assayed the substrate specificity of α-L-fucosidase from different human tissues for fucose containing milk oligosaccharides and for a Fuc α (1→6) GlcNAc oligosaccharide isolated from fucosidosis fibroblasts. The two forms of α-L-fucosidase (α-L-fucosidase I and α-L-fucosidase II) purified by affinity chromatography, hydrolyze to the same extent 2-fucosyllactose, Fuc α(1→2) Gal, Fuc α (1→4) GlcNAc and Fuc α (1→6) GlcNAc linkages with the exception of the enzyme purified from urine which does not hydrolyze Fuc α (1→2) Gal linkages. Liver α-L-fucosidase is most effective in hydrolizing also Fuc α (1→3) GlcNAc linkage with some activity in kidney and brain preparations, but none in spleen, urine or human fibroblast preparations.

Further studies are necessary to determine the precise substrate specificity of human fucosidases and resolve the problem of storage of some substances in fucosidosis.

D. Immunological studies

Purified liver, brain and serum α-L-fucosidase were immunoprecipitated using α-L-fucosidase antiserum obtained by the injection of the purified enzyme into rabbits [55, 56]. All seven isoenzymes of purified liver α-L-fucosidase reacted by double diffusion against the same antiserum suggesting that the presence of sialic acid residues does not affect the antigenicity of multiple forms of the enzyme [98].

Thorpe and Robinson [55] emphasized that much caution is required before assuming monospecificity of the antiserum.

E. α-L-fucosidase uptake

The mechanism of uptake of α-L-fucosidase by cultured skin fibroblasts appears to conform basically to the criteria for mannose-6-phosphate recognition defined by Kaplan et al. [101] and by Von Figura et al. [102] for other lysosomal hydrolases. Human α-L-fucosidase purified from placenta was taken up from the culture medium by skin fibroblasts from deficient individuals [103]. The rate of uptake was low and there was no evidence for saturation of specific cell-surface receptors. Enzyme taken up by the deficient cells was most active in subcellular fractions enriched with lysosomes and was efficient in the catabolism of the material accumulated in fucosidosis fibroblasts.

Recently Chien and Dawson [100] have reported a high and low rate of uptake of α-L-fucosidase II and α-L-fucosidase I, respectively. At present, no studies are available on these specific uptakes of α-L-fucosidase by cells of different tissues. These studies would provide a theoretical basis for enzyme replacement trials in individuals suffering from fucosidosis.

3.7 PROPERTIES OF α-L-FUCOSIDASE IN FUCOSIDOSIS

The low levels of α-L-fucosidase activity in fucosidosis tissues have set considerable limitation to the characterization of the mutant enzyme.

The studies in fucosidosis patients summarized below indicate that the mutant α-L-fucosidase has altered properties (increased Km, decreased enzymatic activity at acidic pH values and increased heat lability), but is antigenically similar to the normal enzyme.

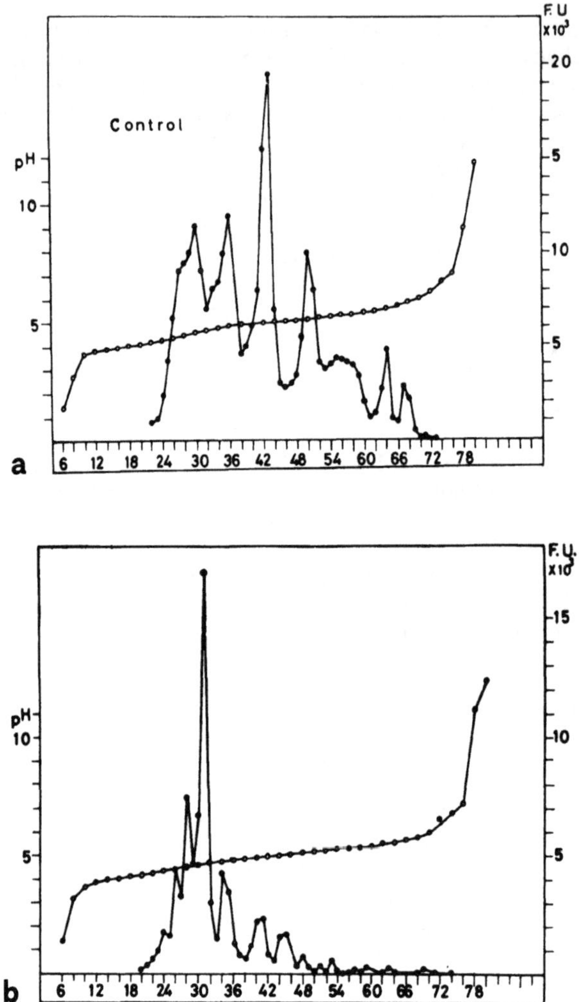

Fig. 3.11 Isoelectric focusing (performed using column 110-LKB equipment) of serum of one control **(a)** and of a fucosidosis patient **(b)**. Fluorescence units (F.U.), ●—●; pH, O—O.

A. Multiple forms

All the electrophoretic forms of α-L-fucosidase separated by starch gel electrophoresis are deficient in crude preparation of fucosidosis liver [67].

Isoelectric profile of partially purified α-L-fucosidase from liver of patients with

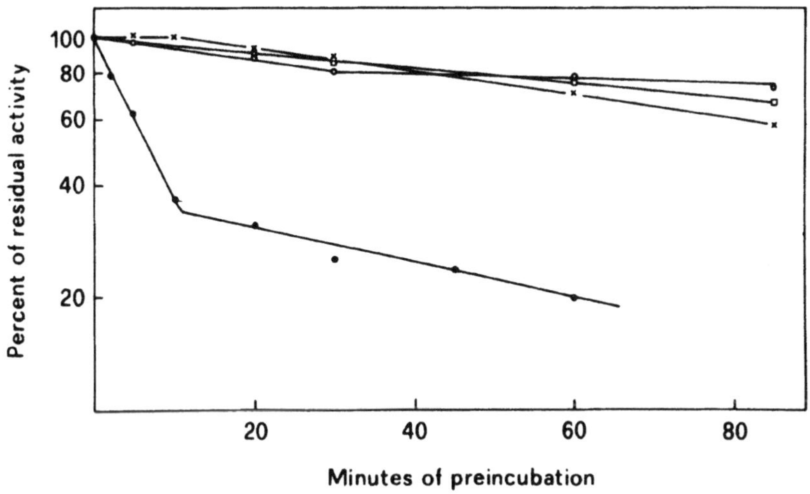

Fig. 3.12 Kinetics of heat inactivation of α-L-fucosidase in extracts of fibroblasts from 3 controls (single rate of decay) and from a fucosidosis patient (double rate of decay).

fucosidosis reveals one major peak of activity with pI 5.8, probably comparable with the peak of pI 5.9 in the normal profile and other minor forms [56].

Similar results have been found for sera of two patients with fucosidosis. Isoelectric focusing of sera shows one major peak of activity with pI of 4.5 and some other peaks [71]. By neuraminidase treatment the single major peak is transformed into several less acidic peaks which do not correspond to the peaks found in normal controls after neuraminidase treatment (Fig. 11).

B. Kinetic and physical properties

The Km value with respect to 4MUB substrate of the single major peak obtained from partially purified mutant liver enzyme [56] and of sera of two related patients [71] is 3-5 times the average values found for normal controls.

Also Km values for the p-nitrophenyl substrate in fucosidosis leukocytes and sera are considerably increased [66].

The pH activity curves of residual activity in the mutant fibroblasts are similar to those of normal cells [47]. The partially purified liver enzyme from a patient with fucosidosis displays no activity in the acidic region (pH 3.0-4.5) of the pH curve, compared with considerable activity of the normal enzyme in the same region [56].

The α-L-fucosidase from fucosidosis fibroblasts [71], sera [20, 71] and liver [56] exhibits decreased thermostability. The rate of thermal inactivation of α-L-fucosidase in sera and in fibroblasts of two fucosidosis patients showed a double rate of decay with respect to the single rate of decay present in control samples [71] (Fig. 12). The partially purified liver enzyme of a patient with fucosidosis loses 100% of its activity after 10 minutes of preincubation at 55°C, while the normal crude or purified enzyme retains approximatively 80% of its initial activity even after 60 minutes of preincubation [56].

C. Immunological studies

Double-immunodiffusion experiments fail to detect cross-reacting material to antibodies prepared against normal α-L-fucosidase in fucosidosis fibroblasts and serum [55] or liver [56].

However, when comparable amounts of α-L-fucosidase from fucosidosis and normal livers are separately preincubated with varying amounts of anti-α-L-fucosidase IgG, both enzymes are immunoprecipitated [56].

3.8 VARIANT(S) OF α-L-FUCOSIDASE CAUSING LOW SERUM AND PLASMA ACTIVITY

The α-L-fucosidase activity is reduced to less than 10% of the mean control value in the serum and plasma of about 6-10% of phenotypically normal individuals in different populations [69, 104, 105]. The variant form of serum α-L-fucosidase is also partially expressed in leukocytes [106, 108, 109].

The kinetic and immunochemical characterization of the low activity serum α-L-fucosidase, shows no significant difference to the normal serum α-L-fucosidase [55, 106, 107], except for a decreased activity at acidic values and an increased heat-lability [106-108].

Gel filtration experiments on sephadex G 200 reveal a reduced peak of low molecular weight (α-L-fucosidase II) and absence of the high molecular weight peak (α-L-fucosidase I) [106].

The α-L-fucosidase in leukocytes from individuals with the low-activity serum α-L-fucosidase is shown to be reduced only when a statistical analysis is performed [108] and has neutrally shifted pH optimum profile, thermolability, increase in the activity of α-L-fucosidase II and decrease in the activity of α-L-fucosidase I, resembling α-L-fucosidase II [106, 108].

The analysis of the mode of inheritance of serum α-L-fucosidase deficiency in different families agrees with the autosomal recessive hypothesis [79, 105, 108, 109].

The distribution of the levels of serum α-L-fucosidase in individuals without the variant serum enzyme is bimodal and separates two classes: the heterozygotes for the allele having intermediate values from the homozygotes for the wild-type allele having high values of α-L-fucosidase activity [108, 109]. Bimodality in the distribution of heterozygotes is more apparent when a ratio of α-L-fucosidase to α-L-mannosidase is calculated [108]. The frequency of the variant allele is estimated at 0.25, 0.29, 0.28 in population samples from North America [109], England [105], Italy [108], respectively.

3.9 NATURE OF THE MOLECULAR DEFECT IN FUCOSIDOSIS AND IN LOW-ACTIVITY SERUM VARIANT

Several investigators have excluded the presence of inhibitors or the lack of activators of α-L-fucosidase in fucosidosis [5, 20, 40, 47, 66,71] and in the low activity serum variant [104].

The characterization of the residual α-L-fucosidase in fucosidosis liver, serum and fibroblasts is consistent with the presence of an antigenically normal enzyme with altered physical and kinetic properties.

These data are suggestive of a structural alteration of the enzymatic molecule. They do not discriminate between the retention of a small portion of a normal enzyme and the synthesis of an altered α-L-fucosidase. The finding of a single major peak of α-L-fucosidase activity with the properties of the most neutral form of normal α-L-fucosidase in the liver and sera of patients would be consistent with the former hypothesis.

Di Matteo et al. [71] proposed that the enzyme found in fucosidosis patients could be structurally altered so that its ability to form the thermostable higher molecular-weight aggregates is impaired.

The findings obtained in family studies by Turner et al. [110], using polymorphism of α-L-fucosidase as a parameter, are consistent with a mutation within the structural locus for α-L-fucosidase resulting in an inactive silent allele.

The similarity of the properties of low activity serum α-L-fucosidase with those of the residual enzyme in fucosidosis suggests that both heritable traits may be related. Since the individuals with deficiency of α-L-fucosidase in plasma have also altered α-L-fucosidase in their leukocytes [106, 108, 109], the hypothesis of a defective sialylation of serum α-L-fucosidase may be excluded.

Further studies need to be done to determine the precise nature of the molecular defect in fucosidosis and in the benign plasma variant.

3.10 PRECURSORS OF THE STORAGE PRODUCTS IN FUCOSIDOSIS

L-fucose is generally located in a terminal, nonreducing position, and is a common constituent of glycoproteins, oligosaccharides and glycolipids [111, 112, 113].

Fucose has not been found in mucopolisaccharides other than keratansulfate.

In brain L-fucose is a common constituent of glycoproteins. No fucose-containing glycolipids or acid mucopolysaccharides were proven to occur in brain [114] but a fucose-containing ganglioside has been identified in mammalian brain [115].

L-fucose is of particular interest in connection with the serological specificity of the blood-group substances [116, 117].

The antigenic specificity appears to be a reflexion of the structural arrangement of certain sugars on the more superficial parts of the carbohydrate moieties of two classes of macromolecules, glycolipids and glycoproteins.

The carbohydrate moiety of the blood-group glycoproteins or glycolipids is thought to be synthesized by the sequential action of a series of glycosyltransferases.

Figure 13 shows the structures of these antigenic determinants. There are two types of backbone structures, type 1 and type 2 chains, termed "precursor substances" devoid of blood group specificity. The presence of fucose on C-2 of the terminal galactose of either type 1 or type 2 chain produces an H determinant. The presence of fucose linked to C-4 of N-acetylglucosamine results in Le^a activity. The simultaneous presence of both fucose substituents on type 1 chain results in the appearance of a new antigenic specificity, Le^b.

The H and Lewis antigens result from the action of genes at two independent

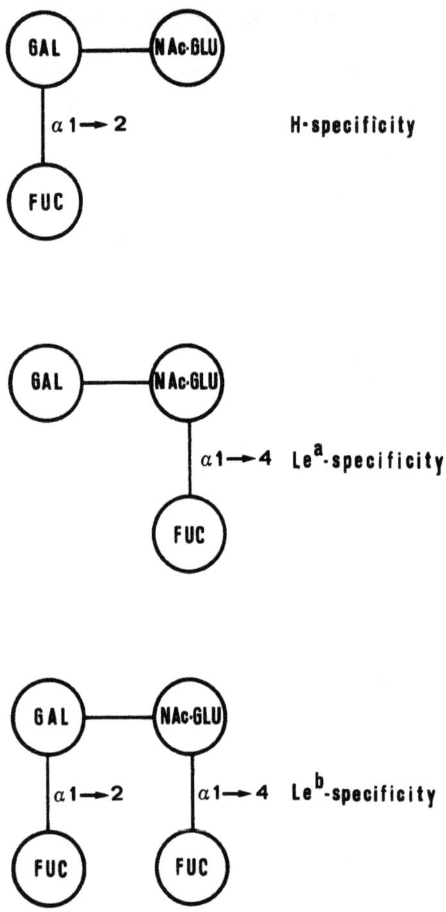

Fig. 3.13 Relationship between fucose and antigenic specificities of blood group substances.

loci, the H and Le loci. The presence of antigens H (and also of A and B) in secretions but not on erythrocytes is due to the action of the independent secretor genes Se and se. The available evidence suggests that most of these genes act by controlling the glycosyltransferase addition of a single sugar to the glycoprotein [118].

The H and Lewis antigens occur in human tissues and secretions in two forms, water soluble (glycoproteins) and alcohol-soluble (sphingoglycolipids).

The H and Lewis substances of erythrocytes are sphingoglycolipids characterized by a high content of fucose. The Le^a and Le^b antigens are not an integral part of red cell membrane but are acquired from the plasma and are primarily glycosphingolipids and not glycoproteins.

Persons with ABH substances in saliva (secretors) have more water soluble substances in their tissues than those lacking the substances in saliva (nonsecretors).

Blood group substances are also detectable in urine [119, 120] and human milk [121].

Only small quantities of fucose-containing glycolipids are present in parenchymatous organs (liver, kidney,. spleen, etc.). This is in striking contrast with the considerable amounts found in glandular epithelial tissues, like intestinal tracts and pancreas. However synthesis of ABH substances in the exocrine acini of the pancreas and the secretory coils of sweat glands is not regulated by the secretor genes.

3.11 STORAGE MATERIAL IN FUCOSIDOSIS

The loss of α-L-fucosidase activity accounts for the accumulation in fucosidosis tissues of a wide variety of fucoglycoconjugates. Nevertheless only a few of the potential storage products have been isolated and characterized.

Structural studies have indicated that the major storage material contains L-fucose in α(1→2) linkage to galactose and α(1→6) linkage to N-acetylglucosamine. Other types of linkages (α(1→3) and α (1→6) to galactose; α (1→3) and α (1→4) to N-acetylglucosamine) are not always found, probably because of experimental procedures used and/or variations in genetically determined availability of substances.

Preliminary biochemical studies indicated the storage of a fucopentaglycosylceramide exhibiting abnormally high titres of H and Lewis specific substances in liver but not in brain, together with storage of fucopolysaccharides in both organs but especially in brain [2, 3, 8].

Dawson and Spranger [12] confirmed that the major stored fucoglycosphingolipid in fucosidosis liver has H antigen specificity with the following structure:

Fuc α(1→2) Gal β (1→3) GlcNAc β (1→4) Gal β (1→4) Glc-ceramide.

Shortly after, Dawson and Tsay [122] and Tsay et al. [123] characterized two further major storage products isolated from fucosidosis tissues: a dekasaccharide (Fig. 14) and a disaccharide of type Fuc α (1→6) GlcNAc.

These authors have reported the relative distribution of the three types of complex carbohydrate storage material in fucosidosis tissues. Fucose-containing ceramide pentahexoside is the major material stored in pancreas and liver. A fucose containing dekasaccharide, structurally related to the oligosaccharide storage material identified in G_{M1} gangliosidosis and Sandhoff's disease, is the major storage compound in fucosidosis brain, together with a lesser amount of the Fuc α (1→6) GlcNAc disaccharide [124].

Kidney contains only a small amount of storage material; in lung and spleen there is a high content of dekasaccharide, whereas lymphonodes contain an equal mixture of dekasaccharide and disaccharide.

Fuc (α, 1→2) Gal (β, 1→4) GlcNAc (β, 1→2) Man (α, 1→3) ↘
Man (β, 1→4) GlcNAc
Fuc (α, 1→2) Gal (β, 1→4) GlcNAc (β, 1→2) Man (α, 1→6) ↗

Fig. 3.14 Structure of fucodekasaccharide isolated from fucosidosis tissues [123].

Cultured skin fibroblasts, in contrast to other fucosidosis tissues, store a glycopeptide of the Fuc α (1→6) GlcNAc-Asn type, similar to the innermost portion of immunoglobulin oligosaccharide chain [125]. The absence of fucooligosaccharide storage material suggests an inability of the cultured fibroblasts to hydrolyze the GlcNAc-Asn linkage.

A. Accumulation of blood group substances

A particular aspect of fucosidosis is the increase of fucose containing specific blood group substances.

The hydrolysis of terminal fucose, by the action of α-L-fucosidase, from the ends of carbohydrate chains in the group specific substances gives rise to the loss of their antigenic specificity [117].

In patients with fucosidosis the impaired degradation of the fucoglycoconjugates causes accumulation of substances with blood group activity. In the red cells and in the biological fluids of patients with fucosidosis an abnormal phenotypic expression of the Lewis blood group system was found [27, 29, 30, 126].

The "secretor" and "non secretor" status of an individual is constant and is genetically determined.

ABH secretors normally have Le (a− b+) red cells, but the cells of fucosidosis patients who are secretors react equally strongly with anti Le^a and Le^b sera.

ABH non secretors normally have (a+ b−) red cells, while the cells of non-secretors with fucosidosis react more strongly with anti-Le^a sera than do control Le (a+ b−) cells and also give distinct, although weak, reaction with anti-Le^b sera. Le (a+ b+) blood group is rare, unless the children are between 3 months and 2 years of age. In fucosidosis several cases of this abnormality have been reported [27, 126].

The increase and abnormal antigenicity of blood specific substances in the cases of a deficiency of α-L-fucosidase clearly show the importance of the enzyme in the catabolism of fucose-containing Le substances.

The inherited blood group differences rather than a specific deficiency of α-L-fucosidase activity for α (1→4) or α (1→2) linkages can explain the differences of accumulation of substances with blood group activity reported in a number of patients with fucosidosis.

B. Urinary excretion of fucose-containing material

Fucosidosis patients excrete in urine an excessive amount of substances related to glycoprotein catabolism [123, 127-132] and a minor amount of keratansulfate [35].

Three abnormal compounds were initially isolated and characterized, the major of which was the same dekasaccharide found in tissues [122, 123].

Subsequently other fucoglycoconjugates were isolated and characterized. To date six major fucooligosaccharides [127, 130] and twenty two fucoglycopeptides [132] have been isolated in urine of fucosidosis patients (Fig. 15). Some of these compounds have not previously been detected in biological fluids.

A detailed study of the fucose-containing material excreted in urine of fucosidosis patients should make the following observations:

1) fucosidosis is unique among the congenital exoglycosidase deficiencies because the excretion of fucoglycopeptides is larger than that of fucooligosaccharides [128, 130, 132].

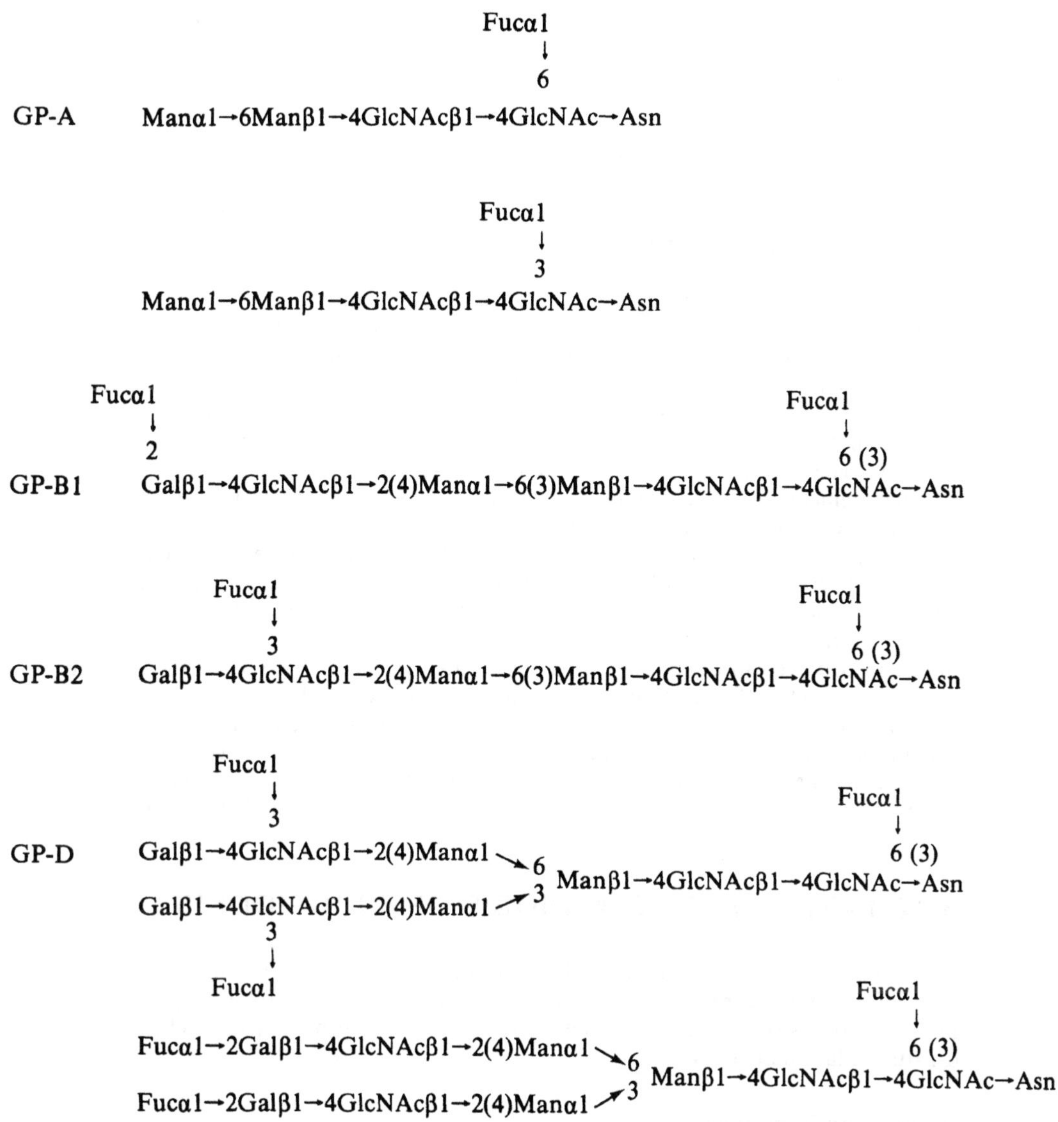

Fig. 3.15 Proposed structures of glycopeptides fractionated by gel permeation chromatography from the urine of fucosidosis patient [132].

2) the differences in excretion patterns of fucose-containing substances could be related to the genetically determined biological substrates of fucosidosis patients and especially to the blood group type and secretor status [127].

3) The exact origin of the urinary fucocompounds is unknown but they are believed to be derived form the impaired catabolism of a large number of different glycoproteins [132].

Other congenital lysosomal exoglycosidase deficiencies result in the accumulation and excretion of high molecular weight oligosaccharides which are also derived from the oligosaccharide side chain of glycoproteins. These oligosaccharides are presumably cleaved from the glycoprotein core by the action of an endo-β-N-acetyl-glucosaminidase and consequently they all terminate in a single N-acetyl-glucosamine residue at the reducing terminus.

The oligosaccharides which accumulate in fucosidosis patients uniquely differ from those found in other lysosomal glycoproteinoses, not only because they contain fucosyl residues at the C-3 or C-6 positions of the N-acetyl-glucosamine residue of the stored oligosaccharide, but also because they are glycopeptides and contain an N,N-dichitobiosyl-asparagine residue in the chain and no reducing terminus. The accumulation of these unique glycopeptides suggests that the endo-β-N-acetylglucosiminidase enzyme is unable to cleave oligosaccharides from the peptide core without the prior action of α-L-fucosidase.

Although the precise sequence of events involved in glycoprotein catabolism has not been determined, it can be inferred that the α-L-fucosidase serves an important, and perhaps limiting, initial step in glycoconjugate degradation.

Greiling et. al. [35] have recently shown an increased excretion of keratan sulfate in two fucosidosis patients. They indicated that the dekasaccharide isolated by Tsay et al. [123] from the urine of a fucosidosis patient closely resembles the structure of keratansulfate. At present, the role of α-L-fucosidase in the degradation of keratansulfate needs to be determined. However, since keratan-sulfate contains α-L-fucose at the non-reducing terminal, increased urinary excretion of long chain keratan sulfate in fucosidosis is a reasonable expectation.

3.12 LABORATORY DIAGNOSIS OF FUCOSIDOSIS

The enzymatic assay is the most direct and precise approach to diagnosis of fucosidosis. Other approaches consisting in abnormal levels of fucocompounds in biological fluids of fucosidosis patients are indirect and probably inconclusive of α-L-fucosidase deficiency.

A. Enzymatic analysis

Source of enzyme The absence or a large deficiency of α-L-fucosidase activity was found in all tissues, types of cells and biological fluids from patients with fucosidosis.

Various tissues are available for diagnostic purposes.

White blood cells [27, 44, 45, 69, 108] isolated mononuclear cells [44, 65], long-term lymphoid cell lines [44, 88], bone marrow fibroblasts [70], cultured skin fibroblasts [47, 71], tears [21] and saliva [73] are the most suitable sources of α-L-fucosidase for diagnosis of fucosidosis. Serum or plasma have proven to be inadequate for diagnosis [69, 104, 108].

Enzyme assay Two artificial substrates for α-L-fucosidase assay are commercially available: the colorimetric p-nitrophenyl-α-L-fucopyranoside and the fluorogenic 4-methylumbelliferyl-α-L-fucopyranoside. No natural substrates for α-L-fucosidase have been used except in special investigations. Milk oligosaccharides and blood group substances containing L-fucose as a terminal non reducing sugar have been used to study the substrate specificity of α-L-fucosidase [78, 100].

The colorimetric and fluorogenic assays are interchangeable [96, 133]. The p-nitrophenyl substrate is relatively inexpensive. The fluorogenic assay is more sensitive than the colorimetric assay and is more useful in studying the tissues with low α-L-fucosidase activity (e.g. fibroblasts) and the residual α-L-fucosidase activity in fucosidosis patients.

Demonstration of a significant difference in residual activity with regard to two synthetic substrates in a patient with fucosidosis has been reported [24]. The patient's liver accumulated fucoglycolipid and retained 25-50% of normal p-nitrophenyl-α-L-fucosidase while the 4-methylumbelliferyl α-L-fucosidase was nearly absent. The patient's spleen had a deficiency of α-L-fucosidase for both substrates. After partial purification the patient's liver residual p-nitrophenyl α-L-fucosidase was found to possess both PNP and 4-MUB activities. The findings suggest that the patient's crude liver supernatant may contain an inhibitor of 4-MUB-α-L-fucosidase activity.

B. Urine analysis

Thin-layer chromatography (TLC) on silica gel is useful as a screening test for fucosidosis [134, 135, 136].

The examination of untreated urine samples from patients with both forms of fucosidosis reveals a distinctive pattern on TLC. Humbel and Dhondt [137] attempted to improve the separation of fucose-containing oligosaccharides using an over-night development in the solvent system n-butanol/acetic-acid/water (50:25:25) followed by a second development in the same direction with a mixture of n-propanol/nitromethane/water (100:80:60). Spraying with a solution of naphtoresorcinol (0,5%) in sulphuric acid/ethanol is the most favoured location procedure. The fucosides develope a red coloration after heating for 10 min. at 100°.

The quantitative and structural analyses of storage products in the urine of fucosidosis patients are purely research methods and are limited to the most specialized laboratories.

3.13 DETECTION OF CARRIERS

Heretozygotes for fucosidosis have α-L-fucosidase levels intermediate between patients and normal individuals (gene dosage effect). Detection of heterozygotes for fucosidosis is, however, not easy.

Serum or plasma, urine, saliva and tears have proven to be inadequate for assessment of carrier status [27, 44, 69, 73].

The use of mixed leukocytes or granulocytes overlaps between normal subjects and obligate heterozygotes [44, 45].

Attempts to correlate α-L-fucosidase to α-D-mannosidase activity in leukocytes [138], failed to clearly separate the two groups [45]. Also cultured skin fibroblasts are not well suited to detect heterozygotes for fucosidosis [68] since α-L-fucosidase activity shows variations according to cell density, growth cycle etc. [139].

The use of purified mononuclear cells, although the isolation requires considerable preparation, has proven to be the best source of α-L-fucosidase assay for detection of heterozygotes [44, 45].

We examined 64 members of two large pedigrees from Calabria in which seven children and a foetus had been diagnosed to be affected with fucosidosis. Of 64 individuals, 23 were diagnosed to be carriers while the values for the remaining subjects were within normal limits indicating a bimodal distribution among individuals at risk.

However, the preliminary results of our screening program for fucosidosis in Calabria suggest that in individual cases it may remain a problem to determine the carrier status unequivocally [140].

The absence of automated α-L-fucosidase assay procedures limits carrier testing for fucosidosis to families at risk or selected populations having a high incidence of this disorder [141].

3.14 PRENATAL DIAGNOSIS

α-L-Fucosidase activity is detectable in the supernatant of amniotic fluid at various stages of pregnancy [80], but the prenatal diagnosis of fucosidosis by assay of cell free amniotic fluid is not recommended.

Since α-L-fucosidase deficiency is manifest in cultured skin fibroblasts, prenatal detection in cultured amniotic cells is feasible. Although no clear-cut correlation exists, there appear to be trends of decreasing specific fucosidase activity with increasing maternal age of donor and increasing gestational time [84].

Considerable variations in α-L-fucosidase activity were found in different types of cultured amniotic fluid cells. The highest values were obtained in primary cultures mainly consisting of epithelioid cells, whereas much lower values were found in fibroblast-like cells in subsequent passages [43, 142, 143].

The marked fluctuations in enzyme levels could make it difficult to determine accurate differentiation between heterozygous and affected.

For instance in the first pregnancy monitored for being "at risk" for fucosidosis, a diagnosis of heterozygosity was erroneously made using primary cultures of epithelial-like amniotic fluid cells, while affected twins were subsequently delivered [144].

Subsequent monitoring of a family at risk for fucosidosis, showed that the foetus was free from the disease [43].

The first example of a prenatal diagnosis of fucosidosis has been described by us [45].

α-L-Fucosidase activity was not detectable in cultured amniotic cells of the monitored pregnancy and prenatal identification of an affected foetus was possible. The results of the α-L-fucosidase assays in lymphocytes, fibroblasts and cultured amniotic cells of the family "at risk" are reported in table 2.

3.15 ANIMAL MODELS OF THE DISEASE

There is no known animal model of fucosidosis.

Table 3.2 Prenatal diagnosis of fucosidosis in a family of Pedigree no. 2 (Fig. 7).

	α-L-fucosidase activity(nM/mg/h) Lymphocytes	Fibroblasts	Cultured amniotic cells*
Controls	38-116	47,9-171,3	68-480
Father	28.4	-	
Mother	19.8	119.8	
Sister	21.9	-	-
Index patient	non detectable activity	1.5	
Monitored pregnancy			non detectable activity
Foetus	non detectable activity		

* The assays were performed in cultured cells after 2 and/or 3 passages. The diagnosis of affected foetus was confirmed, after abortion, by biochemical analysis in purified mononuclear cells. Ultrastructural analysis of the brain, performed by Bugiani [145], showed cytoplasmatic electrolucent inclusions in the cells of the 22-week old foetus.

3.16 β-D-FUCOSIDASE

β-D-fucosidase activity in mammalian tissue was originally described by Levvy [146].

Nothing is known of the presence in animal tissues of D-fucose. Wiederschain et al. [147] in a series of studies on specificity patterns of different types of human fucosidase showed that β-D-fucosidase is able to hydrolyze different substrates with a similar structure.

A comparative study of the splitting of β-D-fucoside, β-D-galactoside and α-L-arabinoside demonstrated that these glycosides are hydrolyzed by the same type of fucosidase: β-D-fucosidase (β-D-galactoside, α-L-arabinoside) hydrolase. This conclusion was confirmed by studies on GM_1 gangliosidosis. It has been shown that in this disease the β-D-fucosidase and α-L-arabinosidase activities are decreased in addition to a specific decrease in β-D-galactosidase activity.

The original works of the present review were supported with grants from the Public Health Service of "Regione Calabria" (Italy).

REFERENCES

1. Durand P., Borrone C., Della Cella G.: *A new mucopolysaccharide lipid-storage disease?* Lancet, *2,* 1313-1314, 1966.

2. Durand P., Borrone C., Philippart M., Della Cella G., Bugiani O.: *Una nuova malattia da accumulo di glicolipidi.* Minerva Pediat. *19,* 2187-2196, 1967.
3. Durand P. Borrone C., Della Cella G.: *Fucosidosis.* J. Pediat. *75,* 665-674, 1969.
4. Loeb H., Tondeur M., Jonniaux G., Mockel-Pohl S., Vamos-Hurwitz E.: *Biochemical and ultrastructural studies in a case of mucopolysaccharidosis "F". Fucosidosis.* Helv. Paediat. Acta *24,* 519-537, 1969.
5. Van Hoof F., Hers H.G.: *The abnormalities of lysosomal enzymes in mucopolysaccharidosis.* Europ. J. Biochem. *7,* 34-44, 1968.
6. Van Hoof F., Hers H.G.: *Mucopolysaccharidosis by absence of α-fucosidase.* Lancet *1,* 1198, 1968.
7. Durand P., Borrone C., Della Cella G., Phillippart M.: *Fucosidosis.* Lancet *1,* 1198, 1968.
8. Phillippart M.: *Fucosidosis: A novel neurovisceral sphingolipidosis.* Neurology *19,* 304, 1969.
9. Patel V., Watanabe I., Zeman W.: *Deficiency of α-L-fucosidase.* Science *176,* 426-427, 1972.
10. Gatti R., Borrone C., Trias X., Durand P.: *Genetic heterogeneity in fucosidosis.* Lancet *2,* 1024, 1973.
11. Kousseff B.G., Beratis N.G., Danesino C., Hirschhorn K.: *Genetic heterogeneity in fucosidosis.* Lancet *2,* 1387-1388, 1973.
12. Dawson G., Spranger J.W.: *Fucosidosis: A glycosphingolipidosis.* New Engl. J. Med. *285,* 122, 1971.
13. Dawson G.: *Glycosphingolipid abnormalities in liver from patients with glycosphingolipid and mucopolysaccharide storage diseases.* In "Sphingolipids, Sphingolipidoses and Allied Disorders" (B.W. Volk and S.M. Aronson Eds.) Plenum Press, New York and London, 395-413, 1972.
14. Van Hoof F.: *Fucosidosis.* In "Lysosomes and storage diseases" (H.G. Hers and F. Van Hoof Eds.) Academic Press, New York and London, 277-290, 1973.
15. Dubesset D.: *La fucosidose.* Thèse Doctorat en Médicine-Université d'Amiens, 1976.
16. Alhadeff J.A., O'Brien J.S.: *Fucosidosis.* In "Practical Enzymology of the Sphingolipidoses" (R.H. Glew and S.P. Peters Eds.) Alan R. Liss, New York, 247-281, 1977.
17. Alhadeff J.A.: *Human α-L-fucosidases and fucosidosis.* In "Lysosomes and Lysosomal storage diseases" (J.W. Callahan and J.A. Lowden Eds.) Raven Press, New York, 299-314, 1981.
18. Voelz C., Tolksdorf M., Freitag F., Spranger J.: *Fucosidose.* Mschr. Kinderheilk. *119,* 352-355, 1971.
19. Matsuda I., Arashima S., Anakura M., Ege A., Hayata I.: *Fucosidosis.* Tohoku J. Exp. Med. *109,* 41-48, 1973.
20. Patel V., Zeman W.: *Variability of expressivity of α-fucosidase deficiency.* In "Current Trends in Sphingolipidoses and Allied Disorders" (B.W. Volk and L. Schneck Eds.) Plenum Press, New York and London, 167-186, 1976.
21. Libert J., Van Hoof F., Tondeur M.: *Fucosidosis: Ultrastructural study of conjunctiva and skin and enzyme analysis of tears.* Invest. Ophthalol. *15,* 626-639, 1976.
22. Troost J., Staal G.E.J., Willemse J., van der Hejden M.C.M.: *Fucosidosis. I. Clinical and enzymological studies.* Neuropädiatrie *8,* 155-162, 1977.
23. Larbrisseau A., Brochu P., Ng Ying Kin N.M.K., Jashmin G., Potier M., Vanasse M., Hausser C.: *Prémière observation canadienne d'une fucosidose.* Union Méd. du Canada, *107,* 968-980, 1978.
24. Alhadeff J.A., Andrews-Smith G.L., O'Brien J.S.: *Biochemical studies on an unusual case of fucosidosis.* Clin. Genet. *14,* 235-240, 1978.
25. Brill P.W., Beratis N.G., Kousseff B.G., Hirschhorn K.: *Roentgenographic findings in fucosidosis type 2.* Am. J. Roentg. Rad. Ther. Nucl. Med. *124,* 75-82, 1975.
26. Lee F.A., Donnell G.N., Gwinn J.L.: *Radiographic features of fucosidosis.* Pediat. Radiol. *5,* 204-208, 1977.
27. Borrone C., Gatti R., Trias X., Durand P.: *Fucosidosis: clinical biochemical*

immunologic and genetic studies in two new cases. J. Pediatr. *84,* 727-730, 1974.
28. Mac Phee G.B., Logan R.W., Primrose D.A.A.: *Fucosidosis: how many cases undetected?* Lancet *2,* 462-463, 1975.
29. Kousseff B.G., Beratis N.G., Strauss L., Brill P.W., Rosenfield R.E., Kaplan B., Hirschhorn K.: *Fucosidosis type 2.* Pediatrics *57,* 205-213, 1976.
30. Landing B.H., Donnell G.N., Alfi O.S., Neustein H.B., Lee F.A., Ng W.G., Bergren W.R., Surgeon P.: *Fucosidosis: Clinical, pathologic, and biochemical studies of five patients.* In "Current Trends in Sphingolipidoses and Allied Disorders" (B.W. Volk and L. Schneck Eds.) Plenum Press, New York and London, 147-165, 1976.
31. Santolaya J.M., Hernandez M., Chabas A., Cabello A., Rivera J.M., Rodriguez M.I.: *Fucosidosis tipo 2. Aportacion de un nuevo caso.* An. Esp. Pediat. *10,* 857-866, 1977.
32. Kornfeld M., Snyder R.D., Wenger D.A.: *Fucosidosis with angiokeratoma. Electron microscopic changes in the skin.* Arch. Pathol. Lab. Med. *101,* 478-485, 1977.
33. Porfiri B., Ricci R., Seminara D., Segni G.: *Ultrastructural studies of type II fucosidosis.* Arch. Dermatol. Res. *270,* 57-66, 1981.
34. Snyder R.D., Carlow T.J., Ledman J. et al.: *Ocular findings in fucosidosis.* In "The Eye and Inborn Errors of Metabolism" (D. Bergsma, A.I. Bron, E. Cotlier Eds.) National Foundation, March of Dimes. Alan R. Liss, New York. Vol. *12,* 241-251, 1976.
35. Greiling H., Stuhlsatz H.W., Cantz M., Gehler J.: *Increased urinary excretion of keratan sulfate in fucosidosis.* J. Clin. Chem. Clin. Biochem. *16,* 329-334, 1978.
36. Gatti R., Costantino G., Borrone C., Durand P.: Unpublished data.
37. Schafer I.A., Powell D.W., Sullivan J.C.: *Lysosomal bone disease.* Pediatr. Res. *5,* 391-392, 1971.
38. Schoonderwaldt H.G., Lamers K.J.B., Kleijnen F.M., Van Der Berg C.J.M.G., De Bruyn C.H.M.M.: *Two patients with an unusual form of type II fucosidosis.* Clin. Genet. *18,* 348-354, 1980.
39. Sovik O., Lie S.O., Fluge G., Van Hoof F.: *Fucosidosis. Severe phenotype with survival to adult age.* Europ. J. Ped. *135,* 211-216, 1980.
40. Liem K.O.: Unpublished data.
41. V. Gemund J.J.: personal communication.
42. Hanefeld F.: personal communication.
43. Poenaru L., Dreyfus J.C., Boué J., Nicolesco H., Ravise N., Bamberger J.: *Prenatal diagnosis of fucosidosis.* Clin. Genet. *10,* 260-264, 1976.
44. Beratis N.G., Turner B.M., Hirschhorn K.: *Fucosidosis: detection of the carrier state in peripheral blood leukocytes.* J. Pediat. *87,* 1193-1198, 1975.
45. Durand P., Gatti R., Borrone C., Costantino G., Cavalieri S., Filocamo M., Romeo G.: *Detection of carriers and prenatal diagnosis for fucosidosis in Calabria.* Hum Genet. *51,* 195-201, 1979.
46. Beratis N.G., Turner B.M., Hirschhorn K.: *Reply to letter by Durand P. et al..* J. Pediat. *89,* 690-691, 1976.
47. Beratis N.G., Turner B.M., Labadie G., Hirschhorn K.: *α-L-fucosidase in cultured skin fibroblasts from normal subjects and fucosidosis patients.* Pediat. Res. *11,* 862-866, 1977.
48. Turner B.M., Smith M., Turner V.S., Kucherlapati R.S., Ruddle F.H., Hirschhorn K.: *Assignement of gene locus for human alpha-fucosidase to chromosome 1 by analysis of somatic cell hybrids.* Somat. Cell Genet. *4,* 45-54, 1978.
49. Corney G., Fisher R.A.F., Cook P.J.L., Noades J., Robson E.B.: *Linkage between α-fucosidase and the Rhesus blood group.* Ann. Hum. Genet. *40,* 403-405, 1977.
50. Van Cong N., Weil D., Finaz C., Rebourchet R., Cochet C., Grouchy J. de, Frezal J.: *Localisation du gene α-Fuc sur 1p chez l'homme et le chimpanzé et sur le 4 chez le cercopithéque.* Ann. Génét. *21,* 41-46, 1978.
51. Mc Kusick V.A.: *Mendelian Inheritance in Man. Catalogs of autosomal dominant, autosomal recessive and X-linked phenotypes.* 5th Ed, The Johns Hopkins University Press., Baltimore and London, 1978.
52. Mc Kusick V.A.: *The anatomy of the human genome.* Am. J. Med. *69,* 267-276, 1980.

53. Goss S.J., Harris H.: *Gene transfer by means of cell fusion. II. The mapping of 8 loci on human chromosome 1 by statistical analysis of gene assortment in somatic cell hybrids.* J. Cell. Sci. *25*, 39-57, 1977.
54. Cook P.J., Noades J.E., Newton M.S., Mey R. de: *On the orientation of the Rh: E1-1 linkage group.* Ann. Human Genet. *41,* 157-162, 1977.
55. Thorpe R., Robinson D.: *Purification and serological studies of human α-L-fucosidase in the normal and fucosidosis states.* Clin. Chim. Acta *86,* 21-30, 1978.
56. Alhadeff J.A., Andrews-Smith G.L.: *Purification and characterization of α-L-fucosidase from the liver of a fucosidosis patient.* Biochem. J. *187,* 45-51, 1980.
57. Bugiani O., Borrone C.: *Fucosidosis: a neuropathological study.* Riv. Pat. Nerv. Ment. *97,* 133-141, 1975.
58. Troost J., Straks W., Willemse J.: *Fucosidosis. II. Ultrastructure.* Neuropädiatrie *8,* 163-171, 1977.
59. Libert I., Martin I.I., Ceuterick C.: Personal communication.
60. Larbrisseau A., Brochu P., Jasmin G.: *Fucosidose de type I. Etude anatomique.* Arch. Franç. Pédiat. *36,* 1013-1023, 1979.
61. Crovato F., Rampini E., Tomatis C., Della Cella G., Borrone C., Durand P.: *Studio istochimico della cute nella fucosidosi.* Pathologica *63,* 269-273, 1971.
62. Epinette W.W., Norins A.L., Drew A.L., Zeman W., Patel V.: *Angiokeratoma corporis diffusum with α-L-fucosidase deficiency.* Arch. Dermatol. *107,* 754-757, 1973.
63. Carpenter S., Karpati G., Larbrisseau A., Charron L.: *Characteristic involvement in biopsies of skeleton muscle, peripheral nerve and skin in alpha-fucosidosis.* Neurology *28,* 365, 1978.
64. Ginsel L.A., Cambier P.H., Daems W. Th.: *Fucosidosis and I-cell disease: A fine-structural and silver staining study of abnormal inclusion bodies in small-intestinal cells.* Virchows Arch. B. Cell. Path. *27,* 99-117, 1978.
65. Bugiani O., Borrone C., Durand P.: Unpublished data.
66. Troost J., van der Heijden M.C.M., Staal G.E.J.: *Characterization of α-L-fucosidase from two different families with fucosidosis.* Clin. Chim. Acta *73,* 329-346, 1976.
67. Alhadeff J.A., Miller A.L., Wenger D.A., O'Brien J.S.: *Electrophoretic forms of human liver α-L-fucosidase and their relationship to fucosidosis (Mucopolysaccharidosis F).* Çlin. Chim. Acta *57,* 307-313, 1974.
68. Zielke K., Veath M.L., O'Brien J.S.: *Fucosidosis: deficiency of α-L- fucosidase in cultured skin fibroblasts.* J. Exp. Med. *136,* 197-199, 1972.
69. Ng W.G., Donnell G.N., Koch R., Bergren W.R.: *Biochemical and genetic studies of plasma leukocyte α-L-fucosidase.* Am. J. Hum. Genet. *28,* 42-50, 1976.
70. Beratis N.G., Hirschhorn K.: *α-L-fucosidase in cultured bone marrow fibroblasts from fucosidosis patients.* Am. J. Med. Genet. *2,* 23-29, 1978.
71. Di Matteo G., Durand P., Gatti R., Maresca A., Orfeo M., Urbano F., Romeo G.: *Human α-fucosidase. Single residual enzymatic form in fucosidosis.* Biochim. Biophys. Acta *429,* 538-545, 1976.
72. Zielke K., Okada S., O'Brien J.S.: *Fucosidosis: Diagnosis by serum assay of α-L-fucosidase.* J. Lab. Clin. Med. *79,* 164-169, 1972.
73. Den Tandt W.R., Jacken J.: *Confirmation of metachromatic leukodystrophy and fucosidosis by enzyme analysis of saliva.* Neuropädiatrie *11,* 189-190, 1980.
74. Aminoff D., Furukawa K.: *Enzymes that destroy blood group specificity. Purification and properties of α-L-fucosidase from clostridium perfrigens.* J. Biol. Chem. *245,* 1659-1669, 1970.
75. Stealey J.R., Watkins W.M.: *Separation of α-L-fucosidases in extracts of trichomonas foetus.* Biochem. J. *126,* 16-17, 1971.
76. Iijima Y., Egami F.; *Purification of α-L-fucosidase from liver of a marine gastropod. Charonia lampas.* J. Biochem. *70,* 75-78, 1971.
77. Levvy G.A., Mc Allan A.: *Mammalian fucosidases. 2. α-L-fucosidase.* Biochem. J. *80,* 435-439, 1961.
78. Wiederschain G.Y.A., Rosenfeld E.L.: *Two forms of α-L-fucosidase from pig kidney and their action on natural oligosaccharides.* Biochem. Biophys. Res. Commun. *44,*

1008-1014, 1971.

79. Wiederschain G.Y.A., Kolibaba L.G., Rosenfeld E.L.: *Human α-L-fucosidases.* Clin. Chim. Acta *46,* 305-310, 1973.
80. Wiederschain G.Y.A., Rosenfeld E.L., Brusilovsky A.I., Kolibaba L.G.: *α-L-fucosidase and other glycosidases in human placenta, foetus and amniotic fluid at various stages of gestation.* Clin. Chim. Acta *35,* 99-107, 1971.
81. Di Matteo G., Orfeo M.A., Romeo G.: *Human α-L-fucosidase: purification and properties.* Biochim. Biophys. Acta *429,* 527-537, 1976.
82. Troost J., van der Heijden M.C.M., Staal G.E.J.: *Human leukocyte α-L-fucosidase.* Clin. Chim. Acta *73,* 321-327, 1976.
83. Alhadeff J.A., Janowsky A.J.: *Human serum α-L-fucosidase.* Clin. Chim. Acta *82,* 133-140, 1978.
84. Alhadeff J.A., Andrews-Smith G.L.: *Human amniotic fluid α-L-fucosidase.* Clin. Genet. *16,* 357-363, 1979.
85. Bosmann H.B.: *Red cell hydrolases: glycosidase activities in human erythrocyte plasma membranes.* J. Membr. Biol. *4,* 113-123, 1971.
86. Avilla J.L., Convit J.: *Studies on human polymorphonuclear leukocyte enzyme. IV Intracellular distribution and properties of α-L-fucosidase.* Biochim. Biophys. Acta *358,* 308-318, 1974.
87. Thorpe R., Robinson D.: *Isoelectric focusing of isoenzymes of human liver α-L-fucosidase.* FEBS Lett. *54,* 89-92, 1975.
88. Turner B.M., Turner V.S., Beratis N.G., Hirschhorn K.: *Polymorphism of human α-L-fucosidase.* Am. J. Hum. Genet. *27,* 651-661, 1975.
89. Alhadeff J.A., Janowsky A.J.: *Purification and properties of human brain α-L-fucosidase.* J. Neurochem. *28,* 423-427, 1977.
90. Turner B.M., Beratis N.G., Turner V.S., Hirschhorn K.: *Isozymes of human α-L-fucosidase detectable by starch gel electrophoresis.* Clin. Chim. Acta *57,* 29-35, 1974.
91. Chester M.A., Hultberg B., Sjöblad S.: *A comparison of the α-L-fucosidase activity of human liver and serum.* Biochim. Biophys. Acta *485,* 147-155, 1977.
92. Alhadeff J.A., Tennant L., O'Brien J.S.: *Isoenzyme patterns of human liver α-L-fucosidase during development.* Dev. Biol. *47,* 319-324, 1975.
93. Robinson D., Thorpe R.: *Human liver α-L-fucosidases.* Clin. Chim. Acta *47,* 403-407, 1973.
94. Robinson D., Thorpe R.: *Affinity chromatography of human liver α-L-fucosidase.* FEBS Lett. *45,* 191-193, 1974.
95. Alhadeff J.A., Miller A.L., O'Brien J.S.: *Purification of human placental α-L-fucosidase by affinity chromatography.* Anal. Biochem. *60,* 424-430, 1974.
96. Robinson D., Thorpe R.: *Fluorescent assay of α-L-fucosidase.* Clin. Chim. Acta *55,* 65-69, 1974.
97. Alhadeff J.A., Andrews-Smith G.L.: *Thermostability of human α-L-fucosidase. Relationship to fucosidosis and low-activity serum α-L-fucosidase.* Biochim. Biophys. Acta *614,* 466-475, 1980.
98. Alhadeff J.A., Cimino G., Janowsky A.: *Isoenzymes of human liver α-L-fucosidase: chemical relationship, kinetic studies and immunochemical characterization.* Molec. Cell. Biochem. *19,* 171-179, 1978.
99. Alhadeff J.A., Andrews-Smith G.L.: *Subunit composition of human liver α-L-fucosidase.* Biochem. J. *177,* 753-756, 1979.
100. Chien S.F., Dawson G.: *Purification and properties of two forms of human α-L-fucosidase.* Biochim. Biophys. Acta *614,* 476-488, 1980.
101. Kaplan A., Achard D.T., Sly W.S.: *Phosphohexosyl components of a lysosomal enzyme are recognized by pinocytosis receptors on human fibroblasts.* Proc. Natl. Acad. Sci. *74,* 2026-2030, 1977.
102. von Figura K., Klein U.: *Isolation and characterization of phosphorylated oligosaccharides from α-N-acetylglucosaminidase that are recognized by cell-surface receptors.* Europ. J. Biochem. *94,* 347-354, 1979.

103. Turner B.M., Turner V.S., Hirschhorn K.: *Metabolic correction of fucosidosis fibroblasts by human α-L-fucosidase.* J. Cell. Physiol. *98*, 225-235, 1979.
104. Ramage P., Cunningham W.L.: *The occurrence of low α-L-fucosidase activities in normal human serum.* Biochim. Biophys. Acta *403,* 473-476, 1975.
105. Playfer J.R., Price Evans D.A.: *Enzyme activity in fucosidosis.* Lancet *2,* 1415-1416, 1976.
106. Wood S.: *Plasma α-L-fucosidase: presence of a low activity variant in some individuals.* J. Lab. Clin. Med. *88*, 469-476, 1976.
107. Alhadeff J.A., Andrews-Smith G.L.: *Kinetic and immunochemical characterization of low-activity serum α-L-fucosidase from a phenotypically normal individual.* Biochem. Med. *20,* 357-363, 1978.
108. Gatti R., Cavalieri S., Romeo G.: *Relationship between α-L-fucosidase deficiency in plasma and α-L-fucosidase activity in leukocytes.* Hum. Genet. *48,* 23-30, 1979.
109. Wood S.: *Human alpha-L-fucosidase: a common polymorphic variant for low serum enzyme activity, studies of serum and leukocyte enzyme.* Hum. Hered. *29,* 226-229, 1979.
110. Turner B.M., Beratis N.G., Turner V.S., Hirschhorn K.: *Silent allele as genetic basis of fucosidosis.* Nature *257,* 391-392, 1975.
111. Spiro R.G.: *Glycoproteins.* Adv. Protein Chem. *27,* 349-467, 1973.
112. Marshall R.D.: *Glycoproteins.* Ann. Rev. Biochem. *41,* 673-702, 1972.
113. Kornfeld R., Kornfeld S.: *Comparative aspects of glycoprotein structure.* Ann. Rev. Biochem. *45*, 217-237, 1976.
114. Margolis R.K., Margolis R.U.: *Disposition of fucose in brain.* J. Neurochem. *19*, 1023-1030, 1972.
115. Ghidoni R., Sonnino S., Tettamanti G., Wiegandt H., Zambotti V.: *On the structure of two new gangliosides from beef brain.* J. Neurochem. *27*, 511-515, 1976.
116. Marcus D.M.: *The ABO and Lewis blood-group system. Immunochemistry, genetics and relation to human disease.* New Engl. J. Med. *280*, 994-1006, 1969.
117. Harris H.: *The blood-group substances.* In "The principles of human biochemical genetics". North Holland Publ. Comp. Amsterdam, London, 191-210, 1971.
118. Schenkel-Brunner H., Chester M.A., Watkins W.M.: *α-L-fucosyltransferases in human serum from donors of different ABO, Secretor and Lewis blood-group phenotypes.* Eur. J. Biochem. *30*, 269-277, 1972.
119. Lundblad A., Berggard I.: *Isolation and characterization of urinary A and B agglutinogen fractions.* Biochim. Biophys. Acta *148,* 146-150, 1967.
120. Strecker G., Montreuil J.:*Isolement et étude de la structure de 16 oligosaccharides isolés de l'urine humaine.* C.R. Acad. Sc. Paris D *277,* 1393-1396, 1973.
121. Denborough M.A., Downing H.J., McCrea M.G.: *ABO system in milk and saliva.* Australasian Ann. Med. *16,* 320-325, 1967.
122. Dawson G., Tsay G.C.: *Fucosidosis.* In "Current Trends in Sphyngolipidoses and Allied Disorders" (B.W. Volk, L. Schneck Eds.), Plenum Press, New York, 187-203, 1976.
123. Tsay G.C., Dawson G., Sung S.S.J.: *Structure of the accumulating oligosaccharide in Fucosidosis.* J. Biol. Chem. *251,* 5852-5859, 1976.
124. Tsay G.C., Dawson G.: *Oligosaccharide storage in brains from patients with fucosidosis,* G_{M1} *gangliosidosis and* G_{M2}*-gangliosidosis (Sandhoff's disease).* J. Neurochem. *27,* 733-740, 1976.
125. Tsay G.C., Dawson G.: *Glycopeptide storage in fibroblasts from patients with inborn errors of glycoprotein and glycosphingolipid catabolism.* Biochem. Biophys. Res. Comm. *63,* 807-814, 1975.
126. Staal G.E.J., van der Heiyden M.C.M., Troost J., Moes M., Borst-Eilers E.: *Fucosidosis and Lewis substances.* Clin. Chim. Acta *76,* 155-157, 1977.
127. Strecker G., Fournet B., Spik G., Montreuil J., Durand P., Tondeur M.: *Structure de 9 oligosaccharides et glycopeptides riches en fucose escrétés dans l'urine de deux sujets atteints de fucosidose.* C.R. Acad. Sc. Paris D *284,* 85-88, 1977.
128. Strecker G., Fournet B., Montreuil J., Dorland L., Haverkamp Y., Vliegenthart

F.G., Dubesset D.: *Structure of the three major fucosyl-glycoasparagines accumulating in the urine of a patient with fucosidosis.* Biochimie *60,* 725-734, 1978.
129. Lundblad A., Lundsten J., Norden N.E., Sjöblad S., Svensson S., Ockermann P.A., Gehloff M.: *Urinary abnormalities in fucosidosis. Characterization of a disaccharide and two glycoasparagines.* Europ. J. Biochem. *83,* 513-521, 1978.
130. Nishigaki M., Yamashita K., Matsuda I., Arashima S., Kobata A.: *Urinary oligasaccharides of fucosidosis. Evidence of the occurence of X-antigenic determinant in serum-type sugar chains of glycoproteins.* J. Biochem. *84,* 823-834, 1978.
131. Ng Ying Kin N.M.K., Wolfe L.S.: *Urinary excretion of a novel hexasaccharide and a glycopeptide analogue in fucosidosis.* Biochem. Biophys. Res. Commun. *88,* 695-705, 1979.
132. Yamashita K., Tachibana Y., Takada S., Matsuda I., Arashima S., Kobata A.: *Urinary glycopeptides of fucosidosis.* J. Biol. Chem. *254,* 4820-4827, 1979.
133. Wood S.: *A sensitive fluorometric assay for α-L-fucosidase.* Clin. Chim. Acta *58,* 251-256, 1975.
134. Humbel R., Collart M.: *Oligosaccharides in the urine of the patients with glycoprotein storage diseases. Rapid detection by thin-layer chromatography.* Clin. Chim. Acta *60,* 143-145, 1975.
135. Sewell A.C.: *An improved thin-layer chromatographic method for urinary oligosaccharide screening.* Clin. Chim. Acta *92,* 411-414, 1979.
136. Tsai M.Y., Marshall J.G.: *Screening for urinary oligosaccharides and simple sugars by thin-layer chromatography.* Med. Lab. Sci. *36,* 85-90, 1979.
137. Humbel R., Dhondt J.L.: *Dépistage des oligosacchariduries.* In "Les Oligosaccharidoses" (Farriaux J.P. Ed.), Crouan et Roques, Lille, 119-122, 1977.
138. Matsuda I., Arashima S., Anakura M., Oka Y.: *α-L-fucosidase and α-D-mannosidase activity in the white blood cells in the disease and carrier state of fucosidosis.* Clin. Chim. Acta *48,* 9-13, 1973.
139. Heukels-Dully M.J., Niermeijer M.F.: *Variation in lysosomal enzyme activity during growth in culture of human fibroblasts and amniotic fluid cells.* Exp. Cell. Res. *97,* 304-312, 1976.
140. Gatti R., Costantino G., Borrone C., Salemi M., Filocamo M., Romeo G., Durand P.: *Detection of carriers for fucosidosis in Calabria.* Helv. Paediat. Acta - Suppl. *45,* 30, 1980.
141. Romeo G., Borrone C., Gatti R., Durand P.: *Fucosidosis in Calabria: founder effect or high gene frequency?* Lancet *1,* 368-369, 1977.
142. Gerbie A.B., Melancon S.B., Ryan C., Nadler H.L.: *Cultivated epithelial-like cells and fibroblasts from amniotic fluid: Their relationship to enzymatic and cytologic analysis.* Am. J. Obstet. Gynec. *114,* 314-318, 1972.
143. Butterworth J., Guy G.J.: *α-L-fucosidase of human skin fibroblasts and amniotic fluid cells in tissue culture.* Clin. Genet. *12,* 297-302, 1977.
144. Matsuda I., Arashima S., Oka Y., Mitsuyama T., Ariga S., Ikeuchi T., Ichida T.: *Prenatal diagnosis of fucosidosis.* Clin. Chim. Acta *63,* 55-60, 1975.
145. Bugiani O.: Unpublished data.
146. Levvy G.A.: *The β-D-fucosidase activity of mammalian tissues and its relation to other glycosidases.* Nature *187,* 1027-1028, 1960.
147. Wiederschain G. Y.A., Beyer E.M., Klyashchitsky B.A., Shashkov A.S.: *Specificity patterns of different types of human fucosidase. Recognition of a certain region of the pyranose ring in sugars by the enzymes.* Biochim. Biophys. Acta *659,* 434-444, 1981.

M. Alan Chester
Arne Lundblad
Per-Arne Öckerman
Seppo Autio

4 Mannosidosis

NAME OF DISEASE	Mannosidosis
PREVALENCE	?
MODE OF INHERITANCE	Autosomal recessive
MUTANT GENE PRODUCT	Acidic α-mannosidases
CHROMOSOMAL LOCATION	19
EXPRESSION	
1. Altered DNA structure	?
2. Disturbed protein function	Enzyme activity absent
3. Disrupted cell and organ function	Accumulation of substrates (α-mannosides) in lysosomes of most cells, including brain, leads to brain dysfunction and somatic changes

4.1 INTRODUCTION

The term Mannosidosis was originally coined to describe a human lysosomal storage disease characterised by the accumulation of α-mannosides caused by a deficiency of α-mannosidase. Subsequently an inherited disease in cattle was also shown to be due to an α-mannosidase defect, and was named bovine Mannosidosis. The disease has also been found in domestic cats. β-Mannosidase is a lysosomal enzyme whose presumed mode of action is the intralysosomal hydrolysis of β-mannoside linkages in glycoproteins. The recent discovery of a deficiency of β-mannosidase in goats means that a distinction must be made between α- and β-Mannosidosis in anticipation of the eventual description of both α- and β-Mannosidosis in the same species. We therefore recommend that the term Mannosidosis is only used when describing the diseases as a group, and that the prefixes α- and β- should be used whenever possible. The major part of this review will deal with human α-Mannosidosis.

A patient with a generalized storage disorder clinically reminiscent of the mucopolysaccharidoses but without mucopolysacchariduria was described in 1967 [1]. A profound deficiency of α-mannosidase was evident. Subsequent characterization of the stored material and the absence of a lysosomal enzyme indicated that the underlying defect impairs the intralysosomal degradation of glycoproteins.

Table 4.1 Published cases of Mannosidosis.

Case no.	Year published	Age at publication (years)	Sex	Nationality	Initials	Reference
1	1967	4[x]	M	Swedish	H.P.	1,2
2	1973	4[1.]	F	Hungarian	—	3
3	1973	7[1.]	F	Hungarian	—	3
4	1973	5[2.]	M	Finnish	A.S.	4, 5
5	1973	12[2.]	M	Finnish	P.S.	4, 5
6	1973	6	M	Finnish	H.K.	4, 5
7	1974	1	M	American	M.G.	6
8	1975	22[3.]	M	Finnish	J.N.	5, 7
9	1975	19[3.]	M	Finnish	P.N.	5, 7
10	1975	15[3.]	M	Finnish	H.N.	5, 7
11	1975	5	M	French	C.F.	8, 9
12	1975	5	F	French	M.G.	8, 9
13	1975	13[4.]	M	French	P.H.	8, 9
14	1975	5[4.]	F	French	B.H.	8, 9
15	1975	10[4.]	M	French	J.H.	8, 9
16	1975	5	M	German	D.C.	10, 11
17	1975	12	M	German	H.D.	10, 11
18	1975	9	F	German	A.S.	10, 11
19	1975	9[5.]	F	German	C.H.	10, 11
20	1975	6[5.]	F	German	E.H.	10, 11
21	1975	12[5.]	F	German	A.H.	10, 11
22	1975	4	?	French	C.P.	12
23	1976	10[6.]	M	European?	P.P.	13, 14
24	1976	19[6.]	F	European?	C.P.	13, 14
25	1976	2	M	American	K.A.	15
26	1976	7	M	American	S.A.	16, 17
27	1976	4[7.]	F	American	—	16, 18
28	1976	5[7.]	F	American	—	16, 18
29	1976	13[8.]	M	American	E.D.	19
30	1976	16[8.]	F	American	J.D.	19
31	1976	26[8.]	F	American	H.D.	19
32	1976	3[x.]	F	American	M.L.	20-23
33	1976	25[9.]	M	American	G.G.	20-23
34	1976	18[9.]	M	American	J.G.	20-23
35	1976	16[9.]	F	American	M.G.	20-23
36	1976	21[9.]	F	American	M.G.	20-23
37	1976	?[10.]	F	French/ N. African?	Z.B.	24
38	1976	?[10.]	F	French/ N. African?	F.B.	24
39	1976	1	F	German	E.Ka.	11
40	1976	6	F	German	A.Ku.	11
41	1976	6	F	German	D.Be.	11
42	1976	10[11.]	M	German	P.Bo.	11
43	1976	11[11.]	F	German	E.Bo.	11
44	1976	13[11.]	F	German	C.Bo.	11
45	1976	20	M	American	—	25
46	1976	15[12.]	F	American	—	25
47	1976	17[12.]	F	American	—	25
48	1976	20[12.]	M	American	—	25
49	1976	24[12.]	M	American	—	25

(continued)

Table 4.1 *(continued)*

Case no.	Year published	Age at publication (years)	Sex	Nationality	Initials	Reference
50	1977	6	M	American	—	26
51	1977	26	M	American	—	26
52	1977	19	M	English	—	27
53	1977	3	M	English	—	27
54	1977	14[13.]	M	American	J.G.	28
55	1977	16[13.]	M	American	L.G.	28
56	1977	18[13.]	M	American	M.G.	28
57	1978	11[14.]	M	Arabic	N.A.S.	29
58	1978	5[14.]	F	Arabic	B.A.S.	29
59	1979	2[x]	F	Italian	R.G.	30
60	1979	4	M	French	C.G.	31
61	1979	?	?	French/ N. African?	N.M.	32
62	1980	18	M	Japanese	—	33
63	1980	10	F	Japanese	—	33
64	1980	4	M	Irish	C.W.	34
65	1980	2	M	Russian	P.A.	35
66	1981	13[15.]	F	English	M.H.	5
67	1981	8[15.]	M	English	N.H.	5

[x.] Deceased.
Suffixes 1-15 indicate siblings.

Five patients with α-Mannosidosis were described in the six years following the discovery of the disease, but during the past seven years almost sixty new cases have been found (Table 1), presumably as a result of increased clinical experience and improvements in biochemical methods of detection. The disease is not localized to any specific ethnic group since patients of Scandinavian, Western and Eastern European, North American, Arabian, African and Japanese origin have been described. Data on 67 published cases are summarized in table 1; we know of 10-15 additional patients with α-Mannosidosis and believe that a number of other cases exist but have not been published. We feel that it is no longer of value to publish new cases of α-Mannosidosis unless the data can improve the understanding of the disease.

4.2 CLINICAL SYMPTOMATOLOGY

A. General problems

The main clinical features of the first patients published were psychomotor retardation and a somewhat coarse face [1-3]. Other symptoms were ataxia, impaired hearing, vacuolized lymphocytes in the peripheral blood and slightly Hurler-like skeletal changes [4]. These features, together with transient hepatomegaly [1-3] and small ocular opacities [3, 4] seem to be the most common clinical characteristics of α-Mannosidosis, based on data from about sixty subsequent cases (Table 2). There are many inherent problems when compiling

Table 4.2 The frequency of some symptoms in α-Mannosidosis.

Symptom	Frequency
Mental retardation	53/55
Delayed motor development	41/45
Hearing loss	39/47
Impaired speech	30/33
Coarse facies	60/61
Dysostosis multiplex	55/59
Ocular changes	21/61*
Hepatomegaly	25/54
Recurrent infections	32/48
Hernia	22/31
Macrocephaly	7/21
Vacuolised lymphocytes	46/52
Bone marrow inclusions	18/18
Mucopolysaccharidia	4/55

* This value is probably not very informative. It is certainly too high if only opacities are considered, but is certainly too low if refraction errors, strabismus and slightly blurred discs are included.

such a table. Since there are without doubt a greater number of interested clinicians than patients with α-Mannosidosis, the estimation of the severity of a particular symptom is bound to be subjective. Nor is the designated presence or absence of a symptom beyond question, since this depends on the experience of the paediatrician, ophthalmologist, psychologist or other clinician. Finally, as more information becomes available and techniques improve, further abnormalities are found which may have been missed or not looked for in earlier case reports.

Perhaps the best way of standardizing the description of clinical symptomatology is to allow an experienced clinical team to examine as many as possible of the known cases of α-Mannosidosis, and compare these data with those in the original publications.

B. Early development and onset of symptoms

The disease does not seem to cause any apparent disturbances during prenatal life. The patients have appeared healthy at birth and their physical and mental development usually seems normal throughout infancy and frequently throughout early childhood. A very common feature, however, is unusually frequent respiratory and ear infections (Table 2) which start during the first two years of life and continue for a few years, before gradually disappearing. This phenomenon has been observed in most patients but does not give cause for alarm since such symptoms are common in childhood. The etiology of this phenomenon is unknown (see also Chapter 5 on Aspartylglucosaminuria).

The first signs of the involvement of the central nervous system usually appear between the end of the first year of life and the fourth year. The onset of these symptoms is gradual and they progress slowly. The first really alarming symptom is the delayed development of speech, mental or motor functions, or a combination of these.

C. Mental retardation

As stated above, early psychomotor development appears normal but mental retardation becomes a common feature in all patients and adult patients are usually moderately or severely mentally retarded [5, 14, 19, 25, 26].

Results from a longitudinal study [25] suggest that the mental retardation is generally static but, occasionally, even improvement in mental performance may occur. The latter statement is based on a somewhat increased mental age in two patients, tested in childhood and at an adult age.

Obvious progressive impairment of mental functions which stabilized during the second decade was recorded in another longitudinal study [5], using conventional IQ-test methods. It was suggested that the previous results [25] also indicate mental regression rather than improvement, although there was some increase in mental age in absolute terms. However, measurement of total mental performance is very complex. There may also be appreciable variation in mental retardation, considering the clinical heterogeneity (see section L). Thus, the suggestion of progressive mental retardation in α-Mannosidosis remains speculative, although the usually normal early development and the poor mental performance of the adults support this speculation.

Special features considered in the psychological tests were confined to the functions of speech and memory [5]. The patients were late in starting to speak, their pronunciation is defective, their vocabulary restricted and their speech is hard to understand. They have difficulties in auditory discrimination, auditory perception and auditory memory. Their speaking ability has deteriorated over the years [5]. The difficulties in linguistic development may be due partly to impaired hearing, found in most of the patients.

D. Motor functions

Most patients have some disturbance of motor function, which has been described as clumsiness [10, 26, 27] or broad based gait [20, 25]. In a longterm study [5], ataxia was the most characteristic motor disturbance as a part of the generally clumsy behaviour. Ataxia has appeared at about the same time as the hearing loss (see section E), from the age of three or four onwards. The generally clumsy behaviour may be associated with mental retardation. However, the disease may particularly affect those areas of the brain which are responsible for fine motor functions and muscular co-ordination. It is also remarkable that none of the patients have been described as having spasticity or rigidity to the degree that is found in some patients with Aspartylglucosaminuria and in practically all patients with Fucosidosis or "Salla"-disease (see Chapters 3, 5, 7). None of them have had any remarkable dyskinesia, which is found, for instance, in "Salla"-disease.

Follow-up observations have also suggested a progressive impairment of motor functions with age [5]. The adult patients could still move independently but their ability to walk was greatly impaired when compared with their movements at a younger age. The progressive nature of these symptoms is more difficult to observe in patients with early severe psychomotor retardation.

E. Sensorineural hearing loss

A loss of hearing ability, due to sensorineural damage, is apparently present in most patients to varying degrees but because of mental retardation it may be sometimes difficult to demonstrate this sign by audiometric methods. It has been noticed to appear during early childhood [5, 25, 27]. This symptom is one of the essential characteristics of α-Mannosidosis and it is also part of the peculiar specific areal involvement of the central nervous system. There is no conclusive evidence for the progressive nature of this clinical phenomenon.

However, the impaired hearing may result partly from the early ear infections. Thus, in many patients, it might be a mixture of conductive and neurosensory elements. Its presence, no doubt, influences the observed disturbances in speech and mental functions.

F. Facial features

The typically coarse face is another clinical characteristic mentioned in almost all clinical reports of α-Mannosidosis (Table 2). The facial features are, most often, described as "slightly Hurler-like", or "gargoyle-like". We have used such description ourselves but we would like to avoid these expressions, although the features have similarities with those seen, for instance, in Hurler's and Hunter's diseases.

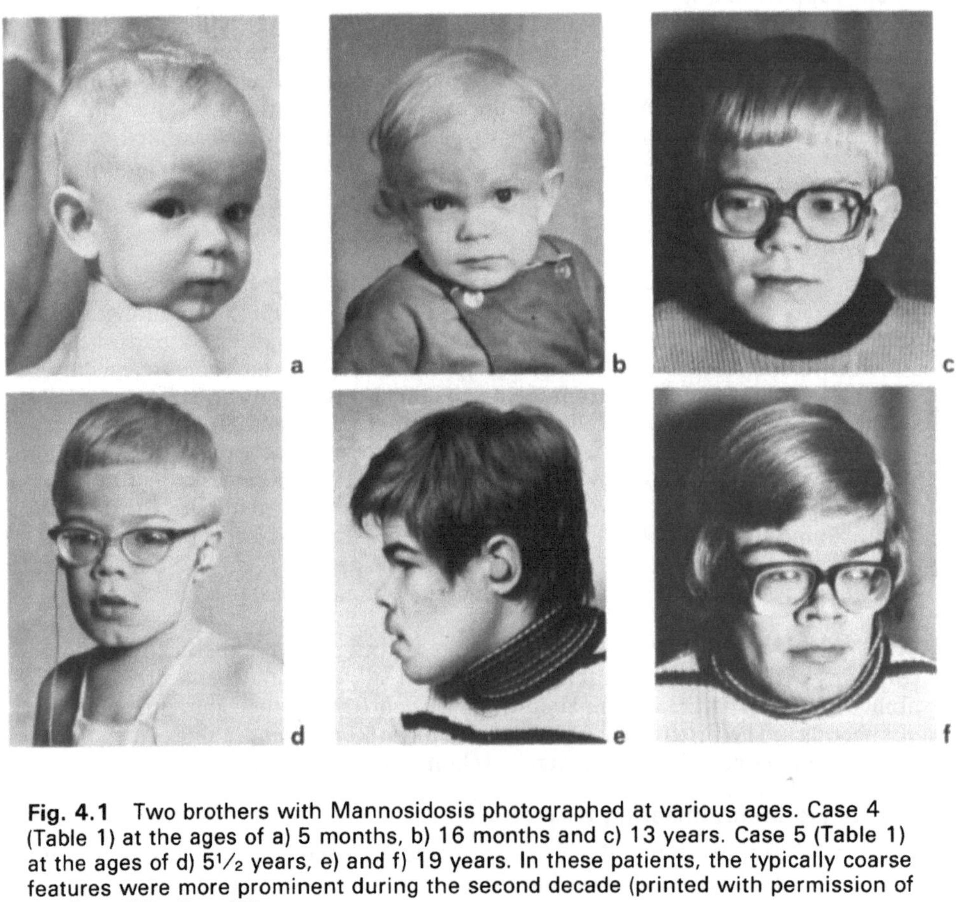

Fig. 4.1 Two brothers with Mannosidosis photographed at various ages. Case 4 (Table 1) at the ages of a) 5 months, b) 16 months and c) 13 years. Case 5 (Table 1) at the ages of d) $5^1/_2$ years, e) and f) 19 years. In these patients, the typically coarse features were more prominent during the second decade (printed with permission of the Ann. Clin. Res. [5]).

Typical facial features are a prominent forehead and a flattened nasal bridge. The ears are often low-set and large, the supraorbital bridges are prominent and there is prognathism. These features are very slight in early childhood but they become more obvious with increasing age. A more evident coarsening of the face appears during the second decade [5]. Examples of the facial features of two patients at various ages are given in figure 1. In spite of the coarse general appearance, the facial features cannot necessarily be called abnormal even at an adult age.

G. Skeletal changes

Osseous changes in several parts of the skeletal system, often called "dysostosis multiplex", have been described in most patients (Table 2). Typical features are a thickened calvarium and vertebral changes, mostly at the thoracolumbar region. Flattened vertebral bodies are most often found in the thoracal region. The ribs may be wider than normal. The inferior portions of the ilia may be hypoplastic and there may be mild expansion in the short tubular bones of the hands. Some examples of these features are shown in figure 2. There is a large variability in these abnormalities, even within the same family. The changes can be observed from the third or fourth year, although they may be mild at a young age [5].

It has been reported that the skeletal changes tend to lessen with age [25], although this tendency was not observed in another follow-up study [5].

H. Ocular changes

Involvement of the eye has been recorded in less than half of all patients (Table 2). The findings reported include lenticular changes characterized by posterior cortical opacities [4, 16, 22, 27], superficial corneal opacities [3, 29] or grey and slightly blurred discs [2]. Minor abnormalities recorded have been severe or moderate myopia, slight or moderate hyperopia and strabismus [5]. Clinically, the ophthalmological findings have no great significance.

I. Hepatomegaly and hepatosplenomegaly

An enlargement of the liver, or liver and spleen has been described in about half of the cases (Table 2). This symptom has been associated with the more severe form of the disease and it has been most pronounced in patients with an early death [2, 20, 30]. In some cases the liver and spleen have been only slightly enlarged [6, 16, 26], and in others the enlargement has been transient [2, 25, 27]. In some patients this organomegaly may be associated with muscular hypotonia and ligamental weakness.

J. General health and physical growth

After the initial incidence of unusually frequent infections, the patients have been described as physically healthy, apart from the apparent malfunction of the central nervous system. However, some patients do succumb to severe infections early in life [2, 20, 30]. It is still unsure whether the patients who die at an early age represent a severe form of the disease or whether the early deaths are only due to an unusually strong liability to infections. This would indicate an abnormal quantitative reaction (unusually frequent infections) and also an abnormal qualitative reaction (unusually severe infections).

The physical growth is often retarded, and the adult height may be about 2 SD below the normal mean [5, 25].

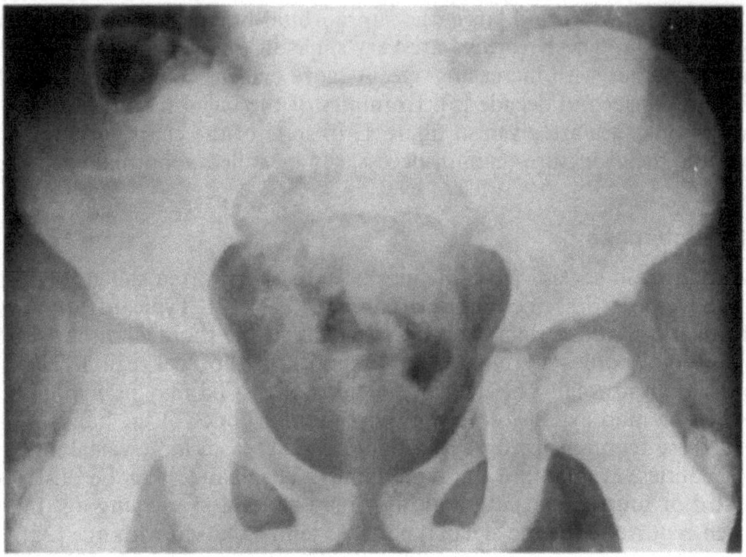

a

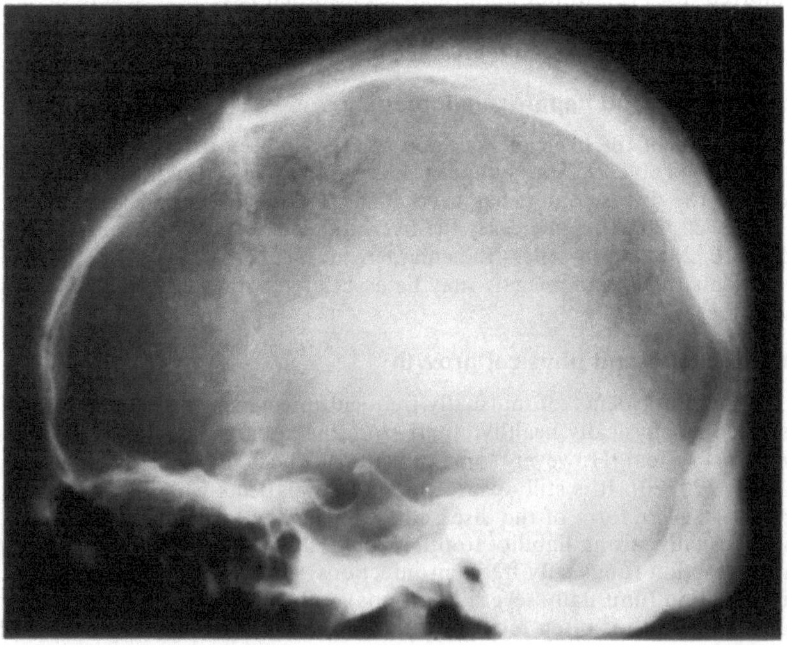

b

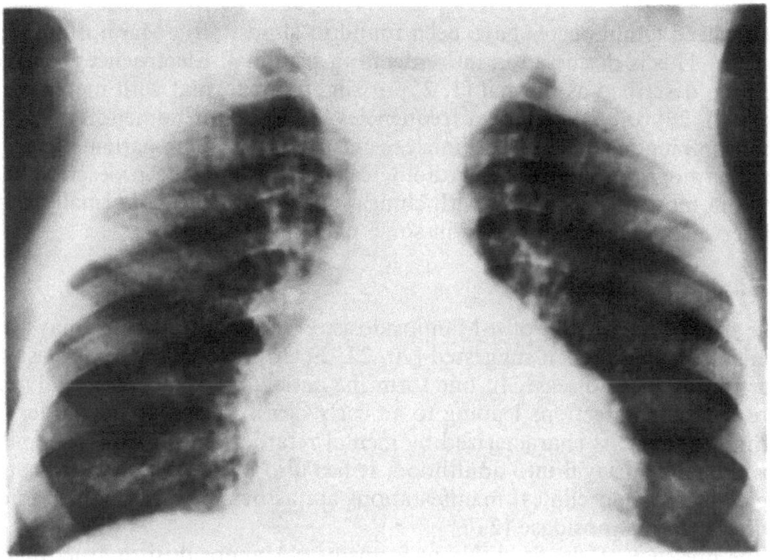

c

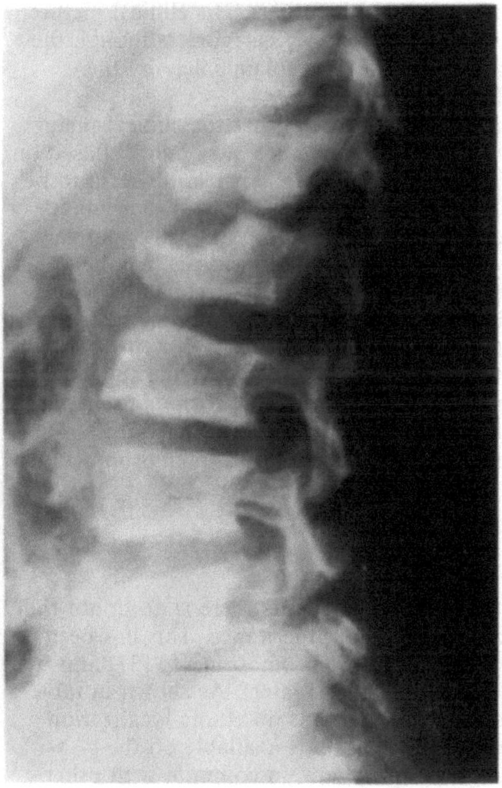

d

Fig. 4.2 Characteristic skeletal abnormalities in three patients with α-Mannosidosis. **a**, slightly hypoplastic ileae in Case 5 (Table 1), age 6 years. **b**, thickened calvarium and the base of skull (the same patient as in a., 16 years). **c**, broad and oar-shaped ribs in Case 4 (Table 1), age 7 years. **d**, marked osteochondrotic vertebral changes in the thoraco-lumbar region. Note also the unusually concave dorsal lines of the vertebral bodies and spondylolisthesis and spondylolysis of the fifth lumbar vertebra. Case 6 (Table 1), age 10 years.

K. Other clinical data

Vacuolized lymphocytes have been found in almost all α-Mannosidosis patients (Table 2). This is discussed in more detail in section 4. Electroencephalograms have been described as normal [1, 25] or slightly abnormal with increased slow activity [5, 26] or excessive beta frequencies [26]. The phenomenon of 14 + 6 c/s positive spiking has been found in sleep recordings of some patients [5]. Electrocardiograms and other cardiological findings have been described as normal in most of the reports with clinical data. However, abnormally short PR-intervals have been recorded in some patients [21].

L. Clinical heterogeneity

The clinical expression of α-Mannosidosis, even among siblings, may vary considerably. It has been suggested [20, 22, 29, 36] that there are at least two major forms of the disease. In one form the patients have hepatomegaly and severe recurrent infections leading to an early death. In the other, milder form, the clinical course is characterized by mental retardation, hearing loss, milder dysostosis and survival into adulthood. It has also been suggested, that at least in some cases the mild clinical manifestations are associated with higher residual activities of α-mannosidase [29].

If the presence of clearly different forms of α-Mannosidosis is true, then the eight patients in the Finnish study group [5] all have the milder form although one of them is clearly more severely affected than the others. His clinical picture, however, is not severe enough to put him in the "severe" category, although the relative severity of his clinical symptoms is difficult to explain solely by the abnormalities in his pre-and perinatal history.

Considering all the clinical data available, it is obvious that the clinical picture of α-Mannosidosis varies from "severe" through "moderate" to "mild". Based on the information published, the patients with a "moderate" expression seem to be in the majority. However, it is difficult to be more precise due to the variability of methods used and the quality of the descriptions in these reports. The relatively small number of patients having the severe form of α-Mannosidosis may be underrepresented, because the patients in this group die early, frequently before the correct diagnosis can be made. The same underrepresentation may be true for the mild form because the clinical manifestation might not be severe enough to arouse the suspicion of a lysosomal storage disease. Consequently the suggestion that α-Mannosidosis can be subdivided into a mild and severe form is premature, if not incorrect. Equally likely is a more uniform variation resulting from polymorphism at the α-mannosidase locus, and other genetic and environmental interactions.

4.3 GENETIC STUDIES

On the basis of the clinical reports on over sixty patients there is no doubt that α-Mannosidosis has an autosomal recessive mode of inheritance. This has been confirmed by localization of the enzyme involved to chromosome 19 [37] and by enzymic studies on parents and unaffected siblings (see later). As shown in table 1, α-Mannosidosis has a widespread occurrence, and has no ethnic localization such as is found in Tay-Sachs disease. No information is available on the frequency of the disease, although it must be quite rare, in common with other lysosomal storage diseases.

4.4 HISTOCHEMISTRY AND ULTRASTRUCTURE

The presence of vacuolized lymphocytes in the peripheral blood strongly indicates a lysosomal storage disease. Almost all patients with α-Mannosidosis have vacuolized lymphocytes (Table 2) although they could not be detected in Cases 7-10 or Case 58, and were only detected after repeated examination in the brother of Case 58 (Case 57). There may be a very large variation in the occurrence of these vacuolized cells over a rather short time, as has been observed in Aspartylglucosaminuria [38]. The presence of false positive (in infections) and

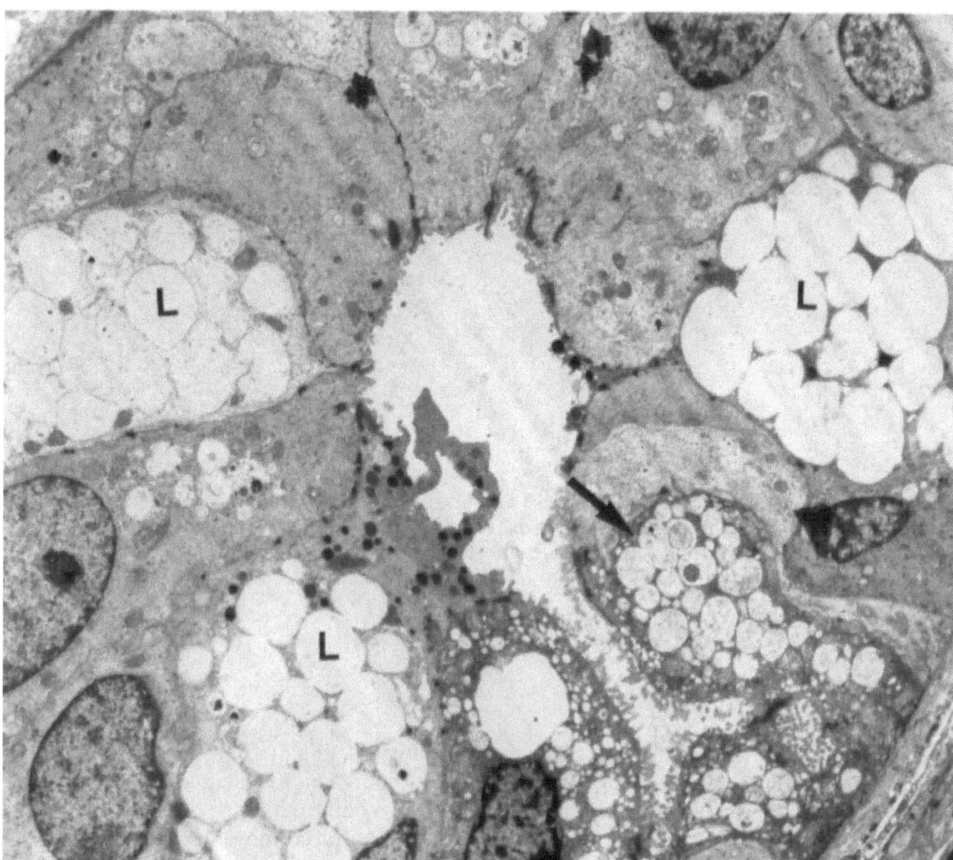

Fig. 4.3 Electron micrograph of the coiled portion of a sweat gland. The cytoplasm of a dark cell is filled with storage lysosomes containing reticulofibrillar material (arrow), whereas the lysosomes (L) in the light cells appear almost empty (x 6,000) [Courtesy of Dr. J. Rapola].

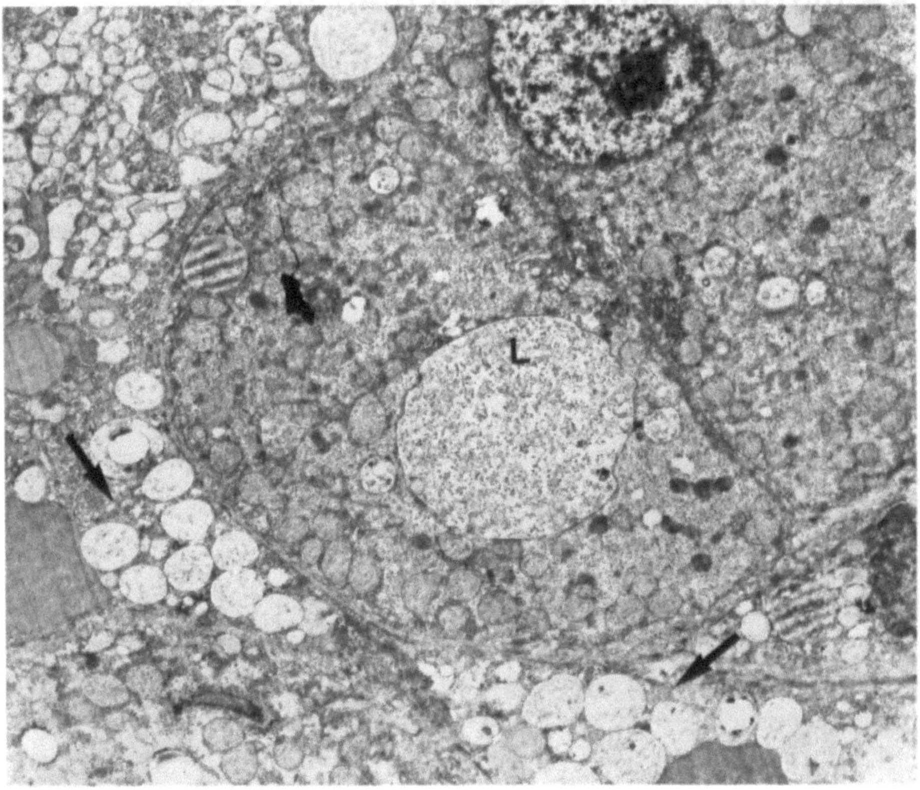

Fig. 4.4 Electron micrograph of liver. The hepatocyte in the middle contains one large storage lysosome (L). It is surrounded by Kupffer cells packed with storage vacuoles (arrows) (x 6,000) [Courtesy of Dr. J. Rapola].

false negative [6, 29] results decreases the reliability of this method for detecting lysosomal storage diseases. Nevertheless the demonstration of vacuolized lymphocytes is useful to the clinician, especially in young patients with mild clinical symptoms. Several cases of Aspartylglucosaminuria and one case of "Salla" disease have been diagnosed after the initial finding of vacuolized lymphocytes in the peripheral blood [Autio, unpublished results].

Several authors have found foamy macrophages containing transparent vacuoles using bone marrow smears [1, 4, 9, 11, 20, 25, 26, 28]. Vacuoles were also found in liver biopsy specimens [4, 14, 20, 34] presumably representing enlarged lysosomes filled with undegradable material. These enlarged vacuoles are present in various types of liver cell. Similar vacuoles have also been found in a conjunctival specimen from an α-Mannosidosis patient [16], and resemble the swollen lysosomes found in other storage diseases. A gingival biopsy from a patient with excessive gum hyperplasia [26] also contained a large number of vacuoles surrounded by a single membrane and containing fine granular material.

In the first *post mortem* study [2] the most striking feature was the presence of a large number of vacuoles in both neural and visceral tissues. A generalized vacuolization of nerve cells was found in the cerebral cortex, brainstem and spinal cord but not in the basal ganglia. There was also cerebellar atrophy and loss of myelin from the centrum semiovale. The optic nerves and retina were normal and evidence of storage was found in the nerve cells in the myenteric plexus. Lymphocytes from the spleen and lymph nodes showed minute cytoplasmic vacuoles. In another case of α-Mannosidosis a complete neuropathological examination was performed within an hour of death from an infection [20, 23] and similar results were found.

Some of the pathological features in α-Mannosidosis are shown in figures 3 and 4. No specific histological, histochemical or ultrastructural methods are available for the diagnosis of α-Mannosidosis, but the presence of a lysosomal storage disease can be diagnosed. Under the electron microscope α-Mannosidosis is very similar to Aspartlyglucosaminuria, and more detailed information is presented in Chapter 5. Perhaps further studies will reveal pathological features characteristic of α-Mannosidosis.

4.5 NATURE OF THE STORAGE MATERIAL

The name Mannosidosis was suggested by Öckerman [39] on the basis of the mannose-rich material stored in liver [1] and grey matter of the brain [39]. In those early reports on α-Mannosidosis the oligosaccharide-nature of the storage material was recognized, and some preliminary analytical data on isolated fractions containing only mannose and glucosamine were given [39]. A part from the studies carried out on tissues from the very first patient discovered most subsequent studies on the storage material have been done on urine, since the majority of patients with diagnosed α-Mannosidosis are still alive. The first detailed characterization of excreted oligosaccharides was done on urine from a group of Finnish patients [40, 41] and the most abundant compound was identified as a trisaccharide

Manα1-3Manβ1-4GlcNAc [33, 40, 42, 43]

In addition a tetra- and pentasaccharide were isolated in somewhat smaller amounts

Manα1-2Manα1-3Manβ1-4GlcNAc [33, 41-43]

Manα1-2Manα1-2Manα1-3Manβ1-4GlcNAc [33, 41-43]

The three oligosaccharides above have also been isolated from urine of other α-Mannosidosis patients [33, 42, 43]. Tsay *et al.* [6] described a branched pentasaccharide and reported that the trisaccharide could not be found in urine from their patient. However, other studies on this particular urine did not reveal any differences in oligosaccharide pattern as compared to the Finnish patients [A. Lundblad, unpublished results].

The isolation and characterization of a hexa-, a hepta-, an octa- and a nonasaccharide in addition to the previously isolated tri-, tetra- and pentasaccharides has also been reported [42]. The suggested structures are:

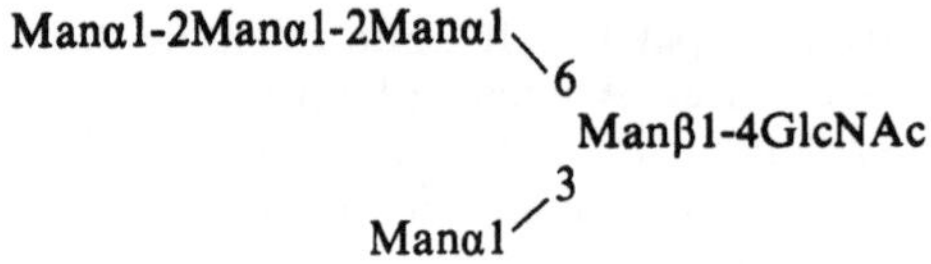

[42]

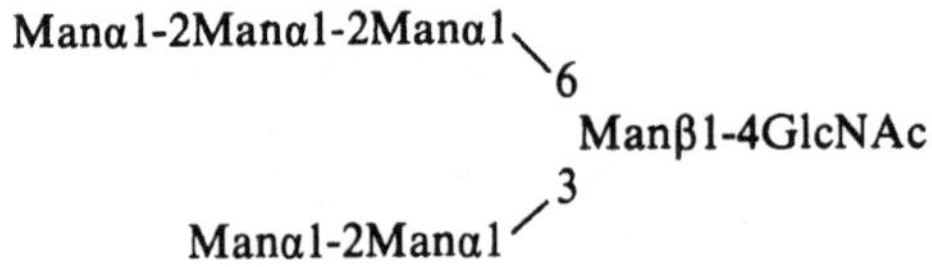

[42]

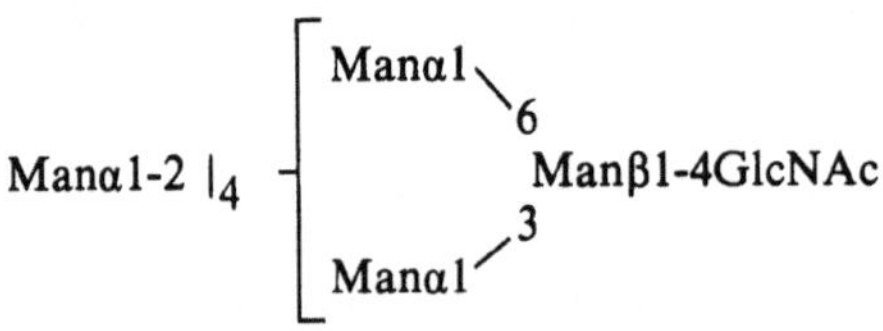

[42]

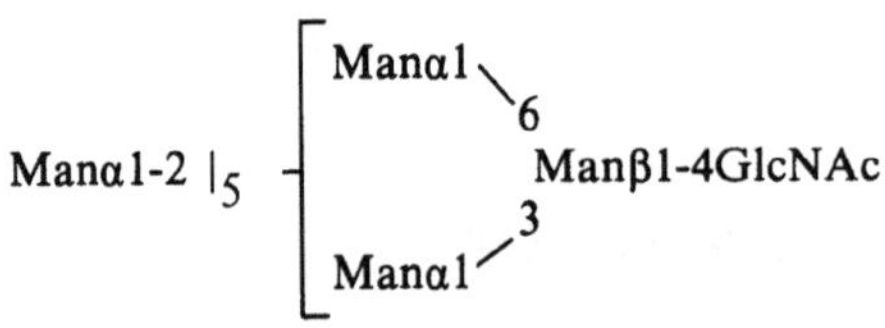

[42]

Two other groups [33, 43] have not been able to confirm these structures but have instead reported the isolation and characterization of thirteen [33] and fourteen [43] new higher molecular weight oligosaccharides from α-Mannosidosis urine. The two groups used similar methodology and characterized the following structures:

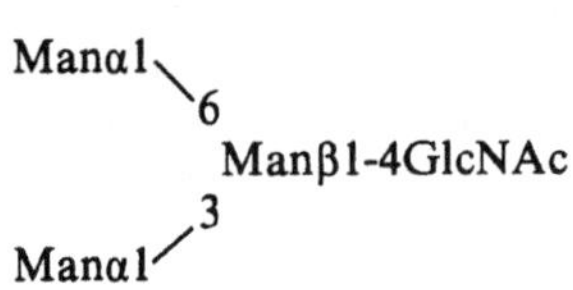

[43]

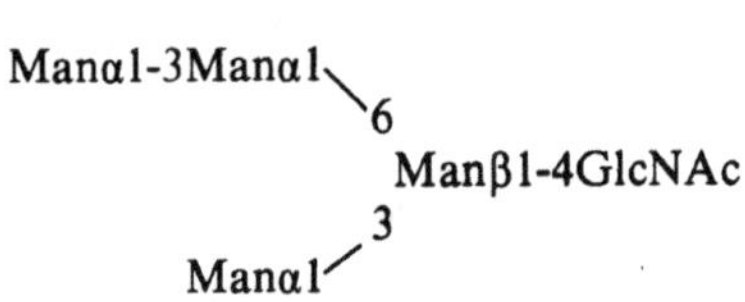

[33, 43]

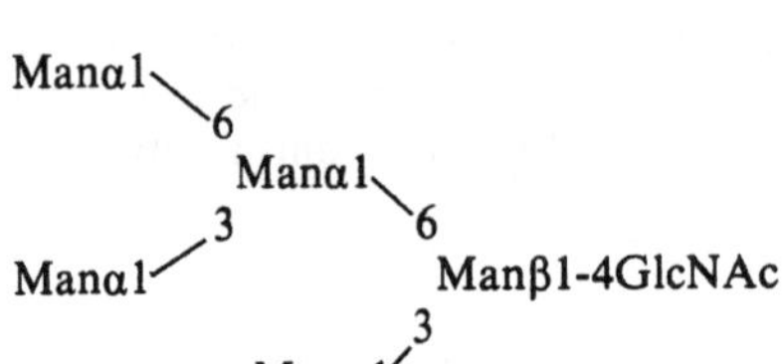

[33, 43]

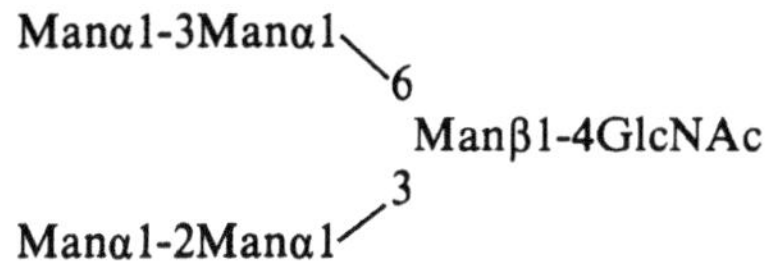

[33, 43]

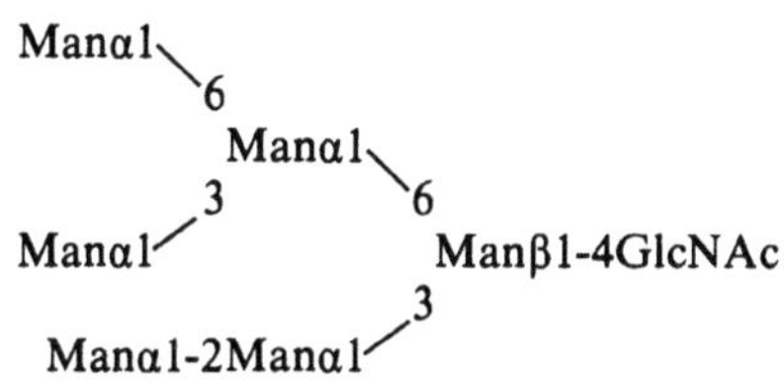

[33, 43]

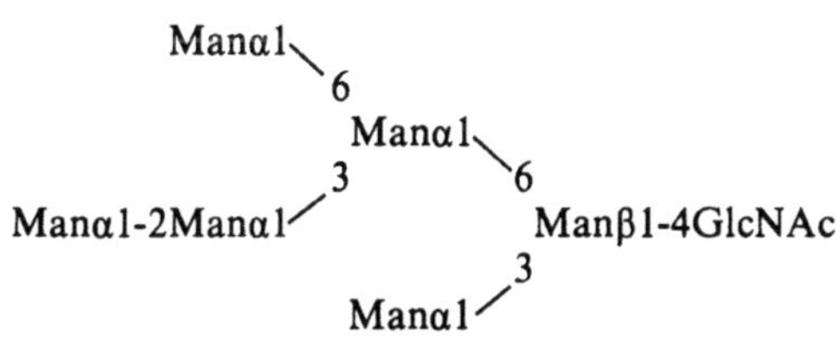

[33, 43]

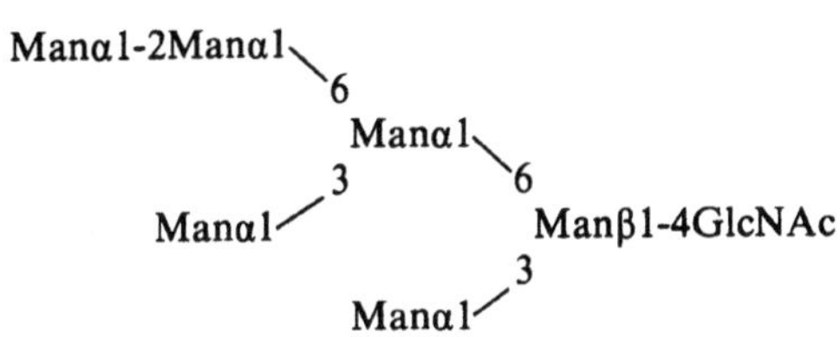

[33, 43]

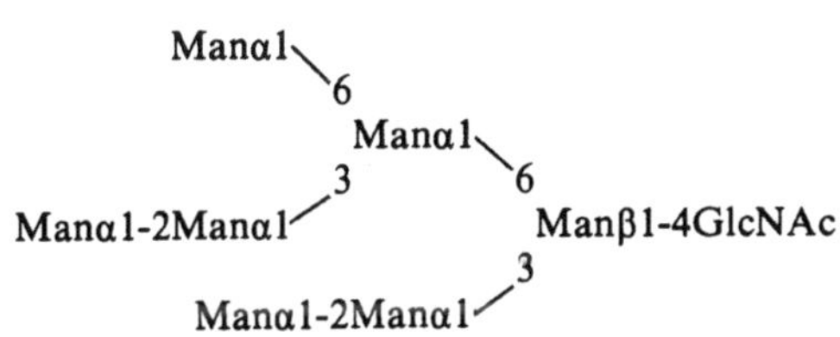

[33, 43]

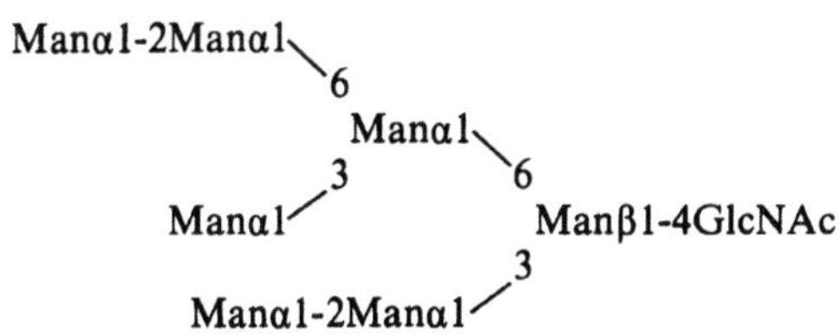

[33, 43]

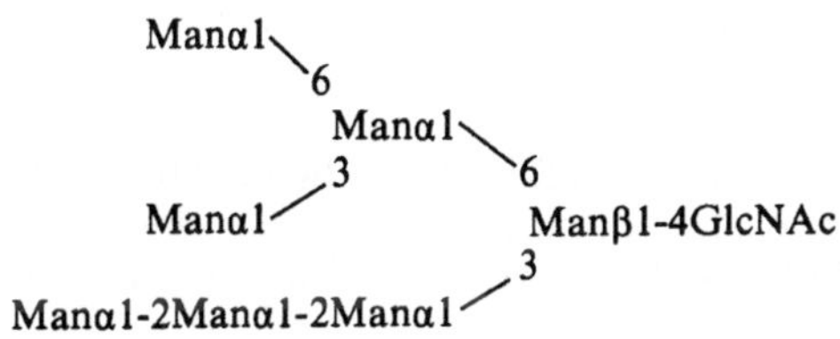

[33, 43]

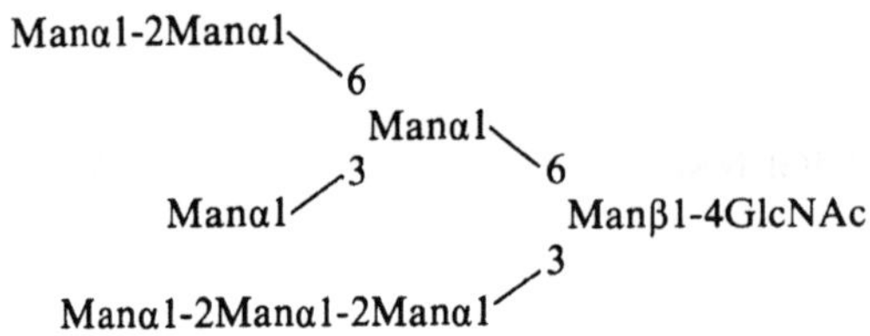

[33, 43]

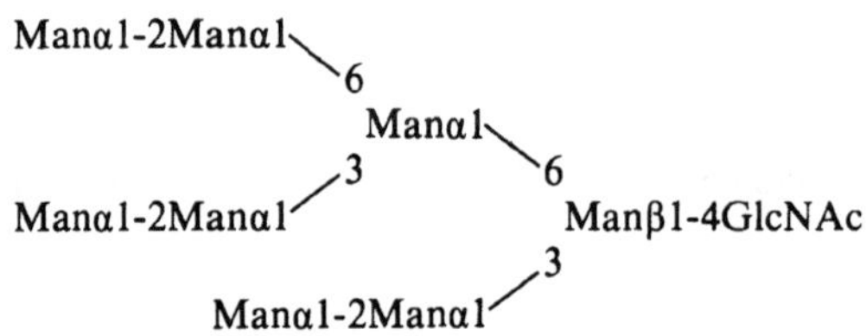

[33, 43]

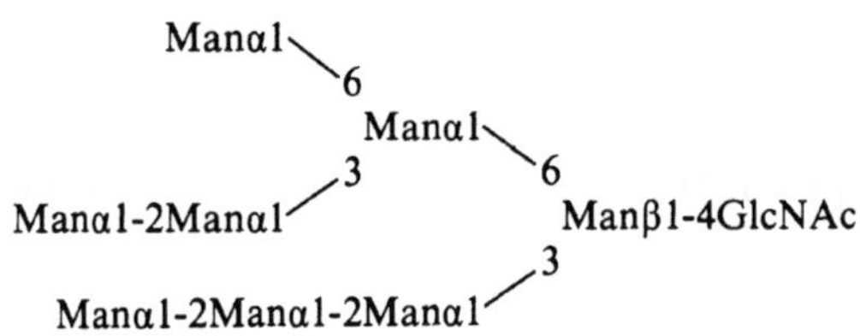

[33, 43]

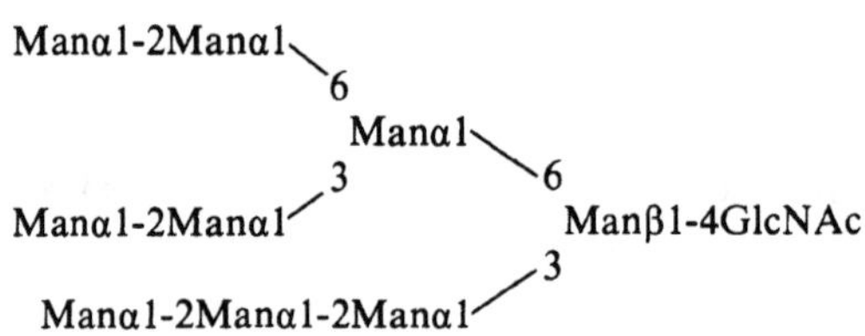

[33, 43]

Nuclear magnetic resonance studies on oligosaccharides from α-Mannosidosis urine have also been reported [43, 44].

The origin of the various oligosaccharides accumulated and excreted in α-Mannosidosis is not known, but various suggestions have been put forward [33, 41-43]. One possibility is that they represent degradation products of various glycoprotein carbohydrate chains. Assuming the presence of an endo-β-N-acetylglucosaminidase [45] which leaves only one N-acetylglucosamine-residue at the reducing end, one would expect accumulation of the tetrasaccharide

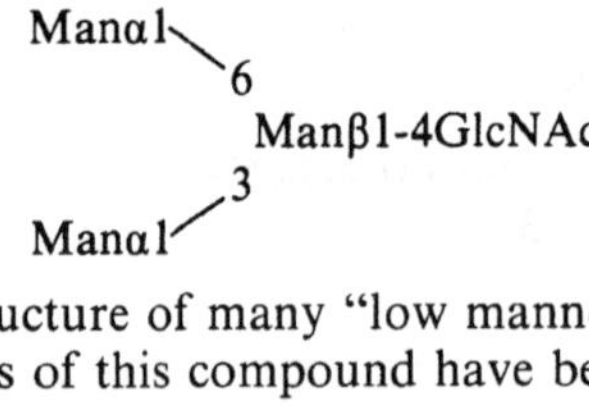

shown to be an integral structure of many "low mannose type" glycoproteins [46-51]. Only trace amounts of this compound have been isolated, however, and the trisaccharide

Manα1-3Manβ1-4GlcNAc

is by far the most predominant compound. This fact raises the possibility of the occurrence of either an endo-α-mannosidase or an α-mannosidase which cleaves only Manα1-6Man linkages. In either case the enzyme would not be deficient in α-Mannosidosis.

The structures suggested by Yamashita *et al.* [33] and Matsuura *et al.* [43] for high molecular weight oligosaccharides are all compatible with products of endo-β-N-acetylglucosaminidase digestion of the sugar chains of high mannose type glycoproteins and their oligosaccharide precursors [52, 53].

4.6 ENZYMOLOGY

A. Introduction

α-Mannosidase (EC. 3.2.1.24) has been studied in a variety of animals and tissues [54-61], but the interest in the enzyme was greatly stimulated by the description of the first patient [1, 39] shown to have a deficiency of α-mannosidase and an accumulation of mannose-rich oligosaccharides in his tissues. The residual α-mannosidase activity in this patient's organs was examined using mannose-containing oligosaccharides isolated from his brain as substrates [62]. The residual activity was abnormal in having a pH optimum of 5.5 compared to 4.0 for normal tissues.

Using rat liver preparations, it was shown [63] that both acidic* and neutral* forms of α-mannosidase existed, having strikingly different properties and being associated with the lysosomal fraction and the cytoplasm, respectively. The use of ion exchange chromatography enabled the separation of the acidic (lysosomal) and neutral (cytoplasmic) human liver α-mannosidases [64] and furthermore indicated the presence of not one, but two acidic enzymes (α-mannosidases A and B), both of which were absent from α-Mannosidosis liver, suggesting they were under a common genetic control. The neutral α-mannosidase in the α-Mannosidosis patient was normal. Thus α-Mannosidosis results from a deficiency of lysosomal α-mannosidase activity which has an acidic pH optimum and presumably is responsible for the normal degradation of α-mannosides as they become available during lysosomal degradation of glycoconjugates. The neutral, cytoplasmic α-mannosidase activity is under independent genetic control and has another, as yet unknown function. The human gene expressing the lysosomal enzymes has been localized to chromosome 19 [37] and that for the cytoplasmic enzyme is on the q 11 → q ter region of chromosome 15 [65]. Other α-mannosidases are located in the Golgi apparatus of normal tissues [66-69] and are presumably under separate genetic control and hence unaffected in α-Mannosidosis. This latter group of enzymes is involved in the complex process of glycoprotein biosynthesis via lipid intermediates [52, 53, 70] and each enzyme has a restricted substrate specificity and a pH optimum near pH 5.5 [66, 67].

In recent years the information on α-mannosidase has increased tremendously as a result of the detailed study of the increasing number of isoenzymes detected in extracts from many sources. More detailed information on the properties of human α-mannosidases are described in the following sections.

B. Molecular weight and subunits

Using Sephadex G-200, Sepharose 6B or sucrose density centrifugation it has been indicated that human tissue α-mannosidase A has a molecular weight in the region of 220,000 - 250,000 [71-74] and α-mannosidase B is slightly larger, around

* "Acidic" and "neutral" are used in this context for the enzymes having an acidic (ca. pH 4) or neutral (ca. pH 6) optimal activity.

300,000 [71, 72, 74]. A value of 97,000 for the acidic activity in α-Mannosidosis liver [73] was intriguing, but was "normalized" on changing the experimental conditions. Polyacrylamide gel electrophoresis [74] indicated that α-mannosidases A and B were of the same size (150,000) but smaller than indicated above. A similar lower value (180,000) was obtained for kidney α-mannosidase A [75] by gel filtration, and this enzyme was shown to consist of two subunits with molecular weights 58,000 and 30,000 indicating a heterotetramer structure. α-Mannosidase C in liver consistently has a molecular weight of 400,000 - 420,000 [71, 73, 74].

In plasma the acidic activity is the expected size (300,000) but the dominant intermediate form (see below) consisted of two components with molecular weights of 130,000 and 480,000 - 600,000 [76] which were not, however, interconvertible.

C. Isoelectric point

Column isoelectricfocussing of homogenates of human liver shows acidic α-mannosidase activity over a broad pI range with optima between pH 5.3 and 6.8 [62, 64, 71]. The residual acidic activity in α-Mannosidosis liver showed a similar pattern [62]. Purified α-mannosidases A and B have pI values of 6.0 and 5.45 respectively [72] which after neuraminidase treatment (α-mannosidase A) increased to pH 6.45. Less clear-cut results were obtained if the purification of α-mannosidase included an ion exchange chromatography stage [71]. Slightly higher pI values were obtained by gel isoelectricfocussing [74].

α-Mannosidase C has a pI of 5.7 when measured in homogenates [64, 71] and 5.45 after purification. This isoenzyme has a normal pI in α-Mannosidosis tissues [64].

D. Separation of isoenzymes

The most common technique used to separate α-mannosidase isoenzymes is chromatography on ion exchange celluloses. The initial observation [64] that human liver α-mannosidases could be separated into three components, two with an acidic pH optimum (α-mannosidases A and B) and one with a neutral pH optimum (α-mannosidase C), has been extended by the detection [71] of a second component in α-mannosidase C having an intermediate optimum (pH 5.5). The subdivision of α-mannosidase B into B_1 and B_2 has also been described [77]. The A and B isoenzymes are absent from α-Mannosidosis liver [20, 64] but α-mannosidase C, including its intermediate component, is unaffected. The structural relationship between α-mannosidases A and B has been demonstrated by showing that rechromatography of the A isoenzyme yields a mixture of A and B [71, 75, 77, 78]. Neuraminidase treatment of the liver sample before ion exchange chromatography results in the conversion of both isoenzymes A and B to another, common form [72, 77, 79].

Separation of α-mannosidase components from skin fibroblasts [34, 80, 81] or cultured amniotic cells [82] on DEAE-cellulose columns shows a qualitatively similar pattern to liver, but the proportions of the A, B and C isoenzymes are quite different. α-Mannosidase A is by far the major component, B is much lower, and C is barely detectable. As expected, the total α-mannosidase activity in α-Mannosidosis fibroblasts is very low [10, 13, 19, 20, 26-28, 34, 80, 81, 83-85].

Mixed leucocytes show an isoenzyme distribution similar to fibroblasts [86] but individual cell populations, for example monocytes and granulocytes, show interesting variations [87]. Human serum can also be shown to contain α-mannosidases A, B_1, B_2 and C by DEAE-cellulose chromatography. In this case the major component is the "B_2" isoenzyme, which in serum consists of the normal acidic enzyme which is absent in α-Mannosidosis [86], plus a high "intermediate'.' activity with an optimum activity around pH 5.5. Other methods for separating proteins dependent on their molecular charge give results generally in agreement with those obtained using DEAE-cellulose chromatography [19, 20, 73, 83, 84].

Several α-mannosidase isoenzymes are glycoproteins [72, 74] and consequently can bind to lectins. Human liver α-mannosidase A and B interact with Concanavalin A whereas the C isoenzyme does not [72, 76]. Fibroblast acidic α-mannosidase, however, can be separated into a binding and non-binding form, both of which are absent in α-Mannosidosis [81]. The acidic α-mannosidase of human plasma also binds to Concanavalin A, but the major plasma isoenzyme, which has an intermediate pH optimum, could be divided into two types, one binding (70%) and one non-binding (30%) [76]. Each of these was further subdivided by gel filtration into a low and high molecular weight form to give a total of four intermediate isoenzymes (I_1 - I_4), one of which (I_1) is probably identical with the neutral α-mannosidase C [76].

Affinity chromatography of human liver α-mannosidases using a ligand terminating in mannosylamine was effective in purifying the enzyme [72, 88]. The C isoenzyme bound preferentially and its prior removal was recommended in order to effectively purify the acidic forms.

E. Immunological properties

Rabbit antibodies raised against human liver α-mannosidase B react equally well with the A isoenzyme but do not react with α-mannosidase C [74, 78]. This antibody also reacted with leucocyte acidic α-mannosidase but not the intermediate plasma form. Similar results were obtained with a rabbit anti-human placenta-acidic-α-mannosidase [73]. Thus the specificity is directed towards the isoenzymes affected in α-Mannosidosis.

Conflicting results have been reported with fibroblasts enzymes. Although the rabbit anti-human-liver-α-mannosidase precipitated normal fibroblast acidic activity, there was no reaction against the residual acidic activity in α-Mannosidosis fibroblasts [81, 89]. In contrast, a rabbit anti-pig-kidney-acidic-α-mannosidase reacted equally well with both normal and α-Mannosidosis fibroblasts [90, 91]. The solution to this dilemma may lie in the type of antibody used. Evidence for this was presented when it was found that the rabbit anti-human-liver-α-mannosidase-B used above did not react with residual activity from three α-Mannosidosis patients, whereas the rabbit anti-human-placenta-acidic-α-mannosidase did react [73].

F. Substrates

α-Mannosidase is usually analyzed using artificial substrates such as p-nitrophenyl α-D-mannopyranoside [54] and 4-methylumbelliferyl α-D-mannopyranoside [58]. The latter compound is used almost exclusively nowadays due to the greater sensitivity of the method. It is assumed that the enzyme requires an α-D-mannopyranosyl structure with the aglycone being of minimal importance.

Early studies on lysosomes [92] indicated that the contents of these organelles could release mannose from ovalbumin. Normal human liver acidic α-mannosidases can degrade the mannose-rich oligosaccharides which were isolated from the brain [62] of an α-Mannosidosis patient, or which accumulated in α-Mannosidosis fibroblasts [93].

G. pH optimum

There is a general consensus of opinion that human liver α-mannosidases A and B have the same optimum, around pH 4 and that α-mannosidase C has an optimum around pH 6 [64, 71, 73, 77, 83, 94]. An "intermediate" α-mannosidase is most active around pH 5.5 [66, 69, 71]. A similar behaviour is observed in other organs [75, 77, 83, 95].

The major α-mannosidase in mixed leucocytes shows optimal activity at pH 4-4.5 [35, 86, 96-99] although the presence of a "shoulder" at pH 5.5 is sometimes observed [86, 99]. The little residual activity remaining in leucocytes from α-Mannosidosis patients has been described as both acidic with a pH optimum of 4.5 [86] and neutral with an optimum at pH 5.5 [35]. This discrepancy may be a result of different assay conditions or leucocyte preparation [100], since the pH profiles of α-mannosidase activity in granulocytes and monocytes show an optimum at pH 4 whereas lymphocytes and platelets show two optima, at pH 3.5-4 and 6.0 [87].

In serum (or plasma) the major α-mannosidase activity has a broad optimum between pH 5.5 and 6.3 [76, 86, 101]. This activity has been shown to have two components, the major one with an optimum between pH 5.2 and 5.8 and the minor one between 6.0 and 6.4. The properties of the latter resemble those of the tissue α-mannosidase C [102]. Some acidic activity is also present although the amounts are low [76, 86, 101] in normal serum, usually lower in α-Mannosidosis heterozygotes, and absent from α-Mannosidosis patients [86, 101].

The pH profiles obtained using skin fibroblast extracts are similar to the mixed leucocyte pattern, usually having maximal activity between pH 3.5 and 4.5 plus a smaller component with its optimum at pH 6 [10, 15, 29, 80, 81, 84, 103]. In α-Mannosidosis fibroblasts the residual activity is almost exclusively at pH 6 [10, 15, 24, 80, 84, 103], although cases with more residual acidic activity are known [26, 29, 35, 81]. Heterozygotes seem to have approximately equal amounts of the two activities [15, 24] although this is not to be recommended as a basis for heterozygote detection. Amniotic fluid cells not surprisingly resemble skin fibroblasts [24, 61, 82].

Human hair roots have optimal activity at pH 4 and no detectable "intermediate" or neutral α-mannosidases [104]. An α-Mannosidosis patient completely lacked this enzyme. A completely different pattern was observed using erythrocytes, in which the enzyme showed a broad optimum between pH 5.5 and 6.0, similar to the serum "intermediate" pattern. No acidic α-mannosidases were apparent and no differences were observed between the normal and α-Mannosidosis erythrocyte enzymes [99].

H. Stability

A method frequently used to differentiate isoenzymes is to study their stability towards preincubation at various temperatures for various times, prior to assay. This type of experiment on α-mannosidases from a variety of sources shows a fairly uniform pattern despite the different conditions frequently used. At temperatures between 55° and 60° C the acidic isoenzymes are stable for upwards of 15 minutes whereas the neutral activity is effectively lost in liver [66, 71, 73, 74, 77], leucocytes [85], serum [101, 102], fibroblasts [15, 29, 35, 80, 81, 85, 103] and amniotic cells [24]. The intermediate form is stable in liver [71, 73] but labile in serum [101, 102]. Purified liver α-mannosidases A and B were both stable when heated at pH 6.5 but only the A isoenzyme was stable when heated at pH 4.4. As expected, α-mannosidase C was labile at both pH values [74].

In α-Mannosidosis the neutral activity behaves normally in being labile and the residual acidic activity shows its mutant properties by also being labile in liver [74], serum [101], leucocytes [85], fibroblasts [34, 80, 81, 85] and amniotic cells [32]. Interestingly, in two reports of patients with high residual acidic activity in fibroblasts [29, 35] this activity may well be more stable than the typical mutant enzyme. In both normal and α-Mannosidosis fibroblasts isoenzyme A is stable to freezing and B is labile [80], again indicating that α-mannosidase B may be more sensitive to its environment than the A isoenzyme.

I. Activators and inhibitors

Native acidic α-mannosidase contains zinc [57, 105-107] and exogenous zinc usually increases the activity of the isoenzymes [20, 26, 29, 35, 72, 78, 81, 95, 97, 101, 103, 108]. An exception is the neutral activity in liver which has been réported to be inhibited [66, 74] or unaffected [71]. This discrepancy may reflect the relative proportions of neutral and intermediate α-mannosidases in liver caused by possible inactivation of the labile neutral enzyme on storage, or alteratively may merely reflect differences in assay conditions. The stimulatory effect of zinc on the residual acidic activity in α-Mannosidosis is frequently slightly greater than that found with the normal enzyme [20, 26, 35, 103, 108].

Cobalt differs from zinc in inhibiting the normal acidic α-mannosidase [20, 26, 74, 80, 81, 95, 97, 103, 108] and strongly activating the mutant enzyme [20, 26, 80, 81, 95, 101-103, 108]. The neutral enzymes are usually activated [20, 74, 95, 103].

The degree of activation or inhibition has been shown to be extremely variable, probably due to the different assay procedures used. However, some attempts at rationalization have been made. The effect of zinc and cobalt over a range of pH values has been studied using purified isoenzymes from normal liver [71], normal and α-Mannosidosis spleen [95], fibroblasts [81] and plasma [101]. Variations in metal ion [20, 81] and substrate [95] concentration, and incubation time [26] have also been examined.

EDTA inhibits α-mannosidase activity [63, 71, 72, 74, 75, 78, 97, 98] although this property is lost after isoenzyme purification [71, 72] which led to the suggestion that the acidic isoenzymes may contain two types of activity with differing zinc requirements. The effect of many other compounds on α-mannosidase activity has been tested [63, 66, 74, 75, 97, 98, 102] but one of the most interesting is swainsonine, a plant alkaloid which causes a toxicosis reminiscent of α-Mannosidosis [109].

J. Michaelis constant

The K_M of the normal liver acidic α-mannosidase activity has been determined under a variety of reaction conditions, but the results have been in good agreement giving values between 0.5 and 2.5 mM for methylumbelliferyl-α-D-mannopyranoside [20, 74, 94] and 1-4.5 mM for the oligosaccharide substrates [72, 94]. α-Mannosidases A and B have the same K_M whereas that of the C isoenzyme is an order of magnitude lower [74]. The residual acidic activity in α-Mannosidosis liver measured using the methylumbelliferyl substrate has a K_M about 15 times higher than the normal enzyme [20].

A similar range of normal values has been found in skin fibroblasts and amniotic fluid cells (0.6-2.2 mM) although the increase in K_M of the mutant enzyme has ranged from 1.5 to 50 times higher than normal [24, 29, 34, 35, 80, 81, 85, 108]. This greater range probably reflects the larger number of publications on the subject. This suggestion is substantiated by the values from different unrelated patients in a given series of experiments which correlate well with each other.

K. Enzyme dynamics

One hitherto unfulfilled aim in studying α-mannosidases is enzyme replacement therapy. Consequently, the possibility that α-Mannosidosis cells can utilize exogenous enzyme to break down the accumulating material has been examined. Cultured fibroblasts [93, 110-112] and rat liver [113] can endocytose some types of α-mannosidase but not others. It is a long step, however, to beneficial therapy for the patients.

The secretion of fibroblast lysosomal enzymes into the culture medium is well known, being extremely pronounced in Mucolipidosis types II and III [114, 115]. Acidic α-mannosidase has been detected in the medium [103, 116], which led to the suggestion [103] that the defect in α-Mannosidosis may be a recognition phenomenon analogous to that postulated for Mucopolidosis II and III, but specific for α-mannosidase. Hybridization of Mucolipidosis II and α-Mannosidosis fibroblasts [117] with subsequent normalization of α-mannosidase activity refuted this suggestion.

α-Mannosidase activity has been detected on the external cell surface of both normal and α-Mannosidosis fibroblasts [91] using an immunological technique. This pericellular activity was also isolated from α-Mannosidosis cells and shown to have properties similar to the mutant intracellular acidic enzyme [108]. The true extracellular acidic activity is independent of the diseased state.

4.7 DIAGNOSIS

A. Clinical

The diagnosis of α-Mannosidosis is first based on the clinical features, although such features may be difficult to notice in the very young. Specific examinations should always be performed when there is inexplicable retardation of mental functions or speech development, often associated with frequent infections during infancy. As the child develops, the typical face, impaired hearing, ataxia and skeletal changes are easier to observe.

B. Morphological

The presence of vacuolized lymphocytes in the peripheral blood has some value at the first diagnostic stage but a lack of these cells does not exclude the possibility of α-Mannosidosis, nor does their presence necessarily indicate a lysosomal storage disease (see Section 4). Electron microscopy of various tissues can unequivocally demonstrate the presence of excessive storage of presumably undegradable material and the subcellular localization can frequently be indicated. This technique is recommended as a "secondary" screening method.

C. Genetic

The presence of an index case in the family strongly implicates the same disease in another member. In most cases, however, no such index case exists but the presence, for example, of a mentally retarded family member in an earlier generation should be regarded as suspicious.

D. Chemical

α-Mannosidosis patients excrete abnormally large amounts of oligosaccharides, the major one being Manα1-3Manβ1-4GlcNAc. Thin layer chromatography of urinary oligosaccharides is a useful screening method for detection of increased oligosaccharide excretion (for further details see Chapter 7 on "Salla" disease), but is inadequate for oligosaccharide characterization. Gas liquid chromatography and, if necessary, mass spectrometry enables the characterization of excreted di- to pentasaccharides by a comparison of retention times and mass spectra with those of authentic samples. The fact that gas chromatography-mass spectrometry is a complicated and expensive technique is no reason to be contented with the less reliable information from screen methods.

E. Enzymic

The ultimate diagnostic proof of α-Mannosidosis is the demonstration of the absence or near absence of lysosomal α-mannosidase activity. Other lysosomal enzyme activities should not be significantly reduced.

Although serum is probably the most easily accessible source of α-mannosidase, its use in diagnosis of α-Mannosidosis is not to be recommended. Serum has a reasonably high α-mannosidase activity but the proportion of acidic (lysosomal) enzyme is very low. Consequently the difference between the activity in α-Mannosidosis and normal serum is slight and the activity at acidic pH can furthermore be influenced by the amount of "intermediate" α-mannosidase activity present, as well as by the amount of true acidic isoenzyme. In leucocytes the major isoenzyme is the acidic component affected in α-Mannosidosis. Thus a great difference in activity is found between normal and α-Mannosidosis leucocytes.

Consequently, leucocytes are a useful source of α-mannosidase and can be recommended for diagnosis, providing it is remembered that the α-mannosidase activity varies, depending on the proportions of the various cell types. These proportions can be quite different in patients suffering from infections.

The acidic α-mannosidase activity also predominates in skin fibroblasts, making these cells useful for diagnosis. Variations in activity can arise, however, as a result of cell culture. Confirmation of the presence of a mutant enzyme can be obtained by testing the activity in the presence and absence of cobalt ions, which inhibits the normal enzyme but strongly activates any residual mutant activity. Care should be taken to use a substrate concentration of approximately 1 mM (using the preferred substrate, 4-methylumbelliferyl-α-D-mannopyranoside) since higher concentrations preferentially increase mutant enzyme activity due to its higher K_M.

F. Definitive diagnosis

All the above factors must give a positive indication of α-Mannosidosis, since no single finding is sufficiently characteristic of the disease. Obviously the more complete the study of a patient, the more reliable is the diagnosis.

4.8 DETECTION OF CARRIERS

Human α-Mannosidosis heterozygote detection has not proved to be easy. As expected, most heterozygotes have acidic α-mannosidase levels intermediate between α-Mannosidosis patients and normal individuals, but the degree of overlap with the latter group is disconcerting. The demonstration, or lack, of overlap in a given study probably merely represents the chance factor when using a small sample. Consequently, a value below the normal range almost certainly indicates a heterozygote but a low normal value does not necessarily preclude the heterozygous state. Attempts to correlate acidic α-mannosidase activity with another enzyme in the same sample may reduce the degree of overlap by increasing the accuracy of the method but no convincing evidence that this can clearly separate the two ranges has been presented.

Analysis of acidic α-mannosidase activity in cultured skin fibroblasts enabled the heterozygote designation of both parents and two siblings of one patient [19] but the parents of a second case [15] showed a low and a low-normal activity. The father of another patient has a clear heterozygote level of acidic α-mannosidase whereas the mother had a below average level [35]. The patient himself has a higher residual fibroblast activity than usual. Interestingly, analysis of leucocytes from this couple indicated that the mother was clearly heterozygous whereas the father had activity only slightly below the lowest control value. The use of more than one enzyme source in borderline cases is therefore to be recommended especially since the use of fibroblasts in heterozygote detection has been criticized [118, 119]. Other studies using leucocytes have shown that less than half of the obligate heterozygotes tested had acidic α-mannosidase activity below the normal range [4, 20, 27, 32, 86]. In one of these reports [86] four heterozygotes had low normal α-mannosidase activities but the ratio between α-mannosidase and β-hexosaminidase clearly indicated their heterozygosity. Serum or plasma acidic α-mannosidase follows the same pattern as above [20, 27, 86, 101], heterozygote detection being helped by correlation with β-hexosaminidase [86] or donor's age [101].

It is apparent that definitive detection of all α-Mannosidosis heterozygotes is difficult, but the efficiency of detection should improve by analyzing more than one enzyme source and increasing the accuracy of the enzyme assay.

Heterozygote detection is only feasible in families-at-risk and possibly highly inbred communities or ethnic groups having a high incidence of a genetic disease [120]. Since neither of the latter situations have been shown to exist for human α-Mannosidosis, only the analysis of families having a member with α-Mannosidosis is recommended. Testing of subsequent marriage partners will give a high proportion of normal values and preclude the necessity of amniocentesis for prenatal diagnosis.

4.9 PRENATAL DIAGNOSIS

The severity of the disease and its frequent slow progression makes prenatal diagnosis and abortion of affected foetuses a socially and economically worthwhile exercise, giving prospective heterozygous parents a relatively risk-free chance of having normal children.

A wide range of α-mannosidase activity has been observed using cultured amniotic fluid cells [60, 82, 84, 121, 123]. Confluent cells have usually been tested which show an even greater variation in activity than non-confluent cells [118]. Extra care must, however, be taken with non-confluent cultures and the use of controls matched for stage of growth and time after last trypsinization is recommended, to avoid variations in activity due to sub-culture shock [119]. No correlation has been found between the length of gestation and enzyme activity [121]. Amniotic fluid cell α-mannosidase has an optimal activity at pH 4.0 with a secondary activity between pH 5.5 and 6.0 [24, 82], the latter being unaffected in α-Mannosidosis [24].

Uncultured amniotic fluid cells also show a wide range of α-mannosidase activity [32, 61, 121] with an optimum at pH 4.0 [61], but the activity is approximately half that found in cultured cells [121]. Amniotic fluid α-mannosidase has an optimum between pH 5.5 and 6.0 [123] and the lack of acidic activity makes it unsuitable for prenatal diagnosis.

Although experience of prenatal diagnosis of α-Mannosidosis is limited, successful analyses have been performed [24, 32, 122] and no insurmountable difficulties encountered. Caution has however been recommended if significant "residual" acidic activity is detected. Differentiation between low normal and high residual mutant enzyme activity should be made by studying the kinetic properties of the enzyme and its response to heat inactivation, and by studying the properties of the enzyme in the index case.

4.10 ANIMAL MODELS

Bovine α-Mannosidosis is without doubt the most studied animal model of a human inborn error of metabolism. A nervous disease of Aberdeen Angus cattle ("pseudolipidosis") was described in 1957 in Australia [124] and was subsequently discovered in New Zealand, from where the disease was introduced.

Clinically, the disease is characterized by progressive ataxia and incoordination, a fine tremor of the head, a slow head-nodding and an aggressive tendency. These symptoms usually appear within a few months of birth and frequently lead to an early death. Some calves, however, have no severe neurological disturbances [125, 126]. Vacuoles found in neuronal cells, macrophages, and endothelial and epithelial cells demonstrate the lysosomal nature of the disease [127, 128]. Fifteen years later, after the description of human α-Mannosidosis, it was shown that these affected cattle lacked lysosomal α-mannosidase and that their lymph nodes contained excessive amounts of a mannose-containing oligosaccharide [129].

Consequently the disease was renamed bovine Mannosidosis, which has also now been diagnosed in Tasmanian Grey and Murray Grey cattle in Australia [130]. Heterozygous Galloway cattle have also been found [131]. As in the human disease the mode of inheritance is autosomal recessive.

Subsequent studies on various tissues [132] have shown that calf α-mannosidase consists of two acidic forms A and B with pH optima between 3.75 and 4.75 and a neutral isoenzyme C (optimum pH 6.6). The presence of an additional isoenzyme with an intermediate pH optimum is indicated by Co^{++} activation studies [133]. In bovine α-Mannosidosis the acidic isoenzymes are selectively reduced and the properties of this residual activity [132-134] indicate that a small amount of a mutant enzyme is present which has a normal activity and cross reacts with the normal enzyme.

A natural example of "enzyme replacement therapy" was studied in an α-Mannosidosis calf which had a normal chimeric twin [135]. The normal lymphocyte acidic α-mannosidase obtained was able to partially normalize the visceral organs but the neurological degeneration was unaltered. An examination of α-Mannosidosis brain, lymph nodes and pancreas showed an increase of storage material with age whereas the liver remained more or less normal, due presumably to the high levels of residual activity (up to 15%) in this organ [136]. An attempt was made to characterize the water soluble carbohydrate material in bovine brain [137] and the results indicated that two tetrasaccharides, both containing mannose and N-acetylglucosamine (molar ratio 1:1) were accumulated. A more detailed study of the most abundant urinary oligosaccharide has also been carried out [138]. The structure

Manα1-6Manβ1-4GlcNAcβ1-4GlcNAcβ1-4GlcNAc

was proposed, which is unusual and distinct from all the oligosaccharides so far isolated from human α-Mannosidosis.

Accurate detection of heterozygotes for bovine α-Mannosidosis and selective mating of normal animals should enable the disease to be controlled and perhaps eventually eradicated. It has been estimated that 8-10% of Angus cattle in New Zealand are heterozygous for α-Mannosidosis [139] and in some herds the value is as high as 25% [126]. The numbers are somewhat lower in Australia [126] and Scotland [140] being 4 and 2% respectively. Cases have also been found in the United States [141]. A mass heterozygote screening programme was suggested in 1974 [142] using heparinized plasma as the enzyme source. The amount of overlap between acidic α-mannosidase activity in normal and heterozygote cattle is small but significant. Overlap can be reduced by using age and sex matched controls [143] and is usually less in a given herd tested at a given time [144], due to environment-induced variation in acidic α-mannosidase activity. The validity of the screening method was demonstrated after testing 8000 cattle [145]. The presence of a small proportion of borderline cases has resulted in attempts to define complementary methods of carrier detection. The use of lymphocytes [146] showed that overlap was reduced when α-mannosidase activity was related to protein rather than reference enzymes. The mean activity in lymphocytes from heterozygotes was 38% of normal which compared well with 37% obtained using plasma [145]. The plasma intermediate α-mannosidase activity is not decreased in the disease [147] but heterozygote detection by relating acidic to intermediate activity was not successful. Although plasma is the enzyme source best suited for heterozygote detection it has been recommended that lymphocytes be used as a supplementary test in borderline cases [148]. Further studies on blood cells [149] have indicated that the comparison of neutrophil acidic α-mannosidase and β-hexosaminidase is an improvement over the lymphocyte assay, whereas another group [150] have recommended the use of granulocyte α-mannosidase: β-hexosaminidase ratios after correction for eosinophil content, and addition of Zn^{++} to eliminate misleading results caused by possible zinc depletion on storage. Both groups emphasized the need to be certain of the blood cell composition of the samples since the glycosidase content of different cell types varied considerably [100]. A recommended method of screening for heterozygotes [126] is acidic α-mannosidase determination in serum or heparinized plasma using age and sex matched controls, followed by neutrophil analysis of the ambigous cases (3-5%). A more general discussion of heterozygote detection with some emphasis on bovine α-mannosidosis has been published [120].

Another animal model for α-Mannosidosis has recently been described [151, 152] in a domestic cat. Urinary excretion of mannose-rich oligosaccharides was detected, combined with a tissue depletion of acidic α-mannosidase. In contrast to the human disease the residual enzyme activity had normal properties, and thus resembles the bovine disease.

An inherited neurodegenerative disease among goats in New South Wales was described in 1973 [153]. Both sexes were affected with a severe neurological disorder presenting with intention tremour, flexion contractures and neurodegenerative muscular atrophy. Pathological examination revealed hypomyelination of the central nervous system, vacuolization of neurons, and axonal lesions. A similar disorder among Nubian goats in the U.S.A. has recently been described [154] and in addition to the findings described above, extensive cytoplasmic vacuolization of renal glomerular and tubular epithelium, vacuolization of visceral parenchymals cellls, and prominent neuroaxonal dystrophy with scattered eosinophilic spheroids in myelinated tracts of the cerebellar white matter were reported.

Chemical characterization of the stored material [154] revealed that the major compound in brain was Manβ1-4GlcNAcβ1-4GlcNAc. This suggested the possibility of a defect in β-mannosidase activity to account for the storage, which was confirmed [155] when a profound deficiency of β-D-mannosidase activity was demonstrated in a number of tissues from affected goats. A partial enzyme deficiency in obligate heterozygotes established the disorder as an autosomal recessive trait.

This is the first description of β-Mannosidosis. The widespread presence of β-mannoside linkages and β-mannosidase in animal tissues makes it most probable that β-Mannosidosis will eventually surface in other species, including Man.

Acknowledgements

We thank Dr. P.F. Benson, Guy's Hospital, London, and Dr. J. Gehler, Johannes Gutenberg-Universität, Mainz for giving us information prior to publication.

REFERENCES

1. Öckerman P.-A.: *A generalized storage disorder resembling Hurler's-syndrome.* Lancet *2*, 239-241, 1967.
2. Kjellman B., Gamstorp I., Brun A., Öckerman P.-A., Palmgren B.: *Mannosidosis: a clinical and histopathologic study.* J. Pediat. *75*, 366-373, 1969.
3. Nordén N.E., Öckerman P.A., Szabo L.: *Urinary mannose in mannosidosis.* J. Pediat. *82*, 686-688, 1973.
4. Autio S., Nordén N.E., Öckerman P.A., Riekkinen P., Rapola J., Louhimo T.: *Mannosidosis: Clinical, fine-structural and biochemical findings in three cases.* Acta Paediat. Scand. *62*, 555-565, 1973.
5. Autio S., Louhimo T., Helenius M.: *The clinical course of mannosidosis.* Ann. Clin. Res. *14*, 93-97, 1982.
6. Tsay G.C., Dawson G., Matalon R.: *Excretion of mannose-rich complex carbohydrates by a patient with α-mannosidase deficiency.* J. Pediat. *84*, 865-868, 1974.
7. Autio S.: *Clinical features of mannosidosis.* J. Pediat. *86*, 314, 1975.
8. Farriaux J.-P., Legouis I., Humbel R., Dhondt J.L., Richard P., Strecker G., Fourmaintraux A.,Ringel J., Fontaine G.: *La mannosidose. À propos de 5 observations.* Nouv. Presse Méd. *4*, 1867-1870, 1975.
9. Farriaux J.-P., Fontaine G.: *La mannosidose: Un diagnostic simple.* Arch. Franç. Péd. *33*, 11-22, 1976.
10. Gehler J., Cantz M., O'Brien J.F., Tolksdorf M., Spranger J.: *Mannosidosis: Clinical and biochemical findings.* Birth Defects. Orig. Art. Ser. *11*, 269-272, 1975.

11. Spranger J., Gehler J., Cantz M.: *The radiographic features of mannosidosis.* Radiology *119*, 401-407, 1976.
12. Perelman R., Nathanson M., Lepastier G., Lesavre P., Plainfosse B., Chirazi S., Seringe P.: *Mannosidose associée à l'absence d'alpha-1-antitrypsine. Présentation d'une observation. Revue de la litterature.* Ann. Pédiat. *22*, 385-396, 1975.
13. Vamos-Hurwitz E., Tondeur M., Humbel R., Philippart M., Hösli P., Loeb H.: *Mannosidosis: Findings in cultured fibroblasts and urine.* Acta Paediatr. Scand. *64*, 865-867, 1975.
14. Loeb H., Tondeur M., Toppet M., Cremer N.: *Clinical biochemical and ultrastructural studies of an atypical form of mucopolysaccharidosis.* Acta Paediatr. Scand. *58*, 220-228, 1969.
15. Aylsworth A.S., Taylor H.A., Stuart C.M., Thomas G.H.: *Mannosidosis: Phenotype of a severely affected child and characterization of α-mannosidase activity in cultured fibroblasts from the patient and his parents.* J. Pediat. *88*, 814-818, 1976.
16. Arbisser A.I., Murphree A.L., Garcia C.A., Howell R.R.: *Ocular findings in mannosidosis.* Amer. J. Ophthalmol. *82*, 465-471, 1976.
17. Murphree A.L., Beaudet A.L., Palmer E.A., Nichols B.L.: *Cataract in mannosidosis.* Birth. Defects. Orig. Art. Ser. *12*, 319-325, 1976.
18. Arbisser A.I., Scott C.I., Howell R.R.: *Mannosidosis and maternal penicillamine therapy.* Lancet *1*, 312-313, 1976.
19. Booth C.W., Chen K.K., Nadler H.L.: *Mannosidosis: Clinical and biochemical studies in a family of affected adolescents and adults.* J. Pediat. *88*, 821-824, 1976.
20. Desnick R.J., Sharp H.L., Grabowski G.A., Brunning R.D., Quie P.G., Sung J.H., Gorlin R.J., Ikonne J.U.: *Mannosidosis: Clinical, morphologic, immunologic and biochemical studies.* Pediat. Res. *10*, 985-996, 1976.
21. Mehta J., Desnick R.J.: *Abbreviated PR interval in mannosidosis.* J. Pediat. *92*, 599-601, 1978.
22. Letson R.D., Desnick R.J.: *Punctate lenticular opacities in Type II mannosidosis.* Amer. J. Ophthalmol. *85*, 218-224, 1978.
23. Sung J.H., Hayano M., Desnick R.J.: *Mannosidosis: Pathology of the nervous system.* J. Neuropath. Exp. Neurol. *36*, 807-820, 1977.
24. Maire I., Zabot M.T., Mathieu M., Cotte J., Hermier M.: *Mannosidosis: Tissue culture studies in relation to prenatal diagnosis.* J. Inher. Metab. Dis. *1*, 19-23, 1978.
25. Yunis J.J., Lewandowski R.C., Sanfilippo S.J., Tsai M.Y., Foni I., Bruhl H.H.: *Clinical manifestations of mannosidosis. A longitudinal study.* Amer. J. Med. *61*, 841-848, 1976.
26. Kistler J.P., Lott I.T., Kolodny E.H., Friedman R.B., Nersasian R., Schnur J., Mihm M.C., Dvorak A.M., Dickersin R.: *Mannosidosis: New clinical presentation, enzyme studies, and carbohydrate analysis.* Arch. Neurol. *34*, 45-51, 1977.
27. Milla P.J., Black I.E., Patrick A.D., Hugh-Jones K., Oberholzer V.: *Mannosidosis: Clinical and biochemical study.* Arch. Dis. Child. *52*, 937-942, 1977.
28. Vidgoff J., Lovrien E.W., Beals R.K., Buist N.R.M.: *Mannosidosis in three brothers. A review of the literature.* Medicine *56*, 335-338, 1977.
29. Bach G., Kohn G., Lasch E.E., El Massri M., Ornoy A., Sekeles E., Legum C., Cohen M.M.: *A new variant of mannosidosis with increased residual enzymatic activity and mild clinical manifestation.* Pediat. Res. *12*, 1010-1015, 1978.
30. Cerruti Mainardi P., Cavalieri S., Battistini E., Gatti R., Borrone C.: *Mannosidosi con grave infiltrazione polmonare.* Minerva Ped. *31*, 235-239, 1979.
31. Krieger D., Bourtel G., Ferrari P.: *La mannosidose: Réflexion à propos d'un cas.* Neuropsychiatr. de l'Enfance *27*, 515-519, 1979.
32. Poenaru L., Girard S., Thepot F., Madelenat P., Huraux-Rendu C., Vinet M.-C., Dreyfus J.-C.: *Antenatal diagnosis in three pregnancies at risk for mannosidosis.* Clin. Gen. *16*, 428-432, 1979.
33. Yamashita K., Tachibana Y., Mihara K., Okada S., Yabuuchi H., Kobata A.: *Urinary oligosaccharides of mannosidosis.* J. Biol. Chem. *255*, 5126-5133, 1980.
34. Gordon B.A., Carson R., Haust M.D.: *Unusual clinical and ultrastructural features in a*

boy with biochemically typical mannosidosis. Acta Paediatr. Scand. *69*, 787-792, 1980.
35. Tsvetkova I.V., Rosenfeld E.L., Prigozina I.G.: *An unusual case of mannosidosis with severe deficiency of acid mannosidase in leucocytes and high residual enzymatic activity in skin fibroblasts.* Clin. Chim. Acta *107*, 37-42, 1980.
36. Grabowski G.A., Welling L.L., Ikonne J.U., Wolfe L.S., Desnick R.J.: *Enzyme manipulation. Evaluation of oral $ZnSO_4$ supplementation in mannosidosis type II.* Pediat. Res. *11*, 456, 1977.
37. Champion M.J., Brown J.A., Shows T.B.: *Studies on the α-mannosidase (MAN_B), peptidase D (PEPD), and glucose phosphate isomerase (GPI) syntenic group on chromosome 19 in man.* Cytogenet. Cell. Genet. *22*, 186-189, 1978.
38. Autio S.: *Aspartylglycosaminuria. Analysis of thirtyfour patients.* J. Mental. Defic. Res., Monogr. Ser. I. 1972.
39. Ockerman P.-A.: *Mannosidosis: Isolation of oligosaccharide storage material from brain.* J. Pediat. *75*, 360-365. 1969.
40. Nordén N.E., Lundblad A., Svensson S., Ockerman P.-A., Autio S.: *A mannose-containing trisaccharide isolated from urines of three patients with mannosidosis.* J. Biol. Chem. *248*, 6210-6215, 1973.
41. Nordén N.E., Lundblad A., Svensson S., Autio S.: *Characterisation of two mannose-containing oligosaccharides isolated from the urine of patients with mannosidosis.* Biochemistry *13*, 871-874, 1974.
42. Strecker G., Fournet B., Bouquelet S., Montreuil J., Dhondt J.L., Farriaux J.-P.: *Etude chimique des mannosides urinaires excrétés au cours de la mannosidose.* Biochimie *58*, 579-586, 1976.
43. Matsuura F., Nunez H.A., Grabowski A.G., Sweeley C.C.: *Structural studies of urinary oligosaccharides from patients with mannosidosis.* Arch. Biochem. Biophys. *207*, 337-353, 1981.
44. Van Halbeek H., Dorland L., Veldink G.A., Vliegenhart J.F.G., Strecker G., Michalski J.-C., Montreuil J., Hull W.E.: *A 500 MHz ^1H NMR study of urinary oligosaccharides from patients with mannosidosis.* FEBS Lett. *121*, 71-77, 1980.
45. Nishigaki M., Muramatsu T., Kobata A.: *Endoglycosidases acting on carbohydrate moieties of glycoproteins: Demonstration in mammalian tissue.* Biochem. Biophys. Res. Commun. *59*, 638-645, 1974.
46. Spik G., Bayard G., Fournet B., Strecker G., Bouquelet S., Montreuil J.: *Studies on glycoconjugates LXIV. Complete structure of two carbohydrate units of human serotransferrin.* FEBS Lett. *50*, 296-299, 1975.
47. Strickland D.K., Hudson B.G.: *Structure of the tryptic glycopeptide isolated from rabbit transferrin.* Biochemistry *18*, 2549-2554, 1979.
48. Kornfeld R., Keller J., Baenziger J., Kornfeld S.: *Structure of the glycopeptide of human γG myeloma proteins.* J. Biol. Chem. *246*, 3259-3268, 1971.
49. Baenziger J., Kornfeld S.: *Structure of the carbohydrate units of IgA_1 immunoglobulin.* J. Biol. Chem. *249*, 7260-7269, 1974.
50. Baenziger J., Kornfeld S., Kochura S.: *Structure of the carbohydrate units of IgE immunoglobulins.* J. Biol. Chem. *249*, 1897-1903, 1974.
51. Zinn A.B., Marshall J.S., Carlsson D.M.: *Carbohydrate structures of thyroxine-binding globulin and their effects on hepatocyte membrane binding.* J. Biol. Chem. *253*, 6768-6773, 1978.
52. Kornfeld S., Li E., Tabas J.: *Synthesis of complex-type oligosaccharides. II. Characterisation of the processing intermediates in the synthesis of the complex oligosaccharide units of the vesicular stomatitis virus G_1 protein.* J. Biol. Chem. *253*, 7771-7778, 1978.
53. Hubbard S.C., Robbins P.W.: *Synthesis and processing of protein-linked oligosaccharides in vivo.* J. Biol. Chem. *254*, 4568-4576, 1979.
54. Conchie J., Findlay J., Levvy G.A.: *Mammalian glycosidases. Distribution in the body.* Biochem. J. *71*, 318-325, 1959.
55. Sellinger O.Z., Beaufay H., Jacques P., Doyen A., De Duve C.: *Tissue fractionation studies. Intracellular distribution and properties of β-N-acetylglucosaminidase*

and β-galactosidase in rat liver. Biochem. J. *74*, 450456, 1960.
56. Conchie J., Hay A.J.: *Mammalian glycosidases. 4. The intracellular localization of β-galactosidase, α-mannosidase, β-N-acetylglucosaminidase and α-L-fucosidase in mammalian tissues*. Biochem. J. *87*, 354-361, 1963.
57. Snaith S.M., Levvy G.A.: *Purification and properties of α-D-mannosidase from rat epididymis*. Biochem. J. *114*, 25-33, 1969.
58. Öckerman P.A.: *Fluorimetric estimation of 4-methylumbelliferyl-α-mannosidase activity in blood plasma*. Clin. Chim. Acta *23*, 479-482, 1969.
59. Wiederschain G.Y., Rosenfeld E.L., Brusilovsky A.I., Kolibaba L.G.: *α-L-fucosidase and other glycosidases in human placenta, foetus liver and amniotic fluid at various stages of gestation*. Clin. Chim. Acta *35*, 99-107, 1971.
60. Gerbie A.B., Melancon S.B., Ryan C., Nadler H.L.: *Cultivated epithelial-like cells and fibroblasts from amniotic fluid: Their relationship to enzymatic and cytologic analysis*. Amer. J. Obstet. Gynecol. *114*, 314-320, 1972.
61. Den Tandt W.R.: *Acid hydrolases in uncultured amniotic fluid cells*. Clin. Chim. Acta *40*, 199-204, 1972.
62. Hultberg B.: *Properties of α-mannosidase in mannosidosis*. Scand. J. Clin. Lab. Invest. *26*, 155-159, 1970.
63. Marsh C.A., Gourlay G.C.: *Evidence for a non lysosomal α-mannosidase in rat liver homogenates*. Biochim. Biophys. Acta *235*, 142-148, 1971.
64. Carroll M., Dance N., Masson P.K., Robinson D., Winchester B.G.: *Human mannosidosis - the enzymic defect*. Biochem. Biophys. Res. Commun. *49*, 579-583, 1972.
65. Champion M.J., Brown J.A., Shows T.B.: *Assignment of cytoplasmic α-mannosidase (MAN_A) and confirmation of mitochondrial isocitrate dehydrogenase (IDH_M) to the q11-qter region of chromosome 15 in man*. Cytogenet. Cell. Genet. *22*, 498-502, 1978.
66. Dewald B., Touster O.: *A new α-D-mannosidase occurring in golgi membranes*. J. Biol. Chem. *248*, 7223-7233, 1973.
67. Tulsiani D.R.P., Opheim D.J., Touster O.: *Purification and characterization of α-D-mannosidase from rat liver golgi membranes*. J. Biol. Chem. *252*, 3227-3233, 1977.
68. Opheim D.J., Touster O.: *Lysosomal α-D-mannosidase of rat liver*. J. Biol. Chem. *253*, 1017-1023, 1978.
69. Tabas I., Kornfeld S.: *Purification and characterization of a rat liver golgi α-mannosidase capable of processing asparagine-linked oligosaccharides*. J. Biol. Chem. *254*, 11655-11663, 1979.
70. Turco S., Robbins P.W.: *The initial stages of processing of protein-bound oligosaccharides in vitro*. J. Biol. Chem. *254*, 4560-4567, 1979.
71. Phillips N.C., Robinson D., Winchester B.G.: *Human liver α-D-mannosidase activity*. Clin. Chim. Acta *55*, 11-19, 1974.
72. Phillips N.C., Robinson D., Winchester B.G.: *Characterization of human liver α-D-mannosidase purified by affinity chromatography*. Biochem. J. *153*, 579-587, 1976.
73. Poenaru L., Miranda C., Dreyfus J.C.: *Residual mannosidase activity in human mannosidosis: Characterisation of the mutant enzyme*. Amer. J. Hum. Genet. *32*, 354-363, 1980.
74. Grabowski G.A., Ikonne J.U., Desnick R.J.: *Comparative physical, kinetic and immunologic properties of the acidic and neutral α-D-mannosidase isoenzymes from human liver*. Enzyme *25*, 13-25, 1980.
75. Marinkovic D.V., Marinkovic J.N.: *Isolation and properties of α-D-mannosidase from human kidney*. Biochem. J. *155*, 217-223, 1976.
76. Hirani S., Winchester B.: *The multiple forms of α-D-mannosidase in human plasma*. Biochem. J. *179*, 583-592, 1979.
77. Chester M.A., Lundblad A., Masson P.K.: *The relationship between different forms of human α-mannosidase*. Biochim. Biophys. Acta *391*, 341-348, 1975.
78. Phillips N., Robinson D., Winchester B.: *Immunological characterization of human liver α-D-mannosidase*. Biochem. J. *151*, 469-475, 1975.
79. Willcox P., Renwick A.G.C.: *Effect of neuraminidase on the chromatographic behaviour*

of eleven acid hydrolases from human liver and plasma. Eur. J. Biochem. *73*, 579-590, 1977.
80. Burton B.K., Nadler H.L.: *Mannosidosis: Separation and characterization of two acid α-mannosidase forms in mutant fibroblasts*. Enzyme *23*, 29-35, 1978.
81. Burditt L., Chotai K., Halley D., Winchester B.G.: *Comparison of the residual acidic α-D-mannosidase in three cases of mannosidosis*. Clin. Chim. Acta *104*, 201-209, 1980.
82. Hultberg B., Lindsten J., Sjöblad S.: *Molecular forms and activities of glycosidases in culture of amniotic-fluid cells*. Biochem J. *155*, 599-605, 1976.
83. Poenaru L., Dreyfus J.-C.: *Electrophoretic heterogenicity of human α-mannosidase*. Biochem. Biophys. Acta *303*, 171-174, 1973.
84. Taylor H.A., Thomas G.H., Aylsworth A., Stevenson R.E., Reynolds L.W.: *Mannosidosis: Deficiency of a specific α-mannosidase component in cultured fibroblasts*. Clin. Chim. Acta *59*, 93-99, 1975.
85. Beaudet A.L., Nichols B.L.: *Residual altered α-mannosidase in human mannosidosis*. Biochem. Biophys. Res. Commun. *68*, 292-298, 1976.
86. Masson P.K., Lundblad A., Autio S.: *Mannosidosis: Detection of the disease and of heterozygotes using serum and leucocytes*. Biochem. Biophys. Res. Commun. *56*, 296-303, 1974.
87. Nakagawa S., Kumin S., Nitowsky H.M.: *Studies on the activities and properties of lysosomal hydrolases in fractionated populations of human peripheral blood cells*. Clin. Chim. Acta *101*, 33-44, 1980.
88. Robinson D., Phillips N.C., Winchester B.: *Affinity chromatography of human liver α-D-mannosidase. FEBS Lett. 53*, 110-112, 1975.
89. Burditt L.J., Chotai K.A., Winchester B.G.: *Evidence that the mutant enzyme in fibroblasts of a patient with mannosidosis does not crossreact with antiserum raised against normal acidic α-D-mannosidase, FEBS Lett. 91*, 186-189, 1978.
90. Mersmann G., Buddecke E.: *Evidence for material from mannosidosis fibroblasts crossreacting with anti-acidic α-mannosidase antibodies*. FEBS Lett. *73*, 123-126, 1977.
91. Von Figura K., Voss B.: *Cell surface-associated lysosomal enzymes in cultured human skin fibroblasts*. Exp. Cell. Res. *121*, 267-276, 1979.
92. Aronson N.N., De Duve C.: *Digestive activity of lysosomes*. J. Biol. Chem. *243*, 4564-4573, 1968.
93. Chester M.A., Hultberg B., Nordén N.E., Szabo L.: *The nature of mannose-containing material which accumulates in cultured fibroblasts from patients with mannosidosis*. Biochim. Biophys. Acta *627*, 244-249, 1980.
94. Hultberg B., Lundblad A., Masson P.K., Öckerman P.-A.: *Specificity studies on α-mannosidases using oligosaccharides from mannosidosis urine as substrates*. Biochim. Biophys. Acta *410*, 156-163, 1975.
95. Hultberg B., Masson P.K.: *Activation of residual acidic α-mannosidosis tissues by metal ions*. Biochem. Biophys. Res. Commun. *67*, 1473-1479, 1975.
96. Hultberg B., Autio S., Berg B., Öckerman P.-A.: *White blood cell acid hydrolases in leukaemias, mucopolysaccharidoses and mannosidosis*. Scand. J. Haemat. *10*, 265-272, 1973.
97. Avila J.L., Convit J.: *Characterization and properties of α-D-mannosidase of human polymorphonuclear leucocytes*. Clin. Chim. Acta *47*, 335-345, 1973.
98. Kolodny E.H., Mumford R.A.: *Human leukocyte acid hydrolases: Characterization of eleven lysosomal enzymes and study of reaction conditions for their automated analysis*. Clin. Chim. Acta *70*, 247-257, 1976.
99. Poenaru L., Dreyfus J.-C.: *α-mannosidase in human red cells*. Biochim. Biophys. Acta *566*, 67-71, 1979.
100. Beutler E., Kuhl W., Matsumoto F., Pangalis G.: *Acid hydrolases in leukocytes and platelets of normal subjects and in patients with Gaucher's and Fabry's disease*. J. Exp. Med. *143*, 975-980, 1976.
101. Hirani S., Winchester B.G., Patrick A.D.: *Measurement of the α-mannosidase activities in human plasma by a differential assay*. Clin. Chim. Acta *81*, 135-144, 1977.
102. Hultberg B., Masson P.K., Sjöblad S.: *Neutral α-mannosidase activity in human serum*.

Biochim. Biophys. Acta *445*, 398-405, 1976.

103. Hultberg B., Masson P.K.: *Normal extracellular excretion of acidic α-mannosidase activity by mannosidosis fibroblast cultures.* Biochim. Biophys. Acta *481*, 573-577, 1977.
104. Phillips N.C., Thorpe R.: *Lysosomal glycosidase activities in human hair roots.* Clin. Chim. Acta *83*, 93-98, 1978.
105. Snaith S.M., Hay A.G., Levvy G.A.: *The relationship between the α-mannosidase activity and the zinc content of mammalian sex organs.* J. Endocrinol. *50*, 659-667, 1971.
106. Snaith S.M.: *Characterization of jack bean α-D-mannosidase as a zinc metalloenzyme.* Biochim. J. *147*, 83-90, 1975.
107. Tabas I., Kornfeld S.: *The synthesis of complex-type oligosaccharides.* J. Biol. Chem. *253*, 7779-7786, 1978.
108. Halley D.J.J., Winchester B.G., Burditt L.J., D'Azzo A., Robinson D., Galjaard H.: *Comparison of the α-mannosidases in fibroblast cultures from patients with mannosidosis and mucolipidosis II and from controls.* Biochem. J. *187*, 541-543, 1980.
109. Dorling P.R., Huxtable C.R., Colegate S.M.: *Inhibition of lysosomal α-mannosidase by swainsonine, an indolizidine alkaloid isolated from Swainsona canescens.* Biochem. J. *191*, 649-651, 1980.
110. Mersmann G., von Figura K., Buddecke E.: *Mannosidosis: Storage of mannose-containing material in cultured human mannosidosis cells and metabolic correction by pig kidney α-mannosidase.* Hoppe-Seyler's Z. Physiol. Chem. *357*, 641-648, 1976.
111. Mersmann G., Wilde F.-J., Buddecke E.: *Differences in intracellular corrective activity in cultured mannosidosis cells of acidic α-mannosidases from various sources.* Hoppe-Seyler's Z. Physiol. Chem. *359*, 421-425, 1978.
112. Mersmann G., Becker R., Buddecke E.: *Amount and properties of uptake forms in preparations of α-mannosidase from pig kidney.* Biochim. Biophys. Acta *525*, 154-161, 1978.
113. Marinković D.V., Petrović S.L., Martinović J.V., Marinković J.N.: *Uptake of iodinated human kidney α-D-mannosidase by rat liver.* Biochem. J. *159*, 729-736, 1976.
114. Wiesmann U.N., Lightbody J., Vassella F., Herschovitz N.N.: *Multiple lysosomal enzyme deficiency due to enzyme leakage?.* N. Engl. J. Med. *284*, 109-110, 1971.
115. Wiesmann U.N., Herschovitz N.N.: *Studies on the pathogenetic mechanism of I-cell disease in cultured fibroblasts.* Pediat. Res. *8*, 865-870, 1974.
116. Hultberg B., Sjöblad S.: *Lysosomal enzymes in medium from cultured skin fibroblasts from normal individuals and patients with lysosomal diseases.* Clin. Chim. Acta *80*, 79-86, 1977.
117. D'Azzo A., Halley D.J.J., Hoogeveen A., Galjaard H.: *Correction of I-cell defect by hybridization with lysosomal enzyme deficient human fibroblasts.* Amer. J. Hum. Genet. *32*, 519-528, 1980.
118. Milunsky A. Spielvogel C., Kanfer J.N.: *Lysosomal enzyme variations in cultured normal skin fibroblasts.* Life Sci. *11*, 1101-1107, 1972.
119. Heukels-Dully M.J., Niermeijer M.F.: *Variation in lysosomal enzyme activity during growth in culture of human fibroblasts and amniotic fluid cells.* Exp. Cell Res. *97*, 304-312, 1976.
120. Jolly R.D., Desnick R.J.: *Inborn errors of lysosomal catabolism. Principles of heterozygote detection.* Amer. J. Med. Gen. *4*, 293-307, 1979.
121. Butterworth J., Sutherland G.R., Broadhead D.M., Bain A.D.: *Lysosomal enzymes of cultured amniotic fluid cells.* Clin. Chim. Acta *44*, 295-299, 1973.
122. Tansley L.R., Fensom A.H., Benson P.F.: Personal communication.
123. Hultberg B., Sjöblad S., Gustavii B.: *Acid hydrolases in amniotic fluid and chorionic villi at various stages of pregnancy.* Clin. Chim. Acta *53*, 221-227, 1974.
124. Whittem J.M., Walker D.: *"Neuronopathy" and "pseudolipidosis" in Aberdeen Angus calves.* J. Path. Bact. *74*, 281-288, 1957.
125. Jolly R.D.: *Diagnosis and control of pseudolipidosis of Angus calves.* N.Z. Vet. J. *13*,

228-229, 1970.
126. Jolly R.D.: *Mannosidosis and its control in Angus and Murray Grey cattle*. N.Z. Vet. J. *26*, 194-198, 1978.
127. Jolly R.D.: *The pathology of the central nervous system in pseudolipidosis of Angus calves*. J. Pathol. *103*, 113-121, 1971.
128. Jolly R.D., Thompson K.G.: *The pathology of bovine mannosidosis*. Vet. Path. *15*, 141-152, 1978.
129. Hocking D.J., Jolly R.D., Batt R.D.: *Deficiency of α-mannosidase in Angus cattle. An inherited lysosomal storage disease*. Biochem. J. *128*, 69-78, 1972.
130. Healy P.J., Cole A.E.: *Heterozygotes for mannosidosis in Angus and Murray Grey cattle*. Aust. Vet. J. *52*, 385-386, 1976.
131. Hart K.G., Healy P.J.: *Galloway cattle heterozygous for mannosidosis*. Aust. Vet. J. *56*, 255-256, 1980.
132. Phillips N.C., Robinson D., Winchester B.G., Jolly R.D.: *Mannosidosis in Angus cattle. The enzymic defect*. Biochem. J. *137*, 363-371, 1974.
133. Burditt L.J., Phillips N.C., Robinson D., Winchester B.G., van De Water N.S., Jolly R.D.: *Characterization of the mutant α-mannosidase in bovine mannosidosis*. Biochem. J. *175*, 1013-1022, 1978.
134. Phillips N.C., Winchester B.G., Jolly R.D.: *A serological investigation into the acidic α-D-mannosidase in normal Angus cattle and in a calf with mannosidosis*. Biochem. J. *163*, 269-277, 1977.
135. Jolly R.D., Thompson K.G., Murphy C.E., Manktelow B.W., Bruere A.N., Winchester B.G.: *Enzyme replacement therapy. An experiment of nature in a chimeric mannosidosis calf*. Pediat. Res. *10*, 219-224, 1976.
136. Jolly R.D., Slack P.M., Winter P.J., Murphy C.E.: *Mannosidosis: Patterns of storage and urinary excretion of oligosaccharides in the bovine model*. Aust. J. Exp. Biol. Med. Sci. *58*, 421-428, 1980.
137. Nordén N.E., Lundblad A., Öckerman P.-A., Jolly R.D.: *Mannosidosis in Angus cattle: Partial characterization of two mannose containing oligosaccharides*. FEBS Lett *35*, 209-212, 1973.
138. Lundblad A., Nilsson B., Nordén N.E., Svensson S., Öckerman P.-A., Jolly R.D.: *A urinary pentasaccharide in bovine mannosidosis*. Eur. J. Biochem. *59*, 601-605, 1975.
139. Jolly R.D., Thompson K.G., Winchester B.G.: *Bovine mannosidosis. A model lysosomal storage disease*. Birth Defects, Orig. Art. Ser. *11*, 273-278, 1975.
140. Jolly R.D., Hartley W.J.: *Storage diseases of domestic animals*. Aust. Vet. J. *53*, 1-8, 1977.
141. Leipold H.W., Smith J.E., Jolly R.D., Eldridge F.E.: *Mannosidosis of Angus calves*. J. Amer. Vet. Med. Assoc. *175*, 457-459, 1979.
142. Jolly R.D., Digby J.G., Rammell C.G.: *A mass screening programme of Angus cattle for the mannosidosis genotype - a prototype programme for control of inherited diseases in animals*. N. Z. Vet. J. *22*, 218-222, 1974.
143. Jolly R.D., Tse C.A., Greenway R.M.: *Plasma α-mannosidase activity as a means of detecting mannosidosis heterozygotes*. N.Z. Vet. J. *21*, 64-69, 1973.
144. Jolly R.D., Thompson K.G., Tse C.A., Munford R.E., Merrall M.: *Identification of mannosidosis heterozygotes-factors affecting normal plasma α-mannosidase levels*. N.Z. Vet. J. *22*, 155-162, 1974.
145. Jolly R.D., Thompson K.G., Tse C.A.: *Evaluation of a screening programme for identification of mannosidosis heterozygotes in Angus cattle*. N.Z. Vet. J. *22*, 185-190, 1974.
146. Thompson K.G., Jolly R.D., Winchester B.G.: *Mannosidosis: Use of reference enzymes in heterozygote detection*. Biochem. Med. *15*, 233-240, 1976.
147. Winchester B.G., Van der Water N.S., Jolly R.D.: *The nature of the residual α-mannosidase in plasma in bovine mannosidosis*. Biochem. J. *157*, 183-188, 1976.
148. Thompson K.G., Jolly R.D., Rammell C.G.: *Lymphocyte α-mannosidase - a supplementary test for determining mannosidosis genotype in cattle*. N.Z. Vet. J. *24*, 167-170, 1976.

149. Jolly R.D., Janmaat A., Van der Water N.S.: *Heterozygote detection: A comparative study using neutrophils, lymphocytes, and two reference parameters in the bovine mannosidosis model.* Biochem. Med. *18*, 402-409, 1977.
150. Healy P.J., Nicholls P.J., Butrej P.: *A granulocyte test for detection of cattle heterozygous for mannosidosis.* Clin. Chim. Acta *88*, 429-435, 1978.
151. Burditt L.J., Chotai K., Hirani S., Nugent P.G., Winchester B.G., Blakemore W.F.: *Biochemical studies on a case of feline mannosidosis.* Biochem. J. *189*, 467-473, 1980.
152. Walkey S.U., Blakemore W.F., Purpura D.P.: *Alterations in neuron morphology in feline mannosidosis.* Acta Neuropathol. *53*, 75-79, 1981.
153. Hartley W.J., Blakemore W.F.: *Neurovisceral storage and dysmyelinogenesis in neonatal goats.* Acta Neuropathol. *25*, 325-333, 1973.
154. Jones M.Z., Laine R.A.: *Caprine oligosaccharide storage disease.* J. Biol. Chem. *256*, 5181-5184, 1981.
155. Jones M.Z., Dawson G.: *Caprine β-mannosidosis.* J. Biol. Chem. *256*, 5185-5188, 1981.

Pertti Aula
Seppo Autio
Kari O. Raivio
Juhani Rapola

5 Aspartylglucosaminuria

NAME OF DISEASE	Aspartylglucosaminuria
PREVALENCE	Finnish people
MODE OF INHERITANCE	Autosomal recessive
MUTANT GENE PRODUCT	1-aspartamido-β-N-acetylglucosamine amidohydrolase
CHROMOSOMAL LOCATION	?
EXPRESSION	
1. Altered DNA structure	?
2. Disturbed protein function	Enzyme activity absent
3. Disrupted cell and organ function	Accumulation of substrate (2-acetamido-I-L-β-aspartamido-1,2-dideoxy-β-D-glucose) in lysosomes of most cells, including brain, leads to brain dysfunction and somatic changes

5.1 INTRODUCTION

Aspartylglucosaminuria (AGU) is the Finnish counterpart of Tay-Sachs disease in Jewish populations as the characteristic and common recessively inherited lysosomal storage disorder.

The disease was first described in England by Pollitt and coworkers 1968 in a sister and brother with severe mental retardation [1]. Both patients excreted large amounts of 2-acetamido-1-(β-L-aspartamido)-1,2-dideoxyglucose (aspartylglucosamine) in the urine. Assays of a sample from the brother revealed deficient activity of aspartylglucosaminidase (AADGase), the enzyme which hydrolyzes the linkage between asparagine and N-acetylglucosamine, splitting the carbohydrate side-chain from the polypeptide chain of glycoproteins.

New cases of AGU were rapidly found in Finland [2-4] and at the present time 138 patients (from 108 families) have been diagnosed. All data from these patients have been collected in our files. In recent years new cases have been detected also outside Finland, suggesting a more widespread distribution of the mutation (see later).

5.2 CLINICAL FEATURES

The clinical characteristics of AGU were first analysed in detail by Autio [3] in a study of 34 patients ranging in age from 1.7 to 37 years (Table 1). Our subsequent experience is in good agreement with this early analysis. The age and sex distribution of the present patients is shown in Table 2.

A. Early history

The preand perinatal histories of AGU-patients are normal. Psychomotor development during the first year of life has not shown any consistent abnormalities.

A history of frequent upper respiratory and ear infections, often associated with enlarged pharyngeal tonsils, is more common than in the general pediatric population. This susceptibility manifests itself during the first months of life and continues for a number of years, but usually decreases after the age of six or seven years. Other infectious diseases are not more frequent in AGU children than

Table 5.1 Clinical characteristics of AGU patients

Findings	Total N=25 (34)[1]
Mental retardation	33/34
Speech disorder	23/25
Motor clumsiness	23/25
Coarse face	21/25
Osteochondrotic vertebral changes	20/25
Vacuolated lymphocytes in the blood	19/25
Mobile joints	19/25
Thin cortex of long bones[2]	17/23
Reduced elasticity of skin	12/25
Hernia	12/25
Macroglossia	12/25
Thick skull[3]	10/23
Systolic cardiac murmur	11/25
Abnormal EEG[4]	11/25
Hoarse tone of voice	9/25
Muscular hypotonia	8/25
Subcapsular, crystal-like lens opacities	8/25
Short stature[5]	10/34
Brachycephaly[6]	7/25
Spasticity of legs	5/25

[1] The number in brackets indicates the total series of 34 patients, data of which have been used for mental retardation and height.
[2] All measurement of combined cortical thickness below the 10th percentile of normal Finnish standards [46].
[3] Mean thickness of calvarium of the skull more than +2SD above the mean of a control sample.
[4] Short episodes of theta activity in eight, focal abnormality in two and bursts with spikes and sharp waves in one of the subjects.
[5] Height more than 2SD below the mean of normal Finnish standards [47, 48].
[6] Cephalic index above the Finnish norms [49].

Table 5.2 The Finnish AGU-patients in 1981[1].

Age-group (years)	Total number of patients	Female/male
Below 10	10	3/7
10 - 19	34	14/20
20 - 29	41	18/23
30 - 39	28	17/11
Over 40	15	8/7
Total	128	60/68

[1] Ten deceased patients are shown separately in table 5.4.

in normal infants. The cause of the unusual susceptibility to infections during infancy and early childhood is unknown. The liability to infections is not a consistent feature at later ages but some AGU-patients may continue to suffer from various infections more so than other mentally retarded patients. Inguinal or umbilical herniae occur in about half of the patients, usually in infancy. The tendency to herniae may be associated with the involvement of the connective tissue.

Peculiar episodes of diarrhea occur in about one third of the patients in infancy. They usually last for one or two days, but some patients are in need of hospital treatment because of more prolonged or severe gastrointestinal symptoms. The pathogenesis of the intestinal symptoms is unknown and they usually disappear after infancy. Clinical symptoms in infancy, mentioned above, are nonspecific but they are the only diagnostic clues at this age.

B. Early neurological manifestation

The first signs of involvement of the central nervous system usually appear between 1 and 4 years. The first anormality observed by most parents is delayed or arrested speech development. Very often, this has initially been ascribed to the respiratory infections.

The retardation of motor functions is not as obvious, although patients during the preschool age tend to be clumsy and may have slight muscular hypotonia. In a careful retrospective analysis of 23 cases of AGU, the first signs of the disease were noted in four patients below one year, in twelve patients between one and five years, and in seven patients between five and nine years [3]. Other early symptoms consist of a poor ability to concentrate and abnormally stubborn behaviour, which are salient features in young children with AGU. These features, combined with motor clumsiness and delayed speech, are also typical of the minimal brain damage syndrome, which is a differential diagnostic possibility.

The facial features are normal early in the course disease (Fig. 3), and skeletal or other somatic changes are not evident.

C. Mental retardation and behaviour

Adult patients with AGU are severely or profoundly mentally retarded. To our knowledge, no adult patient with AGU in Finland or anywere else has a normal intelligence. Patients who are five to ten years of age are mildly to moderately retarded, those between eleven and twenty years are moderately to severely

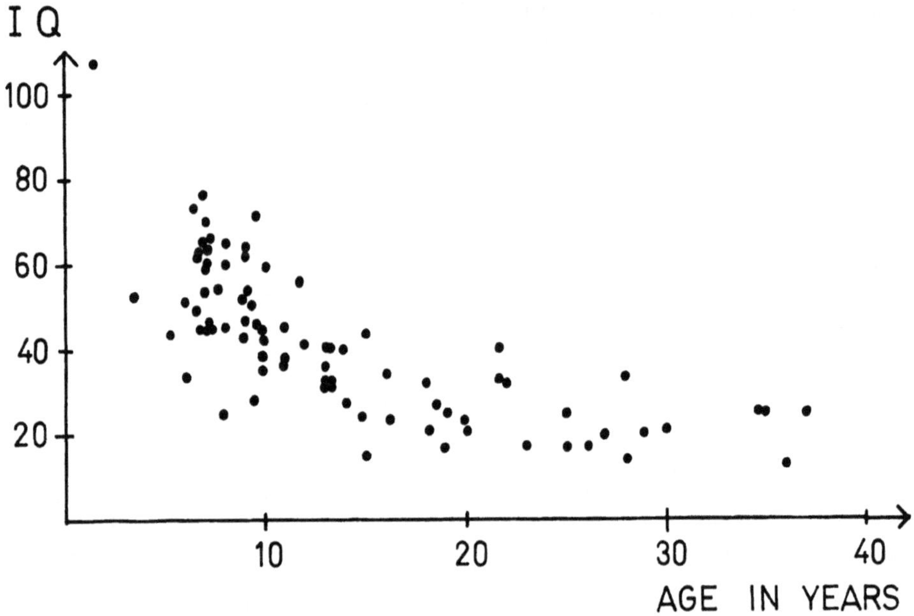

Fig. 5.1 Decline of the mental level in AGU with increasing age. All IQ-tests of 39 AGU-patients plotted on an IQ-age scale. (TML was used in 80 tests, WISC in two and Cattel method in one test, respectively). Source: Autio *et al.* [4].

retarded, and patients over 20 years of age are severely or profoundly mentally retarded.

At approximately of 7-10 years, there is definite progressive impairment of mental function as reflected in IQ-scores (Fig. 1). None of the patients have learned to read or write satisfactorily, and the few who have started normal school have been unable to keep step with normal children.

The abnormally stubborn and restless behaviour apparent in childhood continues in some patients to adult age. About one-fifth of the patients, especially females, develop another cyclic behavioural abnormality at the prepubertal or pubertal age. This can be described as manic, "erethic", or abnormally excitable behaviour lasting from a few weeks to a few months, followed by a more apathetic period. This pattern usually disappears later, but it may continue troughout life in some cases.

About 40 per cent of the Finnish patients are in an institution for the mentally retarded. The others are living at home, although many of them need institutional care from time to time.

D. Motor symptoms and signs

Although the psychomotor development of AGU-patients is usually normal in early childhood, some may show a slight delay in motor milestones at the age of 1-2 years. At the same age muscular hypotonia is observed in one-fifth of the patients.

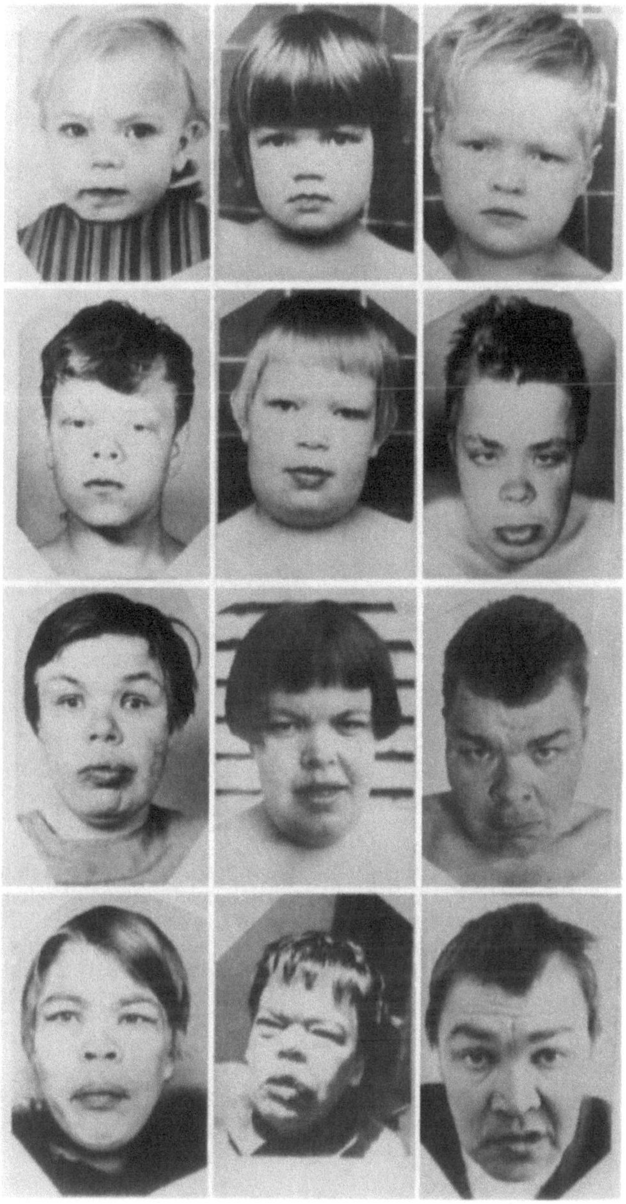

Fig. 5.2 Facial appearance of AGU. Twelve patients with AGU at various age groups. Upper horizontal line: patients below ten years of age. Second line: from ten to twenty years. Third line: from twenty to thirty years. Lower line: thirty years and older. Note the presence of large mouth, thick lips and coarse features in older patients. Source: Autio [3].

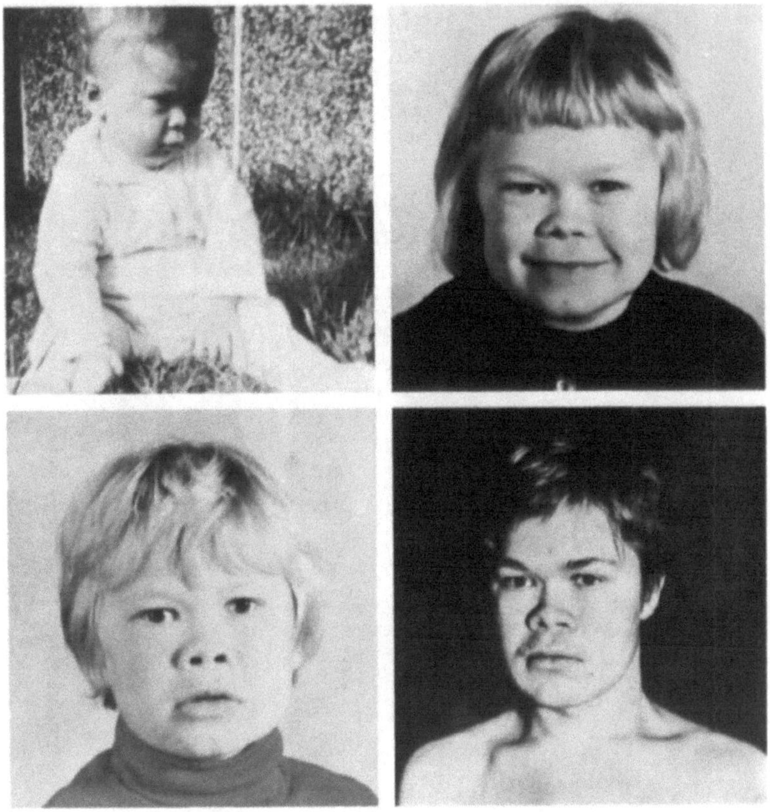

Fig. 5.3 Development of the facial characteristics in AGU in a male patient. Upper-left: the boy at the age of 9 months with normal psychomotor development and without any facial changes. Upper-right: the boy presenting with slight developmental delay at the age of 8 years but no facial changes. Low-left: the patient at the age of 10 years showing the first facial changes in AGU. Lower-right: the moderately retarded patient at the age of 15 years showing typical "coarse" facial features of AGU.

Generally clumsy motor coordination is typical of children with AGU from preschool age on. The clumsiness can be seen, for instance, in the typical way of running in a forward leaning position with swinging hands. Fine motor coordination is also impaired.

The impairment of motor performance in AGU is less pronounced than the deterioration of intellectual ability and the patients are able to move about independently throughout their life. There is, however, some progression of the motor disturbances especially towards the end of the second decade. Most patients walk stiffly, with a wide-based gait; they stumble easily.

Approximately one-fifth of the patients have signs of spasticity of the lower limbs with hyperreflexia, manifesting as increased ankle and knee jerks at puberty or later.

E. Characteristics of face and body

The facial characteristics of a patient with AGU consist of a large mouth, thick lips, coarse facial features, low and broad nasal bridge, and, with increasing age, a generally sagging appearance. Other features often present are persistence of inner epicanthal folds, nasal anteversion, broad face, hypertelorism and a broad nose.

Some typical examples of the facial characteristics are shown in figure 2.

The abnormal facial appearance is not present during the first five or six years (Fig. 3). The coarseness becomes apparent at 7 to 10 years; the typical features are usually obvious from the pubertal age on (Figs. 2 and 3).

In addition to the typical face, the patients have other physical features in common. Most patients have some hypermobility of joints. A protuberant abdomen is seen in about half of the patients. Other features often present are short stature (see below), spinal or thoracic deformities, large tongue, dental diasthema and poorly developed secondary sex characteristics.

F. Skeletal changes

The main skeletal manifestations of AGU are an abnormally thick calvarium, osteochondrotic vertebral changes, and generalized osteoporosis. These changes are similar to those seen in mannosidosis (see Fig. 4.2).

An increased thickness of skull bone was seen in 20 out of 23 patients investigated by Autio [3]. The thick calvarium has usually been observed before the age of ten years.

Spinal changes radiologically similar to juvenile osteochondrosis are found in the majority of patients by 7 to 10 years or during the second decade. Later on, these changes appear as wedge-shaped vertebral bodies and flattened intervertebral spaces, most prominently in the last two thoracic and first two lumbar vertebrae. Vertebral changes are often associated with thoracic or thoraco-lumbar scoliosis or kyphosis after the age of 10-15 years.

The cortex of the long bones is abnormally thin (Fig. 4) and the trabeculation is poor. Combined cortical thickness (CCT) of the second and third metacarpal bones was measured in 23 AGU-patients. All of them have a CCT value below normal and the difference became more pronounced with increasing age [3].

Less commonly encountered skeletal abnormalities include (Fig. 4) a short ulna relative to the radius, small cavities in the long bones, unusually bowed long bones, and thickened, misshapen ribs.

G. Disturbances of physical growth and endocrine functions

The physical growth of AGU-patients proceeds normally until the age of 10-15 years, but the height of the adult patients is usually below normal. The deviation from the normal mean of male patients ranged from -0.5 to -4.0 SD (mean -2.4 SD) and of female patients from -1.0 to -4.5 SD (mean -2.3 SD) in a series of 34 patients.

The growth retardation becomes manifest at the time of the normal pubertal growth spurt. The short stature may also result from the vertebral changes, e.g. flattened vertebral bodies, kyphosis or scoliosis.

The growth retardation may also be associated with endocrine disturbances. Gonadal hypofunction is clinically evident in about one-fourth of the patients; this is evident as scanty or absent pubic or axillary hair or beard, and underdevelopment of the mammary glands. Some females have primary or

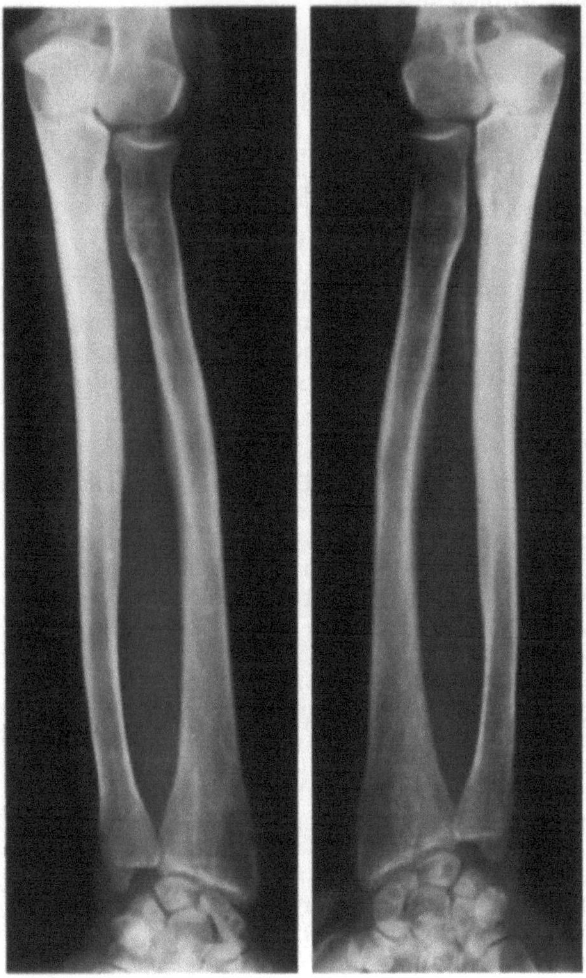

a b

Fig. 5.4 Arm and wrist bones of a 41-years-old woman with AGU. (**a**, left arm; **b**, right arm). The bones are osteoporotic. Ulna and radius are slightly unusually curved, with an uneven thickness of cortex. Ulna is short relative to radius. Note also the small cystic degeneration of the left navicular bone.

secondary amenorrhea.

Except for higher plasma cortisol levels and somewhat inadequate suppression after an oral dose of dexamethasone in some patients, the endocrine functions have been normal. Female patients may be fertile and several male patients have shown heterosexual interest causing secondary social problems (unpublished data).

H. Other clinical data

A systolic murmur at the third or fourth intercostal space or at the apex is present in one third of the patients. The murmur may be heard in childhood. It is usually faint or moderate, but a cardiological evaluation has been deemed necessary in some cases. The pathogenesis of this sign is obscure; the only case autopsied by the authors had granular patches on the ridges of the mitral and tricuspid valves [5].

Clinical laboratory tests are usually normal, except for the presence of vacuolized lymphocytes in peripheral blood. However, there is a tendency to neutropenia from the age of 10-15 years on, and adult patients often have decreased prothrombin levels.

Electroencephalograms are usually normal in children and adults. The most consistent but mild abnormality, recorded in one-fourth of the patients, is the presences of 4-7 c/s theta episodes, lasting for one or two seconds, either generalized or localized to the frontal regions.

Slight or moderate loss of hearing has been documented in some adult patients by recording evoked potentials to auditory stimuli but this finding is not clinically significant. Patients with AGU, in contrast to mannosidosis, have no clinical indications of impaired hearing ability.

Central and cortical brain atrophy varying from mild to moderate, has been observed in most of the adult patients who have been subjected to neuroradiological examinations.

I. Clinical course

The clinical manifestations at various ages are summarized in table 3.

The clinical course, characterized by normal early development progressing to severe psychomotor handicap, with typical features appearing with advancing age, has shown surprisingly little variation in our patients. The uniformity of the clinical findings in the Finnish AGU-patients as well as the lack of variation in the urinary excretion of AADG [6] contrast with the genetic heterogeneity encountered in many other genetic diseases of metabolism.

The clinical data published on the AGU-patients from elsewhere are too scanty for evaluating possible differences in the clinical findings between them and the Finnish patients.

As shown in table 2, the number of patients over 40 years of age is relatively small as compared with other adult age groups. The oldest patient known to us is presently 52 years of age. The Finnish series has included ten AGU-patients who have died at the age of 27-46 years. The autopsy findings and the causes of death are given in table 4.

Except for two cases, the patients died of severe bacterial, purulent, mostly pulmonary infections, such as pneumoniae or pulmonary, pleural and even brain abscesses. In addition to bacterial infections, actinomycosis was diagnosed in one case. Before death most of the patients have experienced a period of poor general condition, fatigue and inactivity.

Table 5.3 Clinical course of Aspartylglucosaminuria.

Age	Somatic and neurological symptoms	Mental and physical symptoms	Laboratory findings	X-ray findings
At birth	Normal	Normal	Not studied	Not studied
0-2 mo	Herniae	Normal	Not studied	Not studied
4-11 mo	Recurrent infections, herniae, diarrhea, motor development occasionally delayed	Normal	AADG-uria, vacuolated lymphocytes, AADG-ase deficiency	Not studied
1-4 y	Recurrent infections, clumsiness, muscular hypotonia	Slight retardation, delayed speech, stubborn behaviour	As above	Hypertrophy of pharyngeal tonsil
5-9 y	Clumsiness increased, slightly coarse face, systolic murmur	Mild to moderate retardation, impaired speech, perceptional disturbance, stubborn behaviour	As above	Somewhat thin cortex of long bones, first signs of vertebral changes
10-15 y	Clumsiness increased, facial abnormality more pronounced	Moderate or severe mental retardation, exitable behaviour	As above	Cortical and vertebral changes more prominent, thick skull
Over 15 y	Impaired motor functions, coarse face, short stature, large tongue and hoarse tone of voice, signs of hypogonadism	Severe retardation, dysarthric speech, exitable behaviour	As above	Thin bone cortex and sequelae from vertebral epiphysitis, thick skull, cerebral atrophy
30-...y	Unusually early death			

Table 5.4 The causes of death and autopsy findings in 10 AGU-patients.

Age at death	Sex	Autopsy findings
27	Female	Purulent abscesses in lungs, liver, pericardium. Osteomyelitis.
27	Male	Pulmonary abscess, pneumoniae.
28	Male	Pleural empyema.
28	Female	Purulent pneumoniae and bronchitis.
29	Male	Asphyxia (drowned in water).
29	Female	Bronchopneumoniae.
31	Male	Cerebral and pulmonary abscesses.
35	Male	Lobar pneumoniae and pulmonary abscesses.
42	Female	Ruptured ulcer with peritonitis.
46	Male	Bronchopneumoniae.

5.3 BIOCHEMISTRY

A. The N-acetylglucosaminyl-asparagine linkage

The oligosaccharide portion of glycoproteins is usually attached to the polypeptide chain by an N-glycosidic linkage from N-acetylglucosamine to the amide group of asparagine. The core region of this type of glycoprotein commonly includes another N-acetylglucosamine and a mannose residue, thereafter chain diversity is attained by branching and attachment of a variable number of mannose, galactose, N-acetylglucosamine, fucose, and sialic acid residues. The N-acetylglucosaminyl-asparagine linkage has been demonstrated in a large number of human glycoproteins, including red cell membrane sialoglycoprotein, thyroglobulin, HCG, IgM, IgE, transferrin, ceruloplasmin, and α2-macroglobulin [7].

The existence of a common core stucture in a large number of glycoproteins is explained by their synthesis from a common oligosaccharide-lipid intermediate, which contains the Man-GlcNAc-GlcNAc-sequence linked to dolichol phosphate [8]. Enzymatic transfer of the oligosaccharide portion from this intermediate to an asparagine residue creates the linkage that patients with aspartylglucosaminuria are unable to hydrolyze.

B. Catabolism of glycoproteins and aspartylglucosaminidase

The breakdown of the oligosaccharide moiety of glycoproteins apparently begins, as a rule, by cleavage of the core structure between the two GlcNAc units, catalyzed by endo-β-N-acetylglucosaminidase [9]. The carbohydrate chain is then degraded from the non-reducing end by the stepwise action of glycosidases, each removing a specific monosaccharide unit. Simultaneously, the polypeptide chain is split by cathepsins and then by amino- and carboxypeptidases. The catabolic enzymes are located primarily, if not exclusively, in lysosomes.

The enzyme responsible for splitting the N-acetylglucosaminyl-asparagine linkage is 1-aspartamido-β-N-acetylglucosamine amidohydrolase (AADGase,

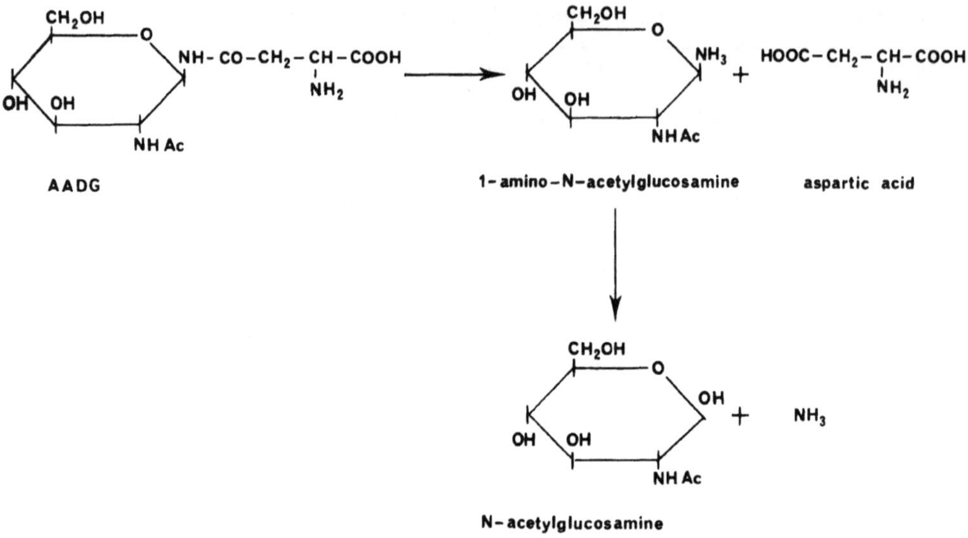

Fig. 5.5 The reaction catalyzed by 1-aspartamido-β-N-acetyl-glucosamine amidohydrolase (AADGase). Ammonia is released from the reaction product, 1-amino-N-acetylglucosamine, nonenzymatically.

E.C.3.5.1.26) (Fig. 5). It is widely distributed in mammalian tissues, with epididymis, kidney, spleen, and liver generally showing the highest specific activities [10-12]. Like the other enzymes of glycoprotein catabolism, AADGase is located in lysosomes [12].

AADGase has been partially purified from a variety of sources, including rat kidney [12], rat liver [12], hog serum [13], hen oviduct [14], and human seminal fluid [15]. Purification to apparent homogeneity has been achieved from hog kidney [11] and human liver [16].

Detailed enzymological studies were reported on the AADGase purified from hog kidney [11]. By chromatography on hydroxylapatite and DEAE-Sephadex A-50, one major and two minor peaks of activity were isolated. The major component was shown to be homogeneous by disc and sodium dodecyl sulfate disc electrophoresis as well as by ultracentrifugation. The protein consisted of a single polypeptide chain with a molecular weight of 70,000 and an isoelectric point of 5.12. Glucosamine was present in the enzyme molecule. The pH-optimum was 5.5, but the enzyme was active over a relatively broad pH range. The pH-activity curve of AADGase from human liver showed an exceptionally high pH-optimum at pH 8-10, but the analytical procedures may have influenced the results [16]. The rat liver enzyme had a rather sharp optimum at approximately pH 7.0 [10]. No activators of the enzyme are known, and the physiological significance of the reported inhibition by bivalent cations appears doubtful [11, 16].

The reported K_M values for the natural substrate, 2-acetamido-1-L-β-aspartamido-1,2-dideoxy-β-D-glucose (AADG) vary between 0.16 and 1.3

mM, depending on the type of preparation, pH, and enzyme concentration [10-13, 16]. The rates of hydrolysis and Michaelis constants of AADGase for ovalbumin glycopeptide were similar to those for synthetic AADG [11].

In the reaction catalysed by AADGase, aspartate and 1-amino-N-acetylglucosamine are first released, followed by nonenzymatic breakdown of the latter to ammonia and N-acetylglucosamine. This second step proceeds rapidly and completely under acidic conditions or at elevated temperatures. The mechanism of the reaction seems to involve an aspartyl-enzyme intermediate [13], which is supported by the fact that both asparagine and aspartic acid act as inhibitors of the enzyme. A number of synthetic aspartyl derivatives of various sugars and other compounds are also hydrolyzed by AADGase.

C. Deficiency of AADGase in aspartylglucosaminuria

Deficient activity of AADGase in a patient with aspartylglucosaminuria was first demonstrated by Pollitt *et al.* [1], who followed the disappearance of AADG upon incubation with a seminal fluid sample. Subsequently, the defect has been shown in many tissues and cells, including brain and liver [17], kidney [18], peripheral blood leukocytes [19], lymphocytes and serum [20], and cultured fibroblasts [21].

Certain discrepancies have been encountered in studies on the enzyme defect in aspartylglucosaminuria. Thus, both normal [17] and reduced [18] activity has been found in kidney samples, and the levels of residual enzyme activity in tissues have also been variable. Such differences are explained, at least in part, by difficulties in the assay of the enzyme. Synthetic substrate has not previously been generally available, and useful fluorescent or radioactive derivatives have not been prepared. The Morgan Elson reaction, used to quantitate the release of the product, N-acetylglucosamine, is relatively insensitive, and since enzyme activity even in normal tissues is low, prolonged incubation times have been necessary. Product disappearance in assays on crude homogenates may also influence the results [18]. Because of the low or unmeasurable residual activity in the patients, kinetic studies on the mutant enzyme have not been possible. Neither is it known, whether material immunologically related to normal AADGase is present in the tissues of patients.

D. Urinary excretion of products of glycoprotein catabolism

The major abnormal metabolite excreted by patients with aspartylglucosaminuria is AADG itself. The compound can be qualitatively identified by methods developed for amino acid analysis, utilizing high-voltage electrophoresis [2, 3] or two-dimensional thin-layer chromatography (see below). AADG is stained by ninhydrin, but the development of the characteristic blue color requires heating of the chromatograms at 120°C for 5 min. With ninhydrin-collidine-acetic acid, AADG and other asparagine-containing oligosaccharides give a typical yellow-brown color [22].

Quantitation of AADG in urine has been performed by measuring the N-acetylglucosamine liberated after acid [23] or enzymatic [24] hydrolysis of the compound, or by gas chromatography-mass spectrometry of methylated AADG [25]. By the latter method, the daily excretion of AADG by patients with aspartylglucosaminuria showed little variation, with a mean of 354 mg/24 h in patients less than 10 yr of age and 441 mg/24 h in older patients [6].

In terms of creatinine excretion, the younger group showed somewhat higher values than the older (0.26 and 0.15 mg/mg creatinine, respectively), but no correlation between the amount of AADG excreted and clinical severity of the disease could be shown. These quantitative results are in good agreement with those obtained using an enzymatic method [24], but are somewhat lower than those found after acid hydrolysis [23]. The concentration of AADG in serum of patients has been unmeasurable, which suggests that the compound is not reabsorbed in the kidney tubuli, in analogy with low-threshold amino acids.

In addition to AADG itself, the urine of patients with aspartylglucosaminuria contains a number of other glycoasparagines, with other carbohydrate units attached to the N-acetylglucosamine end of the molecule. By gel chromatography followed by electrophoresis at pH 2.0 and 5.3, at least five neutral and eleven acidic glycoasparagines were separated, some of them possibly still representing mixtures [22]. Sugar analysis, periodate oxidation, and enzymatic digestion permitted the assignment of tentative structures to four of the neutral and five of the acidic series. Independent verification of some of these structures has been obtained by similar methods complemented with methylation analysis [26, 27].

Next to AADG itself, the compound present in largest amounts in urine from aspartylglucosaminuria patients seems to be a galactosyl derivative of AADG (Gal-GlcNAc-Asn). This structure has not been found as part of any glycoprotein, and it is possibly derived by galactosyl transfer to AADG *in vivo*. Most of the other glycoasparagine tentatively identified contained this same sequence, to which variable numbers of N-acetylglucosaminyl, mannosyl, galactosyl, fucosyl, glucosyl, or N-acetylneuraminyl residues have been added, and in some cases also sulfate [22]. Their origin is unclear but could represent modification of the primary accumulating compound, AADG. Substantial amounts of Man-Man-GlcNAc-GlcNAc-Asn have also been demonstrated in the urine of patients with aspartylglucosaminuria [26, 27]. Since this compound has a structure similar to the core region of glycoproteins, it could represent incomplete catabolism of the oligosaccharide chain rather than secondary modification.

E. Storage material

After water or ethanol extraction of liver and brain tissues from a patient with aspartylglucosaminuria, thin-layer chromatography revealed the presence of free AADG, which was not detectable in samples from normal tissues [28]. Quantitative measurements confirmed substantial accumulation of AADG, with a minor contribution from other glycoasparagines.

A more detailed quantitative study and structural analysis of the water-extractable material in the liver of an AGU patient demonstrated that the major storage compound was indeed AADG, amounting to 2.7 mg/g liver. The other compound detected was Man-Man-GlcNAc-GlcNAc-Asn, its concentration being 0.2 mg/g wet weight [29]. The same storage materials were identified also in neural tissue, spleen, kidney, and thyroid [30]. In lymphocytes isolated from peripheral blood, only AADG but none of the higher glycoasparagines was shown to accumulate in amounts that were suggested to account for the vacuolization phenomenon [31].

Thus it seems that the compound responsible for the storage phenomenon in aspartylglucosaminuria is AADG, although the possibility of minor contributions from other glycoasparagines exists. Such findings are in agreement with the concept of endo-β-N-acetylglucosaminidase having a significant role in initiating the oligosaccharide chain breakdown in glycoprotein catabolism.

5.4 DIAGNOSIS

For all practical purposes, the diagnosis of AGU can be made by demonstration of increased excretion of AADG in the urine. Final proof can be obtained by assay of AADGase in suitable samples from the patient.

A. Detection of AADG in the urine

Because of the constancy of AADG excretion in the patients [6], spot urine samples can be used for analysis, although 24 hr collections are preferable. A preservative, eg. thymol, should be added but freezing of the sample is not necessary. Of the several methods described for the separation and detection of AADG, we have found two-dimensional thin layer chromatography most convenient. A sample (2 μl) of untreated urine is spotted on a 10 x 10 cm cellulose plate and developed first in t-butanol: ethylmethylketone: 25% ammonia: water (125:75:25:50 v/v) and in the second dimension in n-butanol: acetone: acetic acid: water (35:35:10:20 v/v). After drying the plate is sprayed with 0.5% ninhydrin and heated at 120°C for 5 min. Alternatively, one-dimensional development and staining with resorcinol, used for screening for urinary oligosaccharides (Fig. 6), also yields a distinct purple-brown spot in the area usually occupied by tri- and tetrasaccharides. Quantitation of AADG excretion is not necessary for diagnostic purposes.

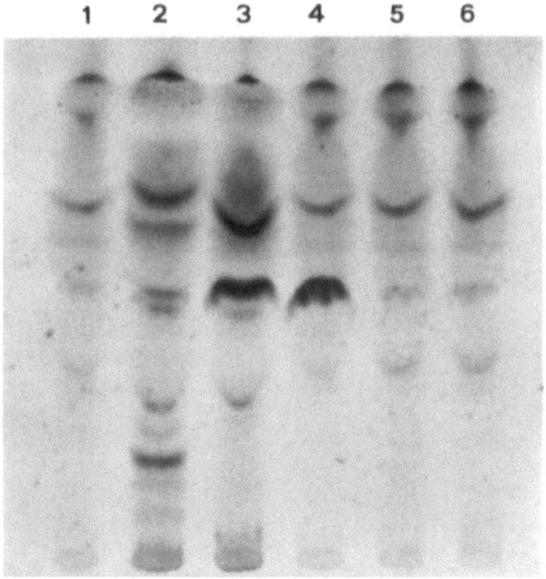

Fig. 5.6 Thin-layer chromatography of urinary oligosaccharides. Samples were concentrated fivefold by lyophilization, spotted on a silica gel plate and developed in n-butanol: acetic acid: water (2:1:1 v/v). The spots were visualized with a resorcinol spray reagent after heating at 120° for 10 min. Lanes 1, 5 and 6: normal urine samples; lane 2: aspartylglucosaminuria; lane 3: Salla disease; lane 4: normal urine + 10 μg/ml sialic acid.

B. Assay of AADGase

Suitable samples that allow the demonstration of deficient AADGase activity in patients include plasma, leukocytes, lymphocytes, and cultured fibroblasts. We have based our assay on the method described by Makino *et al.* [32] with slight modifications [21]. Cells are suspended in 50 mM phosphate buffer pH 7.0 and disrupted by sonication in the presence of Triton X-100. The assay mixture contains a suitable amount of cell extract and 5.5 mM synthetic AADG in 30 mM phosphate buffer, pH 7.0. After incubation at 37° for 20 hr, the reaction is stopped by boiling for 5 min, which also liberates ammonia from the primary reaction product and leaves N-acetylglucosamine to be measured with the Morgan-Elson reaction [33].

In patients with AGU, significant enzyme activity cannot be demonstrated in the samples mentioned, while heterozygotes usually show values intermediate between the patients and normal individuals (see below).

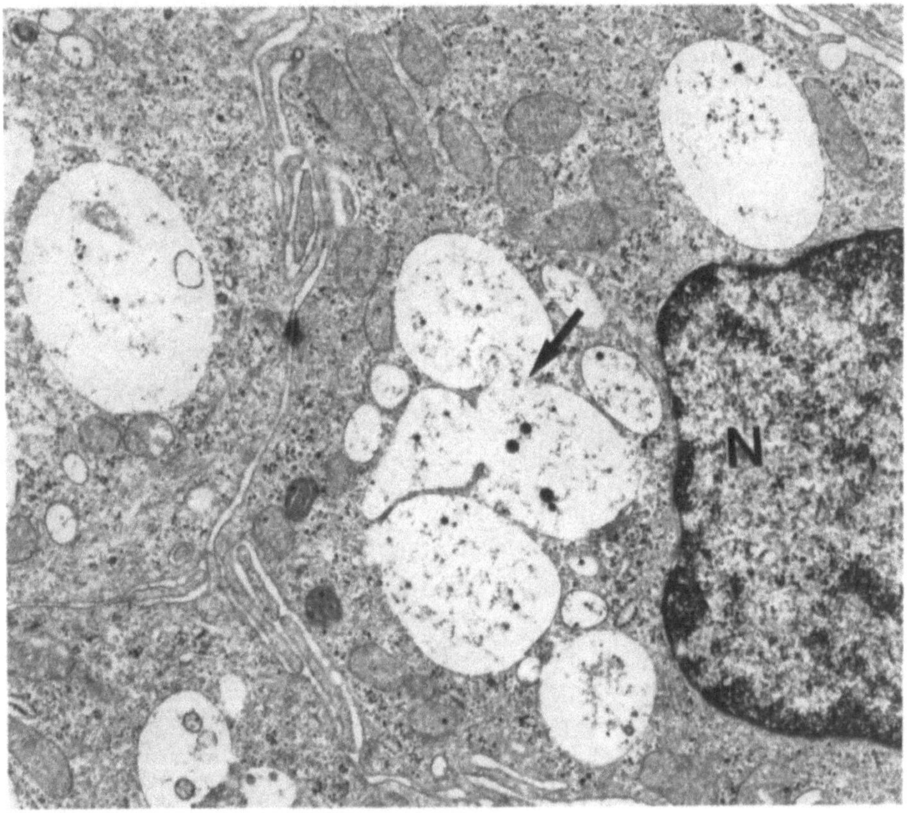

Fig. 5.7 Electron micrograph of sweat gland epithelium of an AGU patient. Several storage lysosomes are present. Three of them are coalescing to form a larger vacuole (arrow). The vacuoles contain small electon dense globules and granular material on clear background. N, nucleus (x 20,000).

5.5 PATHOLOGY

The basic cellular change in AGU, as in many other lysosomal storage disorders [34], is the presence of hypertrophied storage lysosomes. All organs studied by electron microscopy so far have shown storage lysosomes, although some cell types are more often and more extensively affected than others. The storage lysosome in different types of cells is basically similar. It is a cytoplasmic vacuole encircled by a unit membrane containing amorphous fibrillogranular material on a clear background (Fig. 7). Depending on the cell type, the lysosome may also contain electron-dense globules, lipid droplets and membranous structures of various configurations (Fig. 8). The size of the vacuoles varies from about 0.4 to 10 μm in diameter.

The large vacuoles can also be seen by light microscopy under appropriate conditions. The cytoplasmic vacuoles in the lymphocytes of peripheral blood are

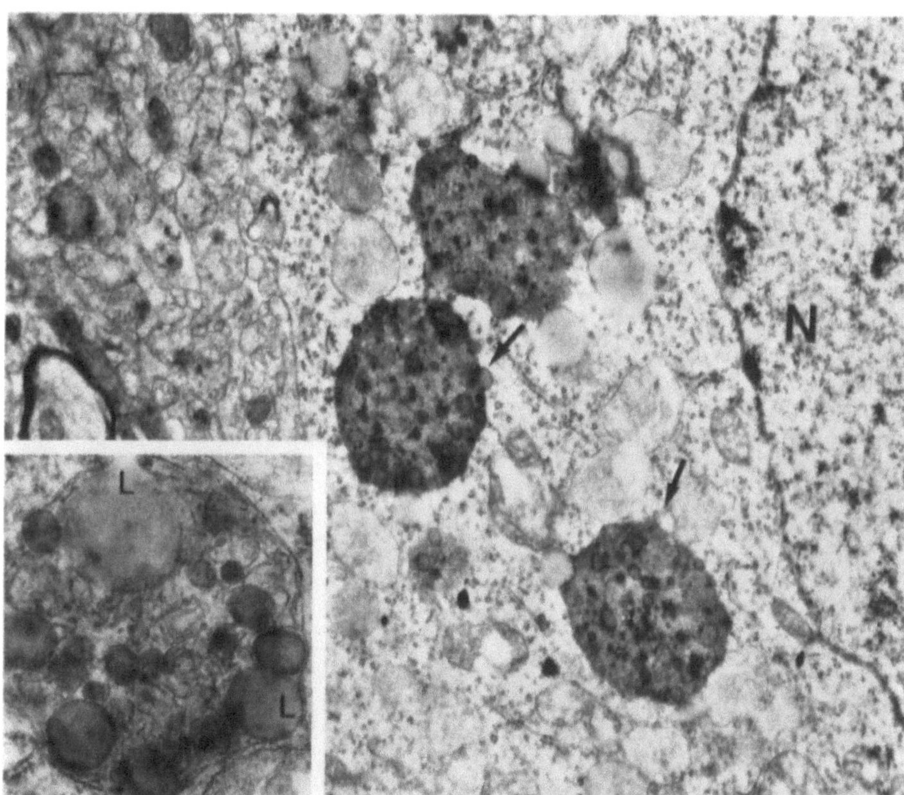

Fig. 5.8 Lipofuscin-like residual bodies (arrows) in an AGU-neuron of cerebral cortex. Insert shows details of a residual body with lipid droplets and membrane fragments. N, nucleus; L, lipid droplets (x 35,000; insert x 80,000) [Courtesy by Dr. M. Haltia].

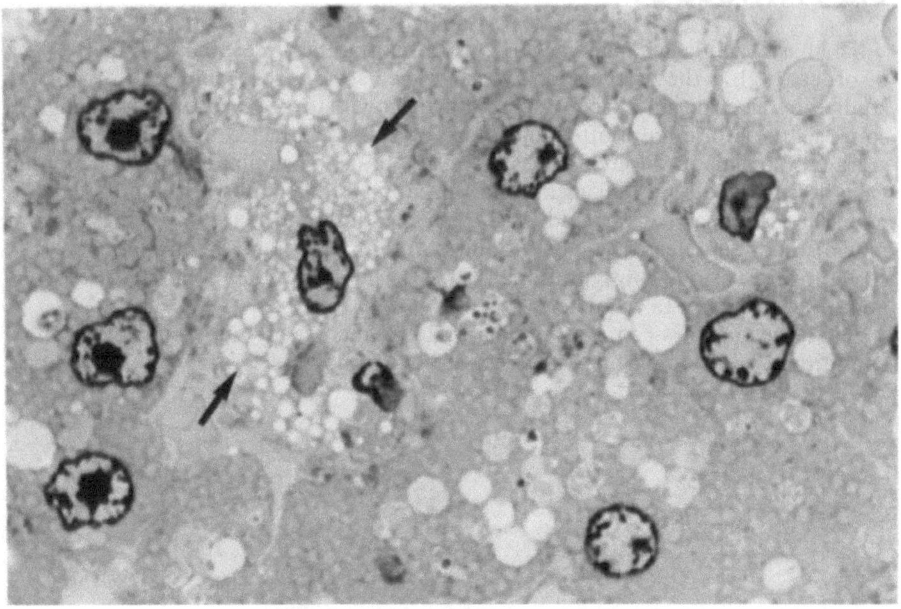

Fig. 5.9 Light micrograph of AGU-liver. Large clear vacuoles in the hepatocytes and Kupffer's cells (indicated by arrows) are packed by smaller vacuoles. 1 μm thick plastic section stained with toluidine blue (x 1,100).

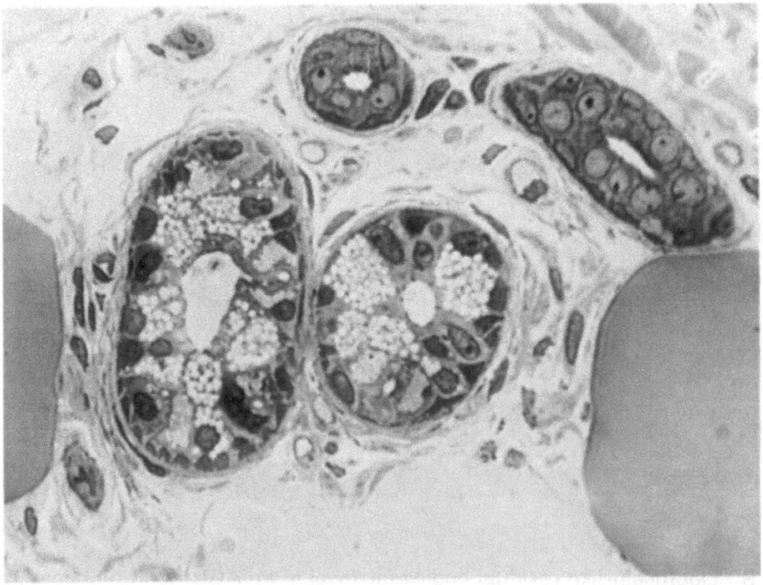

Fig. 5.10 Coiled portion of a sweat gland shows numerous clear vacuoles in the secretory cells. Light microscopic picture as in figure 9 (x 460).

the light microscopic counterpart of the storage lysosomes revealed by electron microscopy [35]. Although some tissues such as liver may appear highly vacuolated in ordinary paraffin sections, the best way to detect the storage lysosomes by light microscopy is to use semithin sections (1 μm) of plastic-embedded tissue. In these sections the vacuoles are easily detected in any tissue (Figs. 9 and 10).

The extensive presence of storage lysosomes, for instance in liver, would suggest histochemical staining methods to be useful in the morphological study of AGU. The few attempts to show the storage material by histochemical methods have generally been negative. Haltia *et al.* [5] performed a careful study of brain and liver biopsies of three patients using both frozen and paraffin sections after various fixations. Several stains to demonstrate lipids, glycosaminoglycans, and proteins were used, but the vacuoles were negative in all stainings, except for the appearance of some lipofuscin-like pigments, probably secondary to cellular destruction. Isenberg and Sharp [19] reported diastase-sensitive PAS-positive globules in hepatocytes, but the vacuoles in other cells did not stain with PAS or Best-carmine. In our experience, frozen and paraffin sections of skin biopsies did not stain with Oil red-O, Sudan black-B, Alcian blue or PAS.

The most plausible explanation for the failure to demonstrate the storage material histochemically has been presented by Haltia *et al.* [5]. The only quantitatively important storage material seems to be AADG itself. This substance of low molecular weight may cause the increase of the lysosomal size by occupying a large volume in isotonic solution. The low concentration of AADG and its high solubility in water and many solvents make the demonstration of the stored material impossible by histochemical methods.

A. Morphological changes in different organs

Brain Only one report on neuropathological changes in the brain of an AGU patient has been published [5]. The brain of a 29-year old female was moderately atrophic with generalized gyral atrophy and wide sulci. There was a slight ventricular dilatation, but the cerebellum and brain stem were normal. The nerve cell bodies of the cerebral cortex were distended, rounded and contained poorly defined vacuoles by light microscopy. All stages of disintegration of the neurons were present and this was most pronounced in the striata which showed spongiform rarefaction of the tissue. Affected cells contained cytoplasmic granules with the staining properties of lipofuscin and coarse lipofuscin-like particles were seen at the sites of destroyed cells. In the cerebellum there was a pronounced loss of Purkinje cells, some proliferation and hypertrophy of Bergmann's glia and the granule cell layer was slightly rarefied. Similar neuronal destruction was also evident in the basal ganglia and thalamus, but the nuclei in the brain stem were much less affected. The white matter appeared normal.

In the electron microscopy the neurons contained storage lysosomes as described before, but in addition there were numerous electron dense granular bodies. These appeared to be residual bodies and consisted of membrane-bound inclusions of granular material associated with lipid droplets (Fig. 8). Similar electron dense bodies were also reported by Arstila *et al.* [36].

Liver Examination of histological sections of liver revealed vacuolization of hepatocytes and Kupffer's cells [5, 19]. The vacuolization may be so extensive that the liver resembles advanced fatty liver, but the staining for lipids is negative [5].

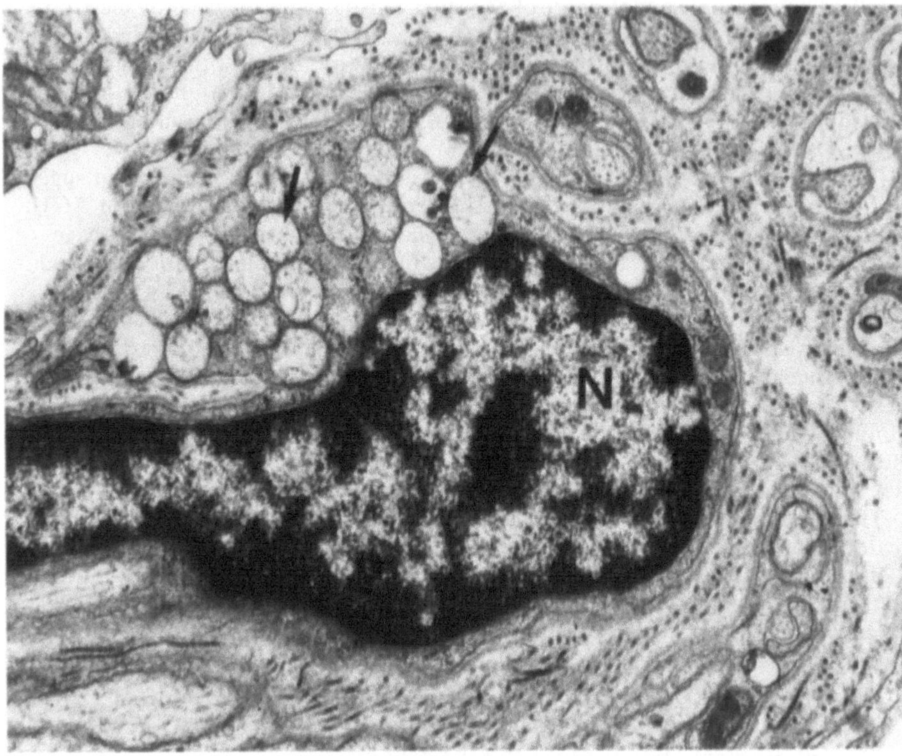

Fig. 5.11

Fig. 5.11-12 Electron micrographs of skin structures in AGU. Schwann cell cytoplasm (Fig. 11) and capillary endothelial cells (Fig. 12) contain numerous storage lysosomes (arrows). N, nucleus; E, red blood cell in the capillary (x 18,000).

By electron microscopy, the hepatocytes contain large numbers of storage lysosomes, some larger than the nucleus [5, 17, 19, 36]. The storage lysosomes contain reticular and fibrillar material, lipid droplets, electron dense globules and concentric membranous arrays, all in relatively small quantities. The Kupffer's cells are often studded with slightly smaller vacuoles than the hepatocytes, and there is often apparent fusion of the vacuoles to form larger ones, as noted also by Haltia *et al.* [5] in hepatocytes. Storage lysosomes are also present in the bile duct epithelium. Despite the extensive vacuolization of the hepatocytes and Kupffer cells, other cellular changes have not been found with the exception of reduced numbers of lysosomes of normal size [5, 19]. The extensive vacuolization of liver cells is easily detected under light microscopy in semithin plastic sections (Fig. 9).

Skin We have used skin biopsies as a screening method for initial diagnosis of various storage disorders, as suggested by O'Brien *et al.* [37]. In five AGU patients both light microscopy of semithin plastic sections and electron

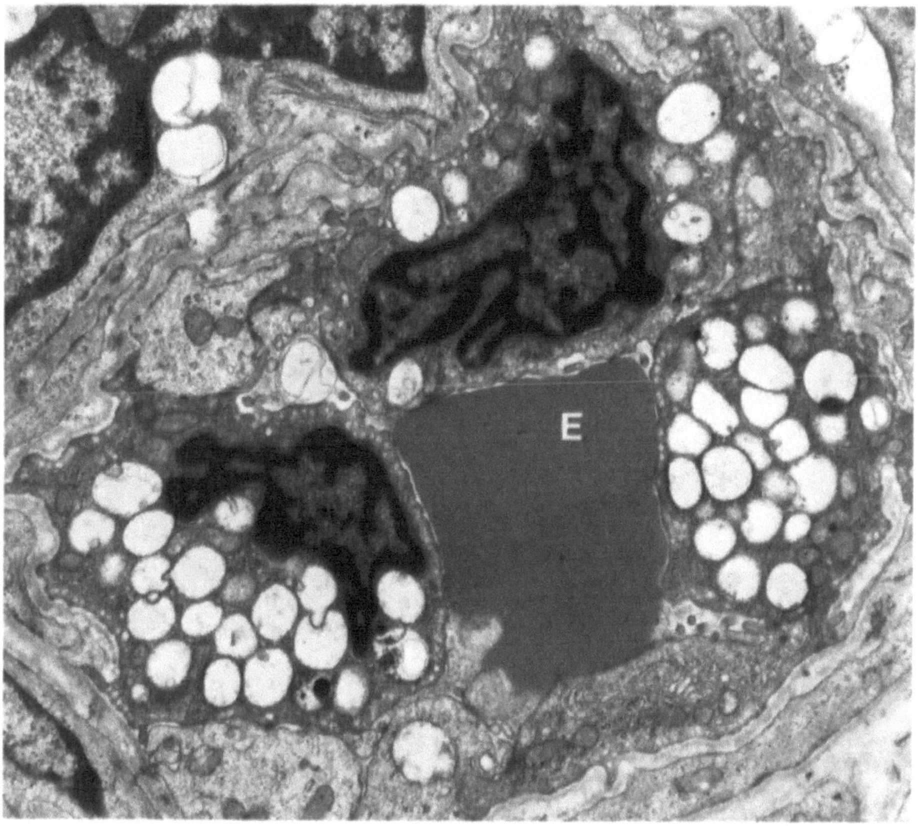

Fig. 5.12

microscopy revealed clear evidence of a lysosomal storage disorder. In light microscopy, the vacuoles in the epithelial cells of the coiled portion of the sweat glands are easily detectable (Fig. 10). Vacuolated skin fibrocytes and Schwann cells could also be found under the light microscope. In electron microscopy, typical storage lysosomes were seen in the sweat gland epithelium, fibrocytes, Schwann cells, endothelial cells, and even in the smooth muscle cells. The only cells not containing storage lysosomes appeared to be the epidermal cells. In the skin the vacuoles contained much less material than in the liver. In the sweat glands there was some reticulogranular material and occasional electron-dense globules, but the vacuoles in other cell types appeared almost "empty" (Figs. 11 and 12).

Lymphocytes The presence of vacuolated lymphocytes in AGU is a constant finding. In electron microscopy the vacuoles appear as membrane-bound storage lysosomes as in other cells (Fig. 13) [19, 35]. Aula *et al.* [35] found that the storage lysosomes were more frequent in T-cells than in B-cells.

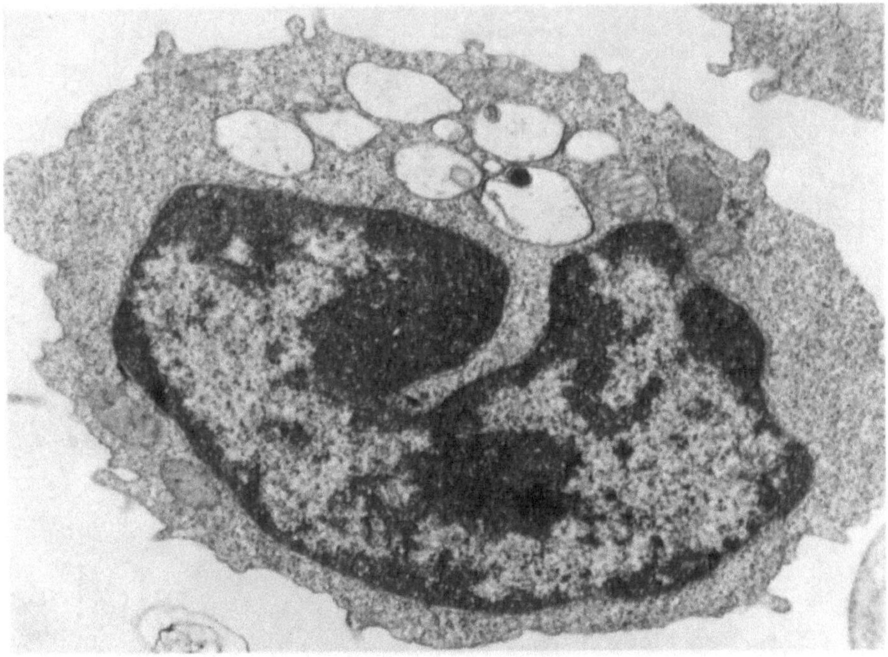

Fig. 5.13 Peripheral blood lymphocyte with several cytoplasmic storage lysosomes (x 18,000).

Other tissues Involvement of other tissues has been briefly mentioned in a few reports of AGU. Arstila *et al.* [36] reported storage lysosomes in the epithelial cells of the renal glomeruli and proximal tubules. Isenberg and Sharp [19] found them in the jejunal epithelium and mesenchymal cells of a lymph node, and we have seen them in a biopsy of the palatinal tonsil.

Cultured fibroblasts A number of fibroblast cultures were studied with the electron microscope, but storage lysosomes were not found in any of them.

B. General pathological findings

No detailed autopsy reports are available. We know of 10 deceased AGU-patients who have been autopsied, but no special investigations have been performed. It seems that the immediate cause of death in 8 cases was purulent infection, usually bronchopneumonia (Table 4). Isenberg and Sharp [19] reported mitral insufficiency and a slightly thickened aortic valve, shown by cardiac chatheterization, and Haltia *et al.* [5] reported nodular thickening of the free edges of the atrioventricular valves of their patient. Diffuse endocardial fibroelastotic changes were not reported in the Finnish autopsy cases.

C. Clinicopathological correlations

The storage phenomenon itself does not necessarily cause severe impairment of cellular functions. The functions of liver and kidney remain normal despite extensive storage. In the brain however, the functional impairment and the pathological changes seem to be concordant. It is not known whether the neuronal destruction is caused by overloading of the cells by the storage lysosomes, or by some other metabolic derangement linked with the AADGase deficiency.

We would emphasize the value of skin biopsy for the demonstration of the storage phenomenon and for the initial screening of a number of storage disorders. Our experience with AGU, mannosidosis and Salla disease fully confirm the results of other studies of skin biopsies in storage disorders [37-39]. The storage lysosomes in AGU as in many other disorders are present before the clinical picture is fully developed. The skin biopsy showed extensive accumulation of storage lysosomes in a two year old boy with AGU. Preliminary results can even be obtained before electron microscopy by studying semithin plastic sections under high power with the light microscope.

The morphological study of skin or even liver biopsies does not, however, distinguish AGU from many other similar storage disorders. The hypertrophied lysosomes in other glycoprotein storage disorders and mucopolysaccharidoses are identical to those of AGU, and the final diagnosis must be made by biochemical investigations.

5.6 EPIDEMIOLOGY

AGU is a prime example of the concept of the Finnish heritage of diseases, enrichments or losses of rare recessive genes [40] in this population. At the present time approximately 30 autosomal recessive disorders and one autosomal dominant disease - lattice corneal dystrophy and amyloidosis [41] - are known to be more common in Finland than in any other part of the world. Interestingly, disorders of glycoprotein degradation, the subject of this book, include three genetic diseases with overrepresentation of Finnish cases. These are aspartylglucosaminuria, Salla disease, and mannosidosis.

The original two AGU-patients were from England [1] but additional new patients outside Finland have been reported only during the last few years. So far, we know of 10 non-Finnish patients with verified AGU. The principal data of these cases are shown in table 5. In addition Borud and coworkers have found 7 AGU-patients in northern Norway [23], but the ancestors of these patients originate in the western part of Finnish Lapland, an area known to have a high frequency of the AGU gene. The limited clinical data available on the AGU-patients outside Finland do not allow a detailed evaluation of the extent of genetic heterogeneity in AGU. It is striking, however, that the patients of Italian (Sicilian) origin seem to have more bony changes on X-ray and less coarse facial features than the Finnish patients.

The existence of the cases mentioned above suggests that the AGU-gene is more widespread than previously thought. The disease should be considered and appropriately looked for in patients with signs of lysosomal storage and mental retardation of unknown origin. Urinary screening with paper chromatography was performed on 1792 mentally retarded patients in Poland, but no AGU-cases were found [42].

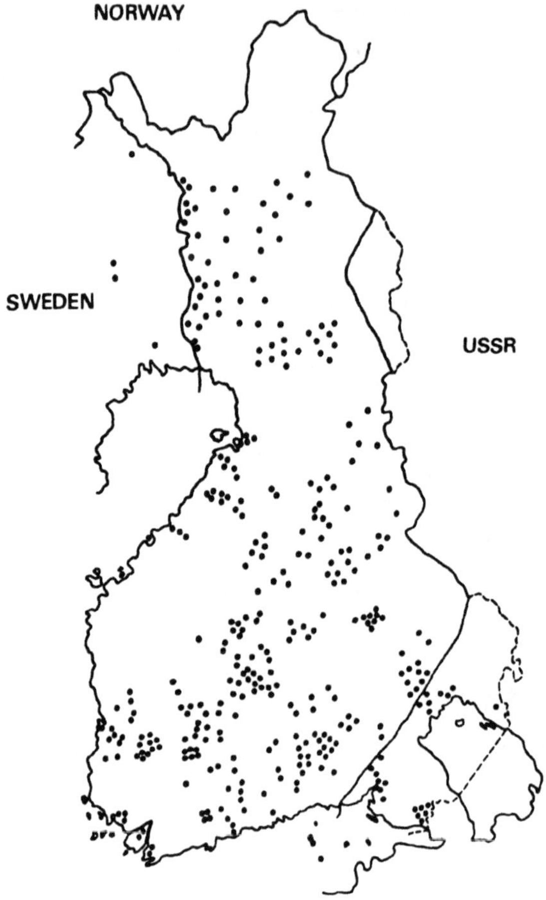

Fig. 5.14 Distribution of AGU gene in Finland. Birthplaces of the grandparents of 74 AGU-patients showing an over-representation of the gene in the northern and eastern parts of the country. The population density is almost a mirrorimage with highest density in South-West Finland.

The distribution of rare genes in the Finnish population is uneven, with local enrichments and deficits. This is illustrated in a map (Fig. 14) dipicting the birthplaces of the grandparents of 74 AGU-patients. The gene frequency of AGU in the Finnish population is unknown but it should be estimated separately for each part of the population. We have recently surveyed the prevalence of AGU and Salla disease (unpublished) among mentally retarded patients over 5 years of age in Lapland, with a total population of 195,000. Urine samples of 1265 mentally retarded patients were assayed using a thin layer chromatographic technique [43]; 26 AGU-patients were found, giving a rough estimate of 1:7000. The corresponding gene frequency would be 1:80. Autio has earlier estimated that the incidence of AGU in newborns is of the order of 1:26,000 [3].

5.7 CARRIER DETECTION

Obligate heterozygotes, i.e. the parents of AGU-patients, do not show any phenotypic manifestations of heterozygosity. Urine of heterozygotes may occasionally contain trace amounts of aspartylglucosamine [24], but normal individuals can also have trace amounts of AADG in urine [44].

In accordance with an autosomal recessive mode of inheritance and in analogy with other lysosomal storage disorders, the heterozygous carriers of the AGU-gene show an intermediate enzyme activity of AADGase. This has been shown with AADGase assay in cultured fibroblasts [21, 45] and in lymphocytes

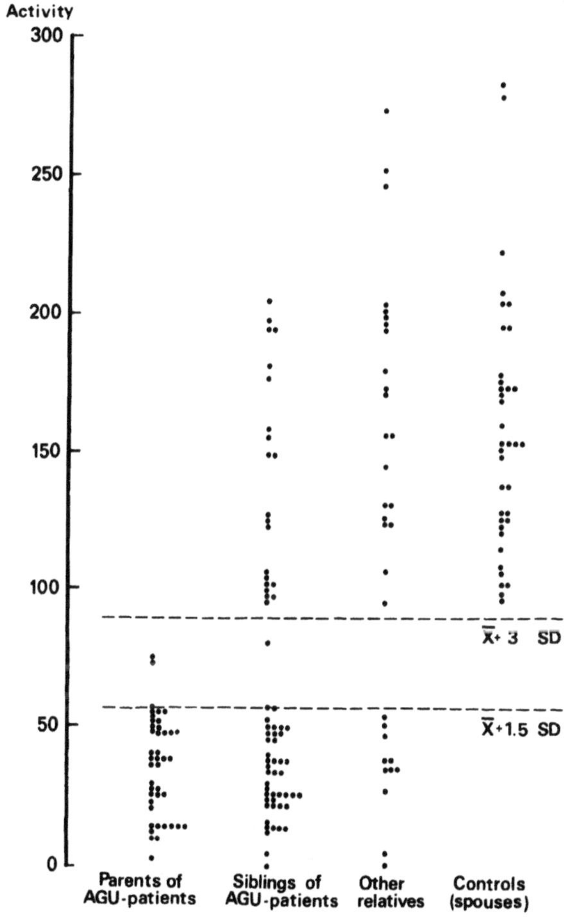

Fig. 5.15 Carrier detection in AGU with assay of AADase activity in peripheral blood lymphocytes. Values above the upper dotted line showing the mean activity +3 SD of obligate heterozygotes indicate normal genotype, values below mean heterozygote activity +1.5 SD indicate carrier genotype. Enzyme activities between the two lines indicate "inconclusive" cases which are subjected to further studies.

Table 5.5 Non-finnish AGU-patients.

Age	Sex	Nationality	Mode of diagnosis	Main clinical characteristics	References
32 y	F	England	AADG in urine	Mental retardation from early years	Pollitt et al. 1968
20 y	M	England	AADG ase assay	Psychotic episodes, coarse features	
3 y	M	England	AAGD in urine	Delayed speech, slight MR, infections and herniae	Rossiter M. (personal comm., 1979)
5 y	M	England	AADG in urine	Delayed speech, MR, coarse features, hoarse voice, infections	
22 y	F	Italy (Sicily)	AADG in urine AADG ase assay	MR (severe), scoliosis, no coarse facial features	Durand P. (personal comm.)
2 y	F	USA (Mexican-Italian)	AADG in urine AADG ase assay	MR (mild), coarse facial features, hepatosplenomegaly, dysostosis multiplex	Isenberg and Sharp 1975
10 y	M	USA (black)	AADG ase assay	MR from 4 years, coarse facial features, hepatomegaly	Thomas et al. (personal comm., 1980)
15 y	M	USA (black)	AADG ase assay	MR from early years, coarse features, limitations of wrist movements	
20 y	F	Italy (Sicily)	AADG ase assay	MR (severe), dysmorphic facial features	Gehler et al. 1980
9 y	M	Italy (Sicily)		Dwarfism, lumbar gibbus, angiokeratoma corporis diff., X-ray abnormality	
7 cases 4-31 y		Northern Norway but Finnish origin	AADG in urine		Borud et al. 1978

separated from peripheral blood [20]. Enzyme assay in serum/plasma or in total leukocyte homogenate does not differentiate the normal and the heterozygous genotype well enough, although the homozygous state can be established. Assay of AADGase in lymphocytes, isolated by Ficoll-Isopaque centrifugation, however, gives a clear distinction between homozygous, heterozygous, and normal AGU-genotypes. The rare individuals with an inconclusive enzyme activity between the upper limit of heterozygotes and lower limit of normal, can be evaluated by retesting or by fibroblast assay.

During the last five years we have carried out a limited AGU-carrier screening among the relatives of verified AGU-patients, to identify carrier-carrier risk couples for possible antenatal studies in a prospective manner.

Up to the present time 64 siblings, 33 other relatives and 37 spouses of siblings have been tested as shown in figure 15. The proportion of heterozygotes and normal genotypes among the tested relatives fits well with the expectation according to the recessive mode of inheritance. The lymphocyte AADGase assay is suitable in limited carrier screening of families of AGU-patients but it is not applicable for screening large populations. The requirement of a freshly taken blood sample for lymphocyte separation sets the limits for the use of the test. Samples cannot be transported except for short distances, and only a limited number of samples can be handled.

5.8 PRENATAL DIAGNOSIS

The basis for the intrauterine diagnosis of fetal AGU is the almost total absence of AADGase activity in cultured fibroblasts from AGU-patients [45] and the normal enzyme activity in amniotic fluid cell cultures from normal mid-term pregnancies [21]. It is not known at which stage of the disease aspartylglucosamine appears in the urine, but the constant amount of AADG in urine of patients ranging from 7 months to 40 years of age strongly suggests an early and probably a before-birth onset. Therefore, determination of amniotic fluid aspartylglucosamine might be another method, although an auxiliary one, for prenatal diagnosis of AGU. Both alternatives have been tested once. Absence of AADG assayed with paper chromatography was found in a pregnancy of a woman whose earlier child had AGU. The product of the pregnancy was a healthy newborn who has had a normal development up to the age of 9 years.

In the second prenatal study of AGU, the enzyme assay on cultured amniotic fluid cells was used. The mother was a 32-year-old III-para whose second child, a 10-year-old girl, has typical AGU. The activity of AADGase in cultured cells was 25 nmol/min/mg of protein, while the control cell lines had 94-110, respectively. The results were interpreted as suggesting a heterozygous fetus and the pregnancy was continued until delivery of a healthy male newborn. The child has had a normal development up to the age of one year and his urine contains no AADG.

Despite the large number of AGU-families in this country, the demand for prenatal diagnosis has remained, and probably will remain, little. The parents have usually already completed their families at the time of final diagnosis of the index patient. The first signs of the disease often are so unremarkable, that several years may elapse before etiological studies are undertaken. This situation will probably not change before a simple serum based carrier test is available for prospective identification of carrier-carrier risk couples in screening of large populations, analogously with Tay-Sachs-carrier screening in Jewish populations.

The original works of the present review were supported with grants from the Association of the Finnish Life Insurance Companies, Sigrid Jusélius Foundation and Foundation for Pediatric Research, Helsinki, Finland.

REFERENCES

1. Pollitt R.J., Jenner F.A., Merskey H.: *Aspartylglucosaminuria. An inborn error of metabolism associated with mental defect.* Lancet *2*, 253-255, 1968.
2. Palo J., Mattsson K.: *Eleven new cases of aspartylglucosaminuria.* J. Ment. Defic. Res. *14*, 168-173, 1970.
3. Autio S.: *Aspartylglucosaminuria. Analysis of thirty-four patients.* J. Ment. Defic. Res. Monograph. Series I, 1-39, 1972.
4. Autio S., Visakorpi J.K., Järvinen H.: *Aspartylglucosaminuria (AGU). Further aspects on its clinical picture, mode of inheritance and epidemiology based on a series of 57 patients.* Ann. Clin. Res. *5*, 149-155, 1973.
5. Haltia M., Palo J., Autio S.: *Aspartylglucosaminuria: A Generalized Storage Disease. Morphological and Histochemical Studies.* Acta Neuropath. (Berl.) *31*, 243-255, 1975.
6. Aula P., Raivio K.O., Maury P.: *Variation of urinary excretion of aspartylglucosamine and associated clinical findings in aspartylglucosaminuria.* J. Inher. Met. Dis. *3*, 159-162, 1980.
7. Kornfeld R., Kornfeld S.: *Comparative aspects of glycoprotein structure.* Ann. Rev. Biochem. *45*, 217-237, 1976.
8. Struck D.K., Lennartz W.J.: *The function of saccharide-lipids in synthesis of glycoproteins.* In "The Biochemistry of Glycoproteins and Proteoglycans". W.J. Lennarz (Ed.) pp. 35-83, Plenum Press, New York, 1980.
9. Tarentino A.L., Maley F.: *Purification and properties of an endo-β-N-acetylglucosaminidase from hen oviduct.* J. Biol. Chem. *251*, 6537-6543, 1976.
10. Conchie J., Strachan I.: *Distribution, purification and properties of 1-aspartamido-β-N-acetylglycosamine amidohydrolase.* Biochem. J. *115*, 709-715, 1969.
11. Kohno M., Yamashina I.: *Purification and properties of 4-L-aspartylglycosylamine amidohydrolase from hog kidney.* Biochim. Biophys. Acta *258*, 600-617, 1972.
12. Mahadevan S., Tappel A.L.: *β-aspartylglucosylamine amidohydrolase of rat liver and kidney.* J. Biol. Chem. *242*, 4568-4576, 1967.
13. Tanaka M., Kohno M., Yamashina I.: *Specificity studies of 4-L-aspartylglycosylamine amidohydrolase.* J. Biochem. *73*, 1285-1289, 1973.
14. Tarentino A.L., Maley F.: *The purification and properties of a β-aspartyl-N-acetylglycosylamine amidohydrolase from hen oviduct.* Arch. Biochim. Biophys. *130*, 295-303, 1969.
15. Pollitt R.J., Jenner F.A.: *Enzymatic cleavage of 2-acetamido-1(β-L-aspartamido)-1,2-dideoxy-β-D-glucose by human plasma and seminal fluid. Failure to detect the heterozygous state for aspartylglucosaminuria.* Clin. Chim. Acta *25*, 413-416, 1969.
16. Dugal B., Stromme J.: *Purification and some properties of 1-aspartamido-β-N-acetylglucosamine amidohydrolase from human liver.* Biochem. J. *165*, 497-502, 1977.
17. Palo J., Riekkinen P., Arstila A.U., Autio S., Kivimaki T.:*Aspartylglucosaminuria II: Biochemical studies on brain, liver, kidney and spleen.* Acta Neuropath. *20*, 217-224, 1972.
18. Dugal B.: *Measurement of 1-aspartamido-β-N-acetylglucosamine amidohydrolase activity in human tissues.* Biochem. J. *163*, 9-14, 1977.
19. Isenberg J.N., Sharp H.L.: *Aspartylglucosaminuria: Unique biochemical and ultrastructural characteristics.* Human Path. *7*, 469-481, 1976.
20. Aula P., Raivio K.O., Autio S.: *Enzymatic diagnosis and carrier detection of aspartylglucosaminuria using blood samples.* Pediatr. Res. *10*, 625-629, 1976.
21. Aula P., Autio S., Raivio K.O., Näntö V.: *Detection of heterozygotes for*

aspartylglucosaminuria (AGU) in cultured fibroblasts. Humangenetik *25*, 307-314, 1974.
22. Pollitt R.J., Pretty K.M.: *The glycosasparagines in urine of a patient with aspartylglucosaminuria.* Biochem. J. *141*, 141-146, 1974.
23. Borud O., Torp K.H., Dahl T.: *Aspartylglucosaminuria. Urinary excretion of aspartylglycosamines related to mental retardation.* Monogr. Hum. Genet. *10*, 23-26, 1978.
24. Sugahara K., Nishimura K., Aula P., Yamashina I.: *Enzymatic determination of urinary aspartylglycosylamine: A rapid and sensitive method to detect aspartylglycosylaminuria (AGU).* Clin. Chim. Acta *72*, 265-267, 1976.
25. Maury P., Kärkkäinen J.: *Identification of 4-N-2-acetamido-2-deoxy-β-D-glucopyranosyl-L-asparagine in biological materials by gas chromatography-mass spectrometry.* Clin. Chim. Acta *91*, 75-79, 1979.
26. Sugahara K., Akasaki M., Funakoshi I., Aula P., Yamashina I.: *Structural studies of glycoasparagines from urine of a patient with aspartylglycosylaminuria (AGU).* FEBS Lett. *78*, 284-286, 1977.
27. Lundblad A., Masson P.K., Nordén N.E., Svensson S., Ockerman P.-A., Palo J.: *Structural determination of three glycoasparagines isolated from the urine of a patient with aspartylglycosaminuria.* Eur. J. Biochem. *67*, 209-214, 1976.
28. Palo J., Savolainen H.: *Biochemical diagnosis of aspartylglycosaminuria.* Ann. Clin. Res. *5*, 156-162, 1973.
29. Maury P.: *Accumulation of two glycoasparagines in the liver in aspartylglycosaminuria.* J. Biol. Chem. *254*, 1513-1515, 1979.
30. Maury C.P.J.: *Accumulation of glycoprotein-derived metabolites in neural and visceral tissues in aspartylglycosaminuria.* J. Lab. Clin. Med. *96*, 838-844, 1980.
31. Maury P., Palo J.: *Characterization of the storage material of peripheral lymphocytes in aspartylglycosaminuria.* Clin. Sci. *58*, 165-168, 1980.
32. Makino M., Kojima T., Yamashina I.: *Enzymatic cleavage of glycoproteins.* Biochem. Biophys. Res. Commun. *24*, 961-966, 1966.
33. Levvy G.A., McAllan A.: *The N-acetylation and estimation of hexosamines.* Biochem. J. *73*, 127-132, 1959.
34. Hers H.G.: *The concept of inborn lysosomal disease.* In "Lysosomes and storage diseases". H.G. Hers, van Hoof (Eds.), New York, Acad. Press, 148-171, 1973.
35. Aula P., Rapola J., Andersson L.C.: *Distribution of cytoplasmic vacuoles in blood T and B lymphocytes in two lysosomal disorders.* Virchows Arch. B Cell Path. *18*, 263-271, 1975.
36. Arstila A.U., Palo J., Haltia M., Riekkinen P., Autio S.: *Aspartylglucosaminuria I. Five structural studies on liver, kidney and brain.* Acta Neuropath. (Berl.) *20*, 207-216, 1972.
37. O'Brien J., Benett I., Veath L., Paa D.: *Lysosomal storage disorders. Diagnosis by ultrastructural examination of skin biopsy specimen.* Arch. Neurol. *32*, 592-599, 1975.
38. Spicer S.S., Garvin A.I., Wöhltmann H. I., Simson J.A.W.: *The ultrastructure of the skin of patients with mucopolysaccharidosis.* Lab. Invest. *31*, 488-502, 1974.
39. Martin J.J., Ceuterick C., Leroy I.G.: *Contribution de la biopsie cutanée au diagnostic des encéphalopathies métaboliques.* Rev. Neurol. (Paris) *132*, 639-651, 1976.
40. Nevanlinna H.R.: *Rare hereditary diseases and markers in Finland: An Introduction.* In "Population Structure and Genetic Disorders". A.W. Eriksson (Ed.), Academic Press, London, 569-576, 1980.
41. Meretoja J.: *Genetic aspects of familial amyloidosis with corneal lattice dystrophy and cranial neuropathy.* Clin. Genet. *4*, 173-185, 1969.
42. Galewicz A., Gorska D., Rodo M., Zaremba J.: *A search for aspartylglycosaminuria in Poland.* J. Ment. Defic. Res. *18*, 135-138, 1974.
43. Humbel R., Collart M.: *Oligosaccharides in the urine of patients with glycoprotein storage diseases.* Clin. Chim. Acta *60*, 143-145, 1975.
44. O'Neil Rowley B., Hamilton P.B.: *Isolation of 2-acetamido-1-(β-L-aspartamido)-1,2-dideoxy-D-glucose from normal urine.* Clin. Chem. *18*, 951-955, 1972.

45. Aula P., Näntö V., Laipio M.-L., Autio S.: *Aspartylglucosaminuria: Deficiency of aspartylglucosaminidase in cultured fibroblasts of patients and their heterozygous parents.* Clin. Genet. *4*, 297-300, 1973.
46. Virtama P., Helelä T.: *Radiographic measurements of cortical bone, variation in normal population between 1 and 90 years.* Acta Radiol. Suppl. *293*, 1970.
47. Takkunen R.-L.: *Anthropometric studies on Finnish children.* Ann. Paediat. Fenn. *10*, Suppl. 24, 1962.
48. Bäckström-Järvinen L.: *Heights and weights of Finnish children and young adults.* Ann. Paediat. Fenn. *10*, Suppl. 23, 1964.
49. Marttila A.: *Uber einige anthropologische Merkmale über die sog. Summenmethode bei Vaterchaftsbestimmungen in einem Material von Finnischen Familien.* Ann. Acad. Sci. Fennicae, Series A.V. Medica. Thesis. P. *49*, 1970.

6

G_{M1} Gangliosidosis and Sandhoff Disease

Thomas G. Warner
John S. O'Brien

NAME OF DISEASE	G_{M1} Gangliosidosis
PREVALENCE	Unknown
MODE OF INHERITANCE	Autosomal recessive
MUTANT GENE	acid β-galactosidase
CHROMOSOMAL LOCATION	3p
EXPRESSION	
1. Altered DNA structure	?
2. Disturbed protein function	Absent enzyme activity
3. Disrupted cell and organ function	G_{M1} ganglioside and galactosyl glycoconjugates) in neurons and bones respectively leads to brain dysfunction and bony abnormalities.

NAME OF DISEASE	Sandhoff Disease
PREVALENCE	Unknown
MODE OF INHERITANCE	Autosomal recessive
MUTANT GENE	β-Chain of Hexosaminidase A and B
CHROMOSOMAL LOCATION	5
EXPRESSION	
1. Altered DNA structure	?
2. Disturbed protein function	Absent enzyme activity
3. Disrupted cell and organ function	Accumulation of substrate (G_{M2} ganglioside) in lysosomes of neurons and N-acetylhexosaminyl-glycoconjugates in viscera, leads to brain dysfunction and visceral storage.

6.1 INTRODUCTION AND GENERAL COMMENTS

The gangliosidoses are inborn errors of metabolism due to a metabolic defect in ganglioside degradation resulting from deficiencies of lysosomal glycohydrolases [1]. Of these disorders only G_{M1} gangliosidosis and G_{M2} gangliosidosis Type 2 or Sandhoff disease involve metabolic defects in both ganglioside and glycoprotein catabolism. G_{M1} gangliosidosis is due to a deficiency of acid β-galactosidase [2]. Sandhoff disease is due to a deficiency of hexosaminidase A and B [3].

In general, G_{M1} gangliosidosis and Sandhoff disease are characterized by: 1. progressive mental and motor deterioration with onset in childhood and fatal outcome; 2. autosomal recessive inheritance; 3. neuronal lipidosis secondary to storage of ganglioside G_{M2} or ganglioside G_{M1}; 4. storage of structurally related glycolipids or oligosaccharides; and 5. absence or severe deficiency of specific lysosomal glycohydrolases. A nomenclature of the G_{M1} and the G_{M2} gangliosidoses based upon their clinical picture is given in table 1.

Table 6.1 Nomenclature of G_{M1} Gangliosidoses and Sandhoff disease.

Common Name	Genetic Classification	Discovered	Patients Known	Inheritance
Infantile G_{M1} gangliosidosis	βGal-2	1964	100	Autosomal recessive
Juvenile G_{M1} gangliosidosis	βGal-3	1968	20	Autosomal recessive
Adult G_{M1} gangliosidosis	βGal-4	1972	10	Autosomal recessive
Sandhoff Disease (G_{M2} gangliosidosis, Type 2)	Hex A-1, Hex A-2	1968	100	Autosomal recessive

The major clinical features of each disorder are given in table 2.

Each gangliosidosis appears to be genotypically distinct, and each phenotype breeds true. Multiple affected children in sibships have been phenotypically similar to probands. No reports have appeared in which more than one ganglioside storage disease has appeared in the same sibship.

6.2. G_{M1} GANGLIOSIDOSIS

A. Phenotypic descriptions

1) *Infantile onset (Type 1)* - In infantile G_{M1} gangliosidosis (generalized gangliosidosis, infantile with bony involvement) symptoms begin at birth or shortly thereafter. Oedema of extremities is often present. Appetite is poor, sucking is weak, and weight gain is subnormal. Ineffective swallowing may necessitate tube feeding. The patient may hold his head up, but usually cannot crawl or sit unsupported. Sensorium is dull. Respirations are labored and irregular, and recurrent bronchopneumonia is frequent. Clonic-tonic convulsions are often

Table 6.2 Major clinical features of ganglioside storage diseases.

	G_{M1} Gangliosidoses			G_{M2} Gangliosidosis
	Infantile	Juvenile	Adult	Sandhoff disease
Age at onset of symptoms	birth	6-20 mon.	teens	3-6 mon.
Age at death	1/2-2 yr.	3-10 yr.	20+ yr.	2-5 yr.
Mental/motor retardation	+	+	+	+
Facial appearance	coarse	normal	normal	doll-like
Edema	+	−	−	−
X-ray changes long bones	+	mild	mild	−
X-ray changes vertebrae	+	mild	mild	−
Vacuolated lymphs	+	+	±	−
Foam cells in marrow	+	+	±	−
Hepatosplenomegaly	+	−	−	−
Splenomegaly	+	−	−	−
Cherry red spot	50%	−	−	+
Retinitis pigmentosa	−	−	−	−
Startle response to sound	+	+	−	+
Macrocephaly	rarely	−	−	+
Macroglossia	+	−	−	−
Siezures	+	+	−	+
Blindness	early	late	−	early
Neuronal lipidosis	+	+	+	+
Visceral histiocytosis	+	+	+	mild
Glomerular epithelial ballooning	+	+	?	−*
Mucopolysaccharidura	±	±	?	−
Dysarthria	+	+	+	−
Spasticity-ataxia	−	+	+	−
Hypotonia	+	−	+	+

* Renal tubular epithelial cell vacuolization.

present. If survival extends beyond the first year the clinical picture is that of decerebrate rigidity with blindness, deafness, spastic quadriplegia, and unresponsiveness. Death due to bronchopneumonia usually occurs by age 2 years.

The appearance is that of a dull-looking, hypoactive, hypotonic infant. Facial abnormalities include frontal bossing, depressed nasal bridge, large low-set ears, and coarse facial features. Gingival hypertrophy and macroglossia are often present. The corneas are usually clear. Cherry-red spots, identical to those seen in Tay-Sachs disease, are present in about half the patients. Hepatosplenomegaly is usually evident within the first few months of life. Macrocephaly may develop, but is not as frequent nor as massive as in TSD. Dorsolumbar kyphoscoliosis and hard, nontender enlargements of the epiphyseal joints, due to cartilaginous hypertrophy, may be prominent.

The radiologic changes are those of dysostosis multiple. Deformities of the vertebral bodies and long bones consist of rarefaction, anterior beaking of vertebral bodies, periosteal cloaking of the long bones early in the course of the disease, spatulate ribs, shoe-shaped sella, and modeling deformities of pelvic, hand and foot bones [4].

Pathologic changes include neuronal lipidosis, visceral histiocytosis, and cytoplasmic ballooning of renal glomerular epithelial cells. Neurons contain numerous cytoplasmic membranous bodies.

Visceral histiocytosis is prominent throughout the reticuloendothelial system. Foamy histiocytes are found in bone marrow, liver, spleen, lymph nodes, and most visceral organs. Renal glomerular epithelial cytoplasmic ballooning is prominent; the cells are filled with a material which is soluble in aqueous fixatives, leaving empty vacuoles after preparation for electron microscopy. Occasional lamellar lipid bodies are also seen in these cells. Similar clear vacuoles, found in epithelial cells in skin biopsies facilitate the diagnosis (Fig. 1) [5].

2) *Juvenile G_{M1} Gangliosidosis (Type 2)* - The phenotype differs from that of the infantile disorder. Onset is later, the course is slower, and bony abnormalities are milder. Mental and motor development is often normal during the first year.

Appetite is good, weight gain is adequate, and developmental milestones are not delayed. Locomotor ataxia beginning at about 1 year of age is usually the initial symptom. Internal strabismus, loss of coordinated manipulative hand movements, loss of speech, and generalized moderate muscular weakness of both upper and lower extremities are early symptoms. Mental and motor deterioration often progress rapidly thereafter. Dulling of sensorium, lack of interest, and lethargy are present. Progressive spasticity of upper and lower extremities develops, and, with time, a state of decerebrate rigidity is reached. Seizures often occur after 16 months of age and may constitute a major problem in medical management. Recurrent infections, especially bronchopheumonia, are constant problems and usually lead to the patient's demise. Thusfar, the average life span has varied between 3 and 10 years [6, 7].

Coarsening of facial features is not evident. Internal strabismus and bilateral nystagmus may be present. Corneas are usually clear, and the retina and macula usually appear normal. Blindness may occur late in the course of the disease. Lymphocytes are vacuolated, and foamy histiocytes occur in the bone marrow. Hepatosplenomegaly is usually absent. Radiographic examination often reveals mild inferior beaking of one or more lumbar vertebral bodies, proximal beaking of metacarpal bones, especially the fifth, and modeling deformities of the pelvic bones. These radiologic changes are mild but can be of considerable diagnostic value. They have been noted as early as age 7 months, prior to neurologic

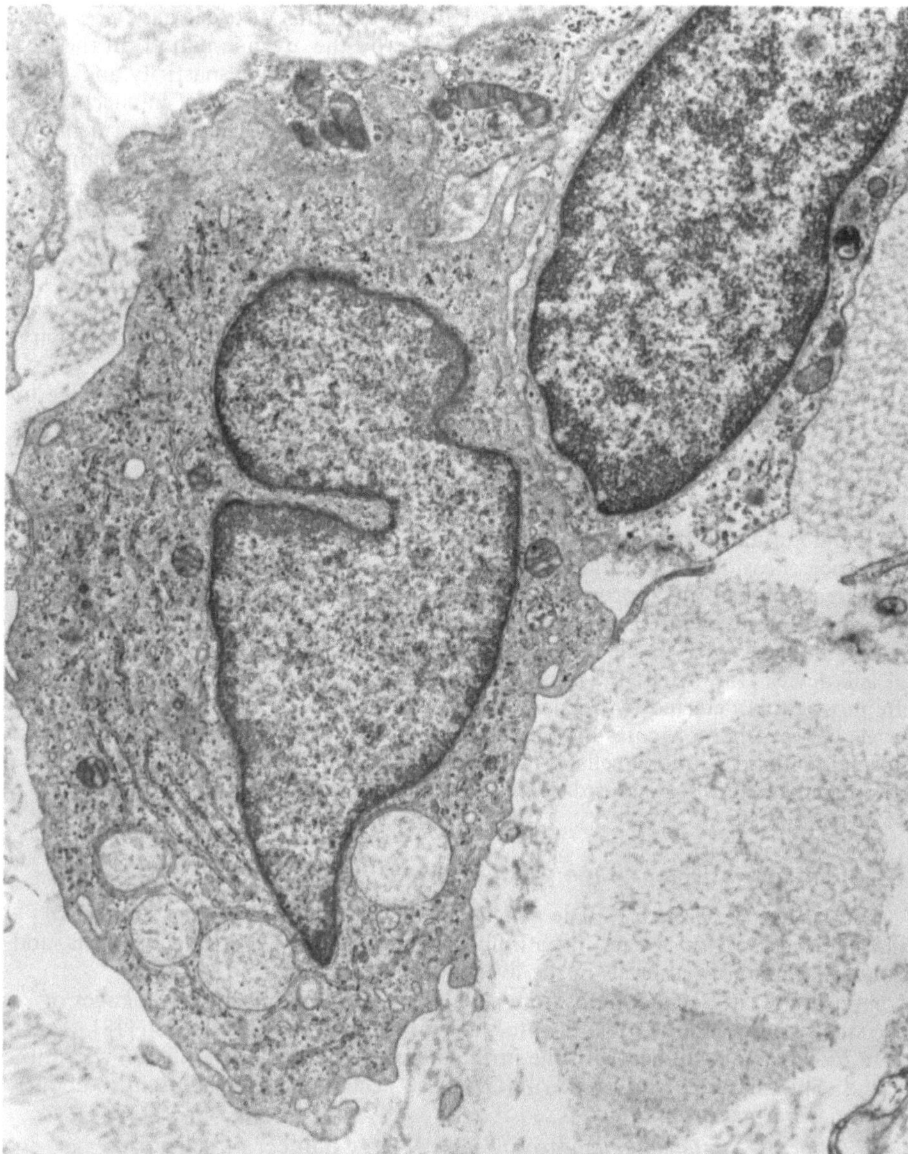

Fig. 6.1 Ultrastructure of dermal fibroblasts from skin biopsy of a patient with G_{M1} gangliosidosis. Type 1. Note: Vacuoles containing finely fibrillar material in perinuclear area.

symptoms [6]. Pathologic changes include neuronal lipidosis which appears identical but less prominent than that in the infantile type. Visceral histiocytosis is usually present but is not as florid as that seen in the infantile type. Renal glomerular cytoplasmic ballooning is similar to that in the infantile type.

3) *Adult G_{M1} Gangliosidosis (Type 3)* - Patients with this disorder have presented with progressive cerebellar dysarthria with onset in the juvenile period. In the best studied patients [8, 9] the picture has been that of progressive spasticity and ataxia. Intellectual impairment has been mild but with time, loss of intellectual function is evident. Progression of symptoms is slow; neurological impairment may only be suspect 10 years after the initial symptoms. Seizures are uncommon and vision remains unimpaired. Thusfar, no patients have expired but neuronal lipidosis should be apparent.

B. The enzyme defect

G_{M1} gangliosidosis is the first ganglioside storage disease shown to be due to the absence of a degradative enzyme [10]. Lysosomal acid β-galactosidase activity is strikingly deficient using ganglioside G_{M1} and galactose-containing glycoproteins [40] as substrates and the deficiency of β-galactosidase is not due to the presence of soluble endogenous inhibitors. Activities of other lysosomal enzymes, including acid phosphatase, β-glucosidase, and β-N-acetylglucosaminidase are either normal or increased. Ganglioside G_{M1} β-galactosidase activity is diminished to less than 0.1% of normal in brain, liver [12], and cultured fibroblasts [13]. Brain β-galactosidase activities for galactocerebroside and ceramide lactoside (ceramide-glucose-galactose) are normal or increased when the appropriate assay is used [14]. The metabolic block in this disorder involves the first step in G_{M1} breakdown (Fig. 2).

The diagnosis of G_{M1} gangliosidosis can be confirmed by β-galactosidase assays of leukocytes [15], urine [16], and cultured skin fibroblasts [13, 17]. Heterozygotes are most easily detected by β-galactosidase assays of leukocytes; intermediate levels between homozygotes and controls are found [15]. The enzyme is present in normal cultured amniotic cells obtained by amniocentesis in the second trimester of pregnancy, and β-galactosidase assay has been used for the prenatal diagnosis of at-risk fetuses [18, 19].

C. Chemistry of the storage products

1) *Lipid storage products* - The ganglioside, G_{M1}, is approximately ten times normal levels in gray matter in both infantile and juvenile G_{M1} gangliosidosis, and 20 to 50 times normal levels in the liver in the infantile type. The sugar composition, fatty acid composition, sugar sequence, and glycosidic linkages of the stored compound are identical to those of normal ganglioside G_{M1} [13].

In four patients studied with the juvenile type, visceral storage of G_{M1} was not found. The asialo-derivative of G_{M1} (G_{A1}) also accumulates in the brain to about ten times normal levels.

2) *Galactose containing oligosaccharides* - A large number of galactose containing oligosaccharides have been described in the liver and urine of G_{M1} gangliosidosis patients, Table 3. The oligosaccharide side chains of many glycoproteins such as α-1-protease inhibitor [25], human ceruloplasmin [26], plasma α-1-acid glycoprotein [27], rat lactalbumin [28] and human plasminogen [29] are partially sialylated or fucosylated derivatives of the major accumulating products in G_{M1} gangliosidosis (compound 1a, 1b, 3 and 4). Thus, the presence of

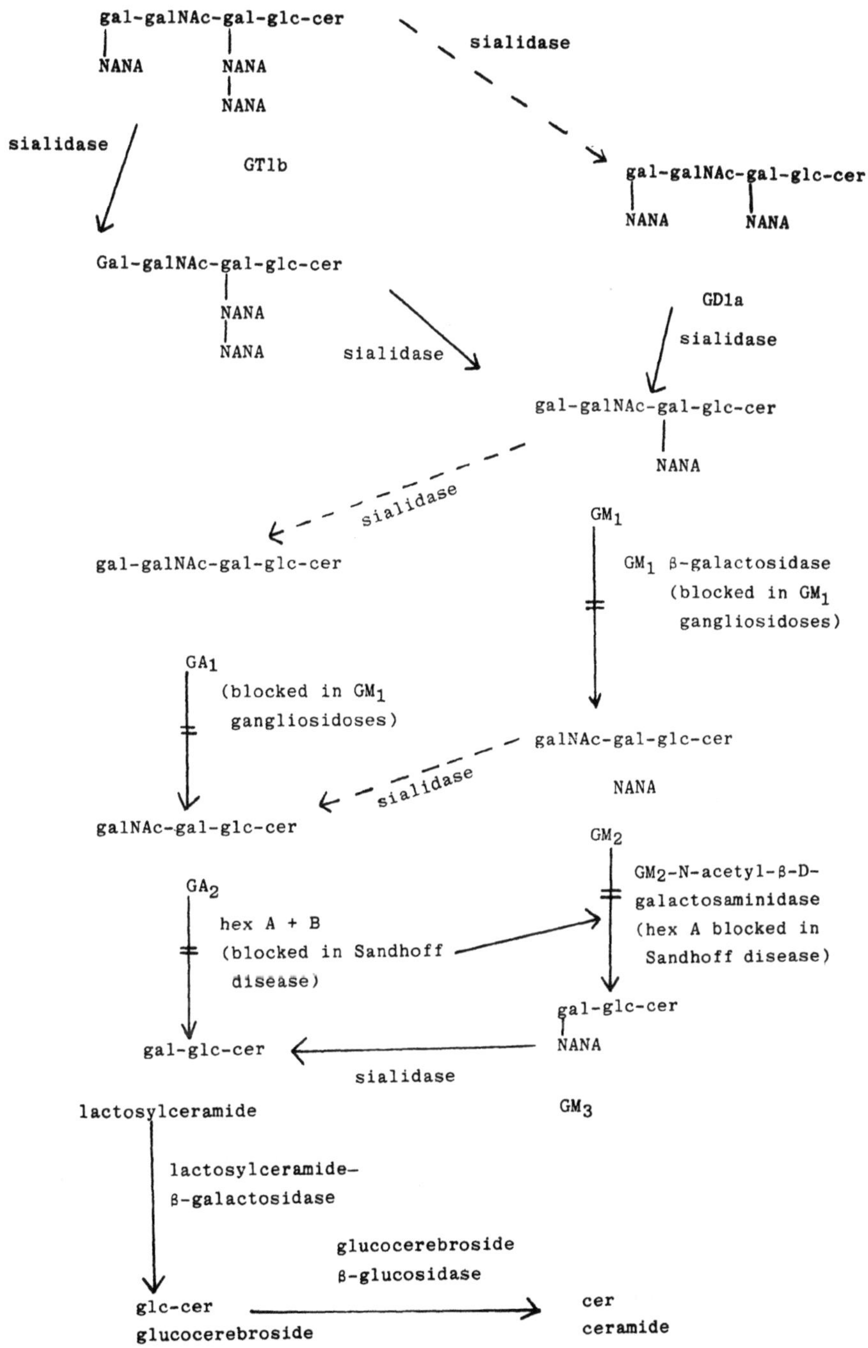

Fig. 6.2 Lysosomal degradation for brain gangliosides. The metabolic blocks involved in G_{M1} and G_{M2} gangliosidosis are indicated.

Table 6.3 Urinary and hepatic oligosaccharides in G_{M1} Gangliosidosis.

	Oligosaccharide	Subtype and tissue source	Reference
1.	a Gal β1-4 GlcNAc β1-2 Man α1-3 Man β1-4 GlcNAc b Gal β1-4 GlcNAc β1-2 Man α1-6 Man β1-4 GlcNAc	Type 1 & 2 liver & urine	21,22,23
2.	Man α 1 | 6 Gal β1-4 GlcNAc β1-2 Man α1-3 Man β1-4 GlcNAc	Type 1&2 liver & urine	21,23
3.	Gal β1-4 GlcNAc β1-2 Man α1 \\ 6 Man β1-4 GlcNAc 3 Gal β1-4 GlcNAc β1-2 Man α1 /	Type 1&2 liver & urine	21,22,23
4.	Gal β1-4 GlcNAc β1-2 Man α1 \\ 6 Gal β1-4 GlcNAc β1 \\ 4 Man α1 / 3 Man β1-4 GlcNAc 2 Gal β1-4 GlcNAc β1 /	Type 1&2 liver & urine	21,23
5.	Gal β1-4 GlcNAc β1 \\ 6 Man α1 \\ 6 2 Man β1-4 GlcNAc Gal β1-4 GlcNAc β1 / 3 Gal β1-4 GlcNAc β1-2 Man α1 /	Type 1&2 liver & urine	21,23
6.	Gal β1-4 GlcNAc β1-2 Man α1 \\ 6 Man β1-4 GlcNAc 3 Gal β1-4 GlcNAc β1-3 Gal β1-4 GlcNAc β1-2 Man α1 /	Type 1 urine	23
7.	Gal β1-4 GlcNAc β1-3 Gal β1-4 GlcNAc β1-2 Man α1 \\ 6 Man β1-4 GlcNAc 3 Gal β1-4 GlcNAc β1-2 Man α1 /	Type 1 urine	23
8.	Gal β1-4 GlcNac β1 \\ 6 Man α1 \\ 2 Gal β1-4 GlcNAc β1 / 6 Man β1-4 GlcNAc Gal β1-4 GlcNAc β1 \\ 3 4 Man α1 / 2 Gal β1-4 GlcNAc β1 /	Type 1&2 liver & urine	21,23

(continued)

Table 6.3 *(continued)*

Oligosaccharide	Subtype and tissue source	Reference

9.

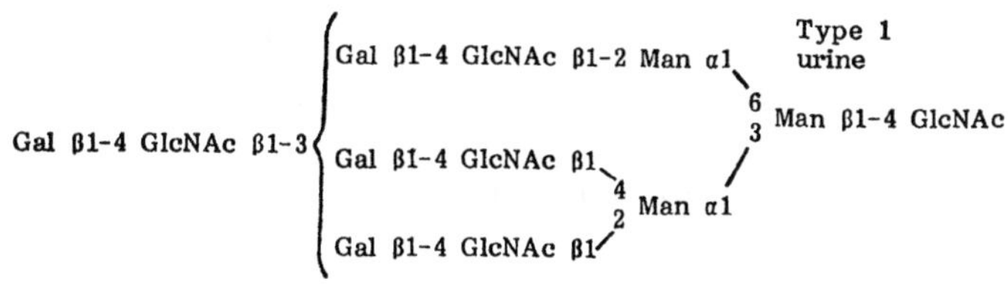

Type 1 urine — 23

10.

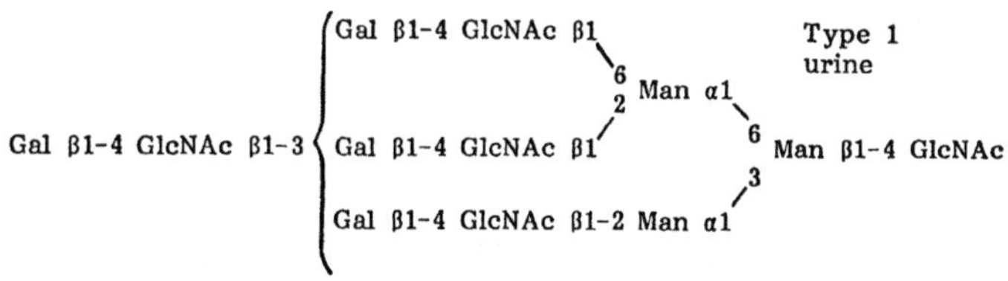

Type 1 urine — 23

11.

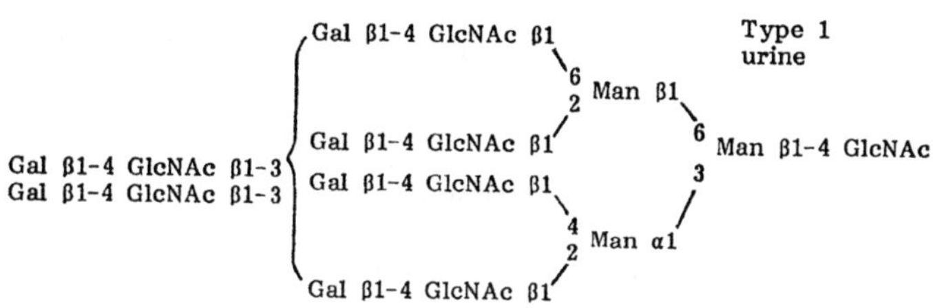

Type 1 urine — 24

12.

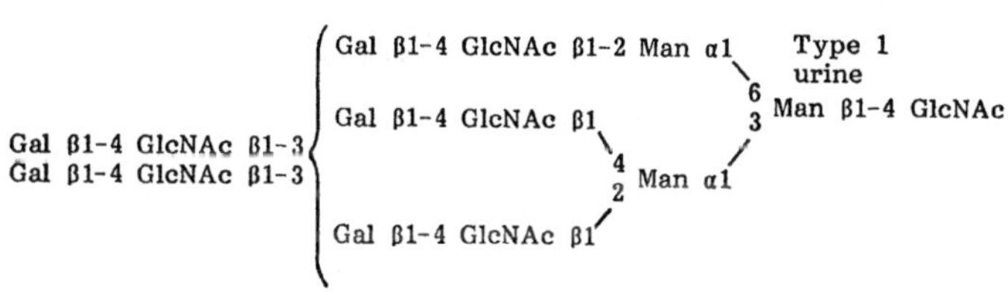

Type 1 urine — 24

13.

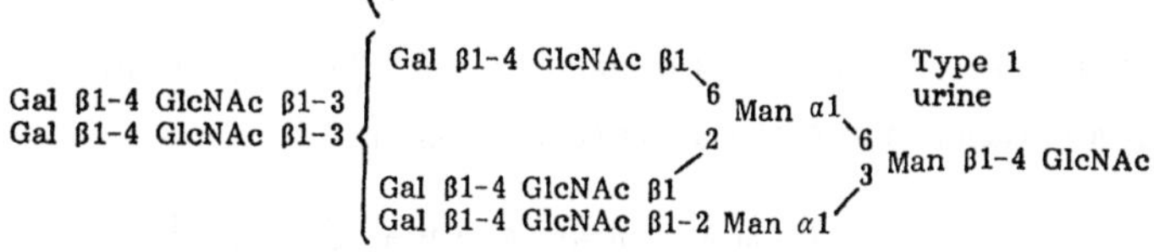

Type 1 urine — 24

14.

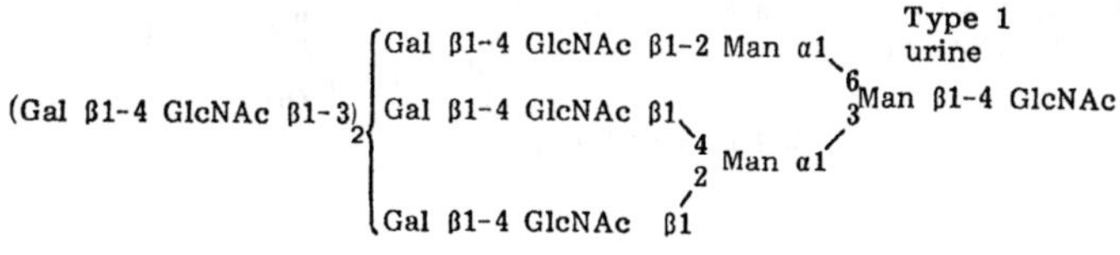

Type 1 urine — 24

(continued)

Table 6.3 *(continued)*

Oligosaccharide	Subtype and tissue source	Reference
15. 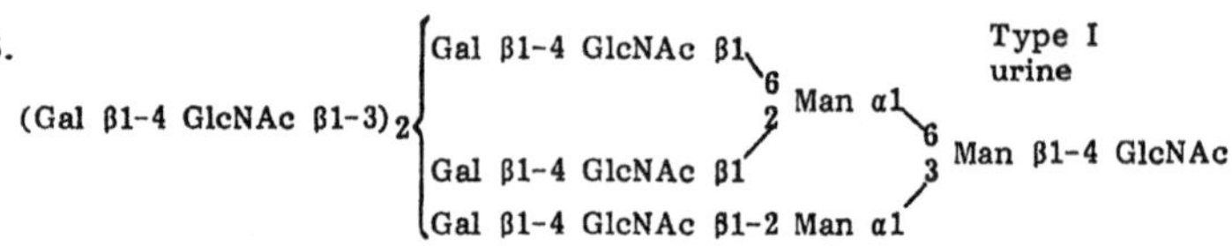	Type I urine	24
16. 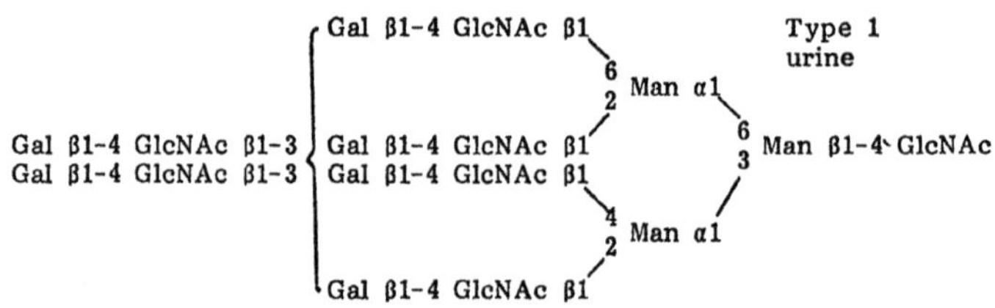	Type 1 urine	24
17. 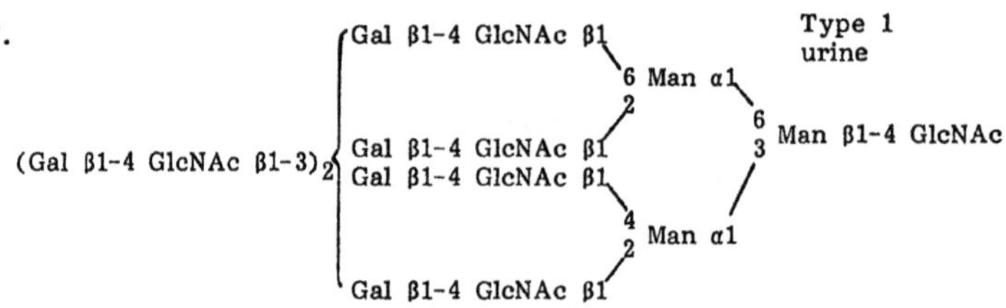	Type 1 urine	24
18. 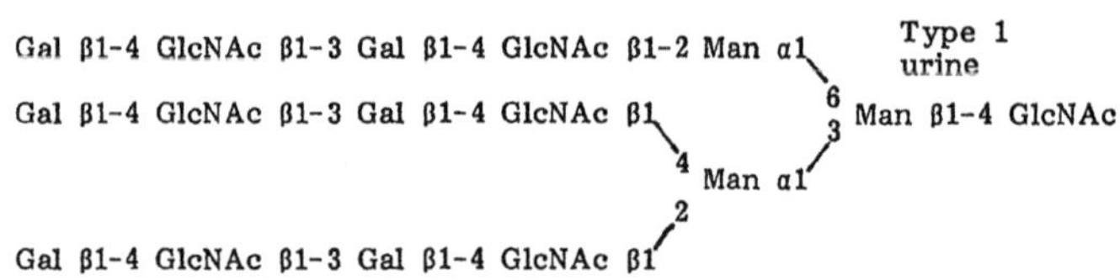	Type 1 urine	24
19. 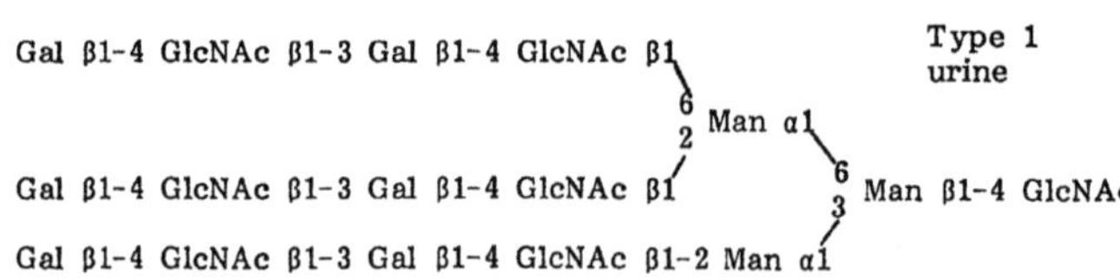	Type 1 urine	24
20. 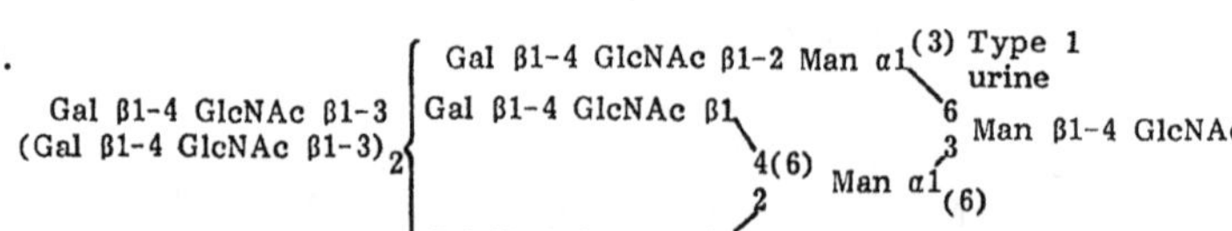	Type 1 urine	24

(continued)

Table 6.3 *(continued)*

Oligosaccharide	Subtype and tissue source	Reference
21. (Gal β1-4 GlcNAc β1-3)$_3$ { Gal β1-4 GlcNAc β1-2 Man α1(3) Gal β1-4 GlcNAc β1 4(6) Man α1(6) 6,3 Man β1-4 GlcNAc Gal β1-4 GlcNAc β1 2	Type 1 urine	24
22. Gal β1-4 GlcNAc β1-3 / Gal β1-4 GlcNAc β1-3 / Gal β1-4 GlcNAc β1-3 { Gal β1-4 GlcNAc β1 6 Gal β1-4 GlcNAc β1 2 Man α1 6 Man β1-4 GlcNAc Gal β1-4 GlcNAc β1 4 Gal β1-4 GlcNAc β1 2 Man α1 3	Type 1 urine	24
23. Gal β1-4 GlcNAc β1-3 / (Gal β1-4 GlcNAc β1-3)$_2$ { Gal β1-4 GlcNAc β1 6 Gal β1-4 GlcNAc β1 2 Man α1 6 Man β1-4 GlcNAc Gal β1-4 GlcNAc β1 4 Gal β1-4 GlcNAc β1 2 Man α1 3	Type 1 urine	24
24. (Gal β1-4 GlcNAc β1-3)$_3$ { Gal β1-4 GlcNAc β1 6 Gal β1-4 GlcNAc β1 2 Man α1 6 Man β1-4 GlcNAc Gal β1-4 GlcNAc β1 4 Gal β1-4 GlcNAc β1 2 Man α1 3	Type 1 urine	24
25. Gal β1-4 GlcNAc β1-3 / Gal β1-4 GlcNAc β1-3 / (Gal β1-4 GlcNAc β1-3)$_2$ { Gal β1-4 GlcNAc β1-2 Man α1(3) 6 Gal β1-4 GlcNAc β1 4(6) Man α1(6) 3 Man β1-4 GlcNAc Gal β1-4 GlcNAc β1 2	Type 1 urine	24
26. (Gal β1-4 GlcNAc β1-3)$_2$ / (Gal β1-4 GlcNAc β1-3)$_2$ { Gal β1-4 GlcNAc β1-2 Man α1(3) 6 Gal β1-4 GlcNAc β1 4(6) Man α1(6) 3 Man β1-4 GlcNAc Gal β1-4 GlcNAc β1 2	Type 1 urine	24

(continued)

Table 6.3 *(continued)*

Oligosaccharide	Subtype and tissue source	Reference
27. 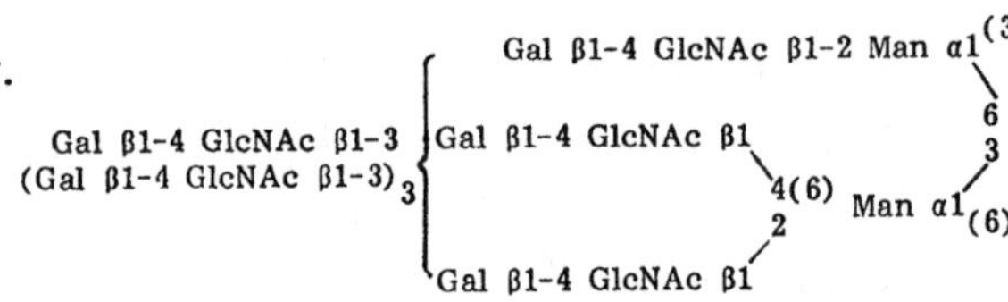	Type 1 urine	24
28. 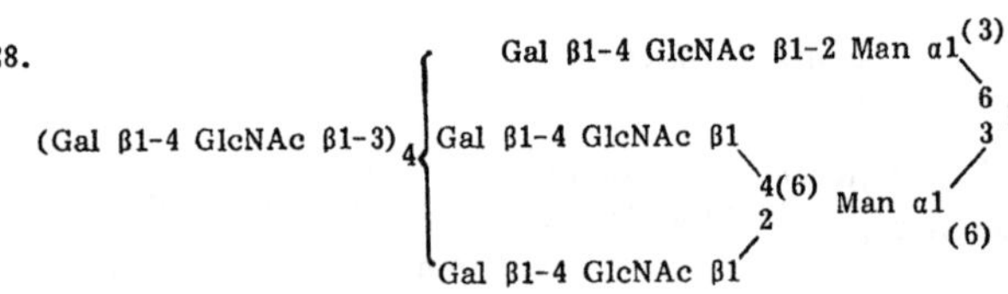	Type 1 urine	24

these oligosaccharides in the tissue and in urine of these patients can be explained by a block in glycoprotein catabolism at a step subsequent to cleavage of the oligosaccharide unit from the peptide chain by an endo-N-acetylglucosaminidase which cleaves the dichitobiosyl (N-acetylglucosaminyl-N-acetylglucosamine) linkages and after action of neuraminidase and fucosidase cleavage of sialic acid and fucose residues respectively (Fig. 3).

A recent report comparing urinary oligosaccharides from the infantile and juvenile forms of G_{M1} gangliosidosis has revealed that some high molecular weight oligosaccharides are not present in the urine of patients with the juvenile form of the disease [24]. It would be of interest to know if these additional storage products could be correlated with specific clinical symptoms such as bony abnormalities, edema, seizures, etc. which differ in severity among different patients.

An additional series of complex oligosaccharides differing from those presented here has been reported in brief [30]. However, details of this analysis have not been reported, making it difficult to assess the validity of the structural assignments, especially since they differ from those found by others.

Recently we have employed high performance liquid chromatography for the detection and analysis of urinary oligosaccharides (Fig. 4) [31]. This highly sensitive method may be a valuable tool for prenatal diagnosis if the presence of these oligosaccharides can be demonstrated in amniotic fluid of an affected fetus.

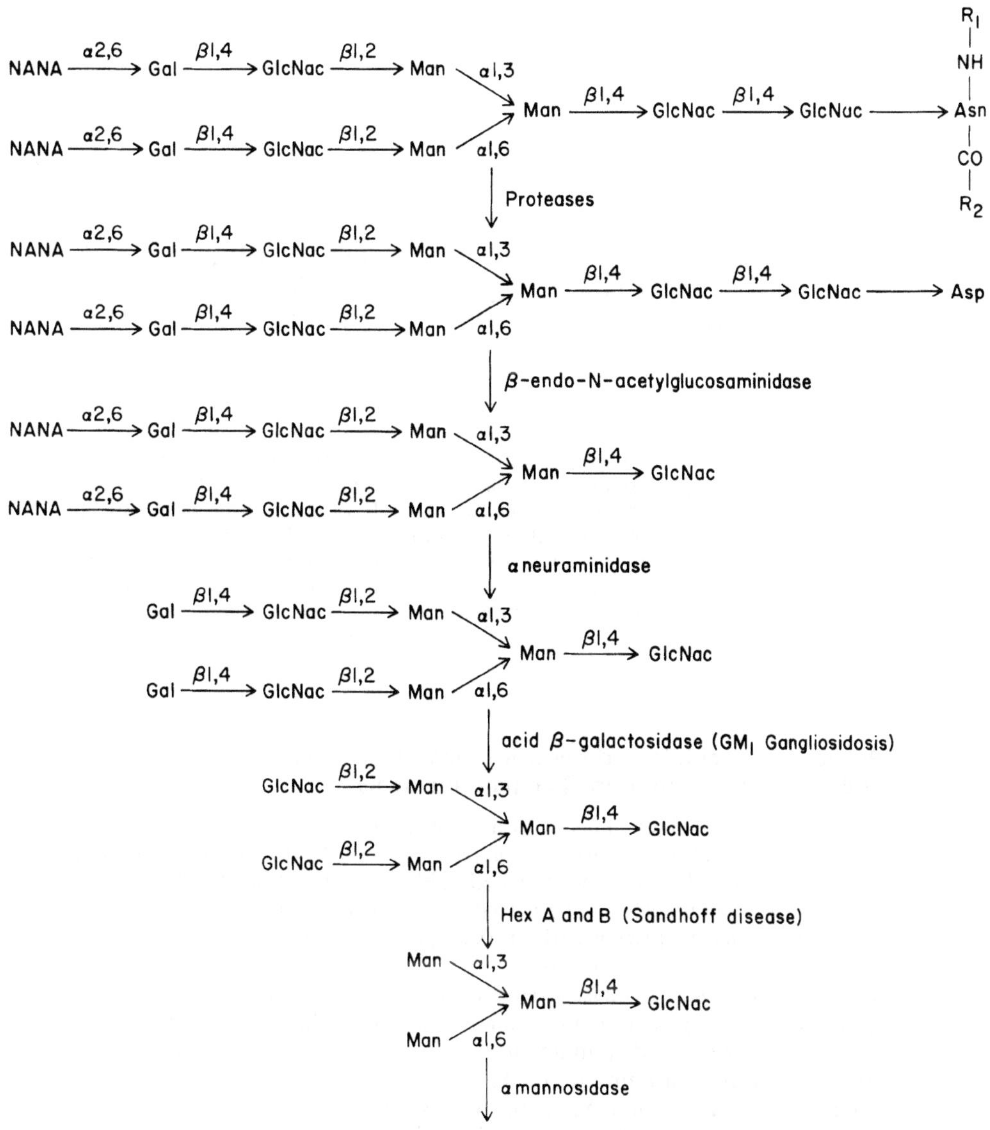

Fig. 6.3 Lysosomal degradation pathway for typical asparagine-linked glycoproteins. The metabolic blocks giving rise to G_{M1} gangliosidosis and G_{M2} gangliosidosis, Sandhoff disease are indicated.

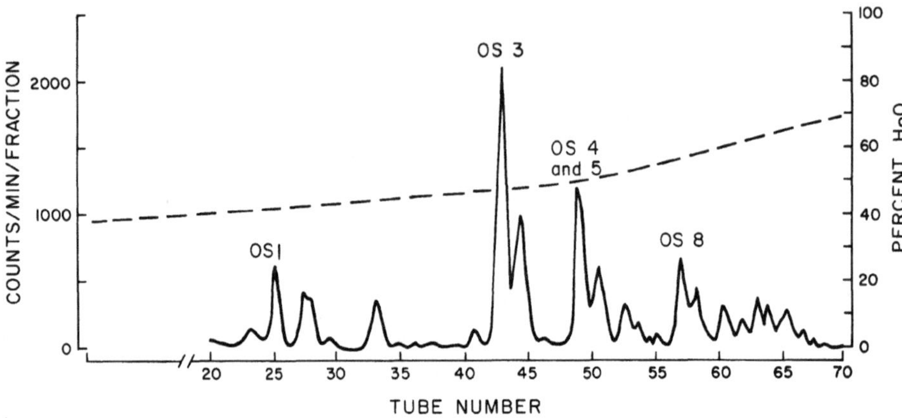

Fig. 6.4 High performance liquid chromatography of urinary oligosaccharides from a G_{M1} gangliosidosis, Type I patient. The oligosaccharides are first reduced with NaB^3H_4 and separated on a column of Micro pak Ax-5 in acetonitrile-water 65:35 v/v, increasing the water content at 0.3%/min for 50 min followed by a 0.6%/min increase for 20 min. The flow rate was 1.0 ml/min and 0.3 ml fractions collected, see table 3 for structure of OS in each fraction. Identification of peaks of the high molecular weight compounds (greater than OS 3) is tentative.

D. Properties of acidic and neutral β-galactosidases and their involvement in G_{M1} gangliosidosis

Several different β-galactosidases, assayed using synthetic substrates occur in human tissues [32]. Three of these, *isoenzymes* A_1 and A_2 (formerly called acid β-galactosidase A) and A_3 (formerly called acid β-galactosidase B) [50] are deficient in G_{M1} gangliosidosis. The activity of a fourth β-galactosidase, called *neutral β-galactosidase*, varies greatly in G_{M1} gangliosidosis, being sometimes elevated, sometimes diminished [13].

The major acid β-galactosidase isoenzymes, β-galactosidase A_1 and A_2 have been purified to homogeneity from human liver and studied in detail [34-35]. Form A_1 is monomeric and spontaneously dimerizes to A_2 under appropriate conditions. Form A_3 is a multimer of A_1 [35].

β-Galactosidase A_1, A_2 and A_3 (Table 4) possess pH optima (4.2 to 4.4) in the acid region; thus their designation as *acid β-galactosidases* [36]. Each is stimulated and stabilized by sodium chloride [57], are thermolabile, and cleave aryl-β-D-glucoside linkages but not aryl-β-D-glucoside linkages. Each hydrolyzes β-linked galactose residues from G_{M1} [34, 38]. β-Galactosidase A_3 has a molecular weight (660,000) about ten times higher than that of A_1 (65,000) and under appropriate conditions. A_1 can be generated from A_3 [35]. Recent experiments with pure A_3 reveal it is comprised solely of multimers of A_1 [39]. Monospecific antibodies prepared against homogeneous A_1 selectively precipitate A_2 and A_3 [34, 35]. Thus all the A forms possess a common polypeptide, and appear to be coded by a common locus.

Table 6.4 Properties of β-galactosidases in human liver.

Isoenzyme	Molecular Weight	pH Optimum (4MUβGal)	Heath Stability*	Effect of Chloride	Cleaves 4MUβGal	Cleaves 4MUβGal
A_1	62,000	4.35	Labile	Stimulates	+	0
A_3	600,000–800,000	4.20	Labile	Stimulates	+	0
Neutral	57,000	5.80	Stable	Inhibits	+	+

Isoenzyme	Cleaves G_{M1}	Cleaves Asialofetuin	Activity in G_{M1} Gangliosidosis	Binds to Con A-Sepharose	Cross Reacts with Anti-A Antibodies	% of Total Activity **
A_1	+	+	Deficient	+	+	85
A_3	+	+	Deficient	+	+	10
Neutral	0	0	Variable	0	0	5

* Heating at 37°C.
** Values are approximate; assayed at pH 4.3 with 4MU-β-galactoside.

Neutral β-galactosidase (table 4) possesses a more neutral pH optimum than the A forms and a lower molecular weight (57,000). It is inhibited by sodium chloride, is more heat stable than β-galactosidase A [37] and cleaves both aryl-β-D-galactoside and aryl-β-D-glucoside linkages, in contrast to the A forms. Neutral β-galactosidase does not cleave galactose from ganglioside G_{M1} or from asialofetuin [34, 38]. Neutral β-galactosidase from human liver does not adsorb to concanavalin A-Sepharose columns and is readily separated from A_1 and A_3 by this means [40]. Monospecific anti-A-antibodies contains polypeptide different from the A forms and is coded by a separate genetic locus.

In normal human liver, using 4-methylumbelliferyl or p-nitrophenyl-β-galactosides as substrates, approximately 85 percent of the activity assayed at pH 4.5 is due to A_1 (plus A_2), 10 percent to A_3, and 5 percent to neutral β-galactosidase. Variations from these values are obtained from individual to individual or when the chloride concentration or pH of the assay is varied. Neutral β-galactosidase has some β-galactosidase at the pH optimum of acid β-galactosidase and if one equates 4-methylumbelliferyl-β-galactosidase activity with G_{M1} β-galactosidase activity, an error will result if neutral β-galactosidase is present. Neutral β-galactosidase must be removed prior to assay to avoid this error, or ganglioside G_{M1} must be used as the susbtrate.

Initially, patients with infantile G_{M1} gangliosidosis (Type 1) appeared to have much lower activities of neutral β-galactosidase than those with juvenile G_{M1} gangliosidosis (Type 2) [13]. This finding was not confirmed when larger numbers of patients were studied [41, 42]. It is now evident that the activity of neutral β-galactosidase in liver varies over a seven-fold range both in patients with G_{M1} gangliosidosis and in normal controls, without apparent relationship to clinical status or subtype [43]. The reason for this large variability is not known. Unfortunately, since neutral β-galactosidase comigrates with β-galactosidase A_1 on gel electrophoresis, they can be confused. The proposal that Type 2 patients may be A (+) and Type 1 patients A (-) resulted from this overlap. It now appears that neutral β-galactosidase is not involved in the pathogenesis of G_{M1} gangliosidosis.

Neutral β-galactosidase has been found to be strikingly deficient in liver tissue of subjects who expired from causes unrelated to β-galactosidase deficiency [44]. The benign deficiency of this enzyme suggests it is not involved in storage disease.

Competitive inhibition studies indicate that the same catalytic site(s) on purified β-galactosidase A_1 cleaves G_{M1} and 4MU-β-D-galactoside [34]. Isoenzyme A_1 also cleaves lactose, N-acetyllactosamine, 4MU-α-L-arabinoside, and p-nitrophenyl-β-D-fucoside [34]. The latter two substrates are isomers of β-D-galactose, and the explanation for depressed hydrolysis of these substrates in G_{M1} gangliosidosis [45-46], is apparent. β-Galactosidase A_1 can cleave β-D, 1-3 linkages which occur in G_{M1}, and β-D-, 1-4 linkages which appear in glycoproteins [34].

Substrates cleaved poorly or not at all include galactosylhydroxylysine, galactosyl-hydroxylysine-peptides from a proteolytic digestion of collagen, galactocerebroside, lactosylceramide, and monogalactosyldiglyceride [34].

If one measures the activities of galactocerebroside β-galactosidase, lactosylceramide β-galactosidase, and monogalactosyldiglyceride β-galactosidase during the purification of β-galactoside A_1 these three activities do not copurify with A_1. Monospecific anti-A_1 antibodies do not precipitate galactocerebroside β-galactosidase, lactosylceramide β-galactosidase, or monogalactosyldiglyceride β-galactosidase from solution. Accordingly these enzymatic activities are the properties of protein(s) different from β-galactosidase A_1 [34].

Thus the mutation in Krabbe's disease, wherein there is a deficiency of galactocerebroside β-galactosidase [47] (and lactosylceramide β-galactosidase and monogalactosyldiglyceride β-galactosidase [48], is not allelic with the G_{M1} gangliosidosis mutation. Earlier findings that galactocerebroside β-galactosidase activities are not deficient in G_{M1} gangliosidosis [48] and that acid β-galactosidase activity is not deficient in Krabbe's disease [47] pointed to the same conclusion.

Two groups have reported partial or severe deficiencies of lactosylceramide β-galactosidase activity in G_{M1} gangliosidosis [49, 50]. The apparent explanation for these resulsts is that the assay used, in which crude taurocholate is employed instead of purified taurocholate, stimulates acid β-galactosidase to cleave lactosylceramide [49]. With this assay, one finds a deficiency of lactosylceramide β-galactosidase in G_{M1} gangliosidosis. When purified taurocholate is the detergent, no deficiency is found.

The important conclusion is that the deficiency of lactosylceramide β-galactosidase activity in G_{M1} gangliosidosis appears to be due to methodologic peculiarities, not to production of the two enzymes by a common locus.

Human liver acid β-galactosidase appears to exist in vivo as a single monomeric polypeptide chain of 65,000 molecular weight or as multimers of A_1 (35). Since acid β-galactosidase is monomeric, its polypeptide chain is apparently the production of a single autosomal locus.

E. Nature of the mutation in G_{M1} gangliosidosis

The genetic defect in G_{M1} gangliosidosis appears to be a mutation of the structural gene which codes for β-galactosidase A, located on the short arm (p12-q21) of human chromosome 3 [51]. Liver samples from 11 patients with G_{M1} gangliosidosis (five with infantile, three with juvenile, and three with "intermediate" phenotypes) contain normal or supranormal quantities of crossreacting material to anti-β-galactosidase A_1 antibodies. This indicates that structurally altered β-galactosidase A was produced in all [52]. A similar result was obtained in a study of another patient [53].

Quantitative analyses of mutant β-galactosidase, determined immunologically, in infantile, juvenile and adult G_{M1} gangliosidosis fibroblasts have given normal amounts of kinetically defective enzyme in each type and higher enzyme activity per unit of immunoreactivity in the later onset patients [54].

One unusual patient (C.E.) with juvenile onset has an electrophoretic variant of β-galactosidase A_1 [43] with a more positive charge than normal. The enzyme from C.E. has a Km that was five times normal with ganglioside G_{M1}. The mutant enzyme crossreacted immunologically with normal β-galactosidase A antibodies, but possessed only 1 percent of the normal catalytic activity per unit of antigenic activity. This is the first electrophoretic variant reported for a patient with a lysosomal hydrolase deficiency disease.

Patient C.E. is proven kinetic mutant, and many other patients are likely to be also.

Since a single locus appears to code for the synthesis of β-galactosidase A_1 polypeptide, the mutations in all types of G_{M1} gangliosidosis studied thusfar appear to be allelic, since structurally altered A protein is produced in nearly normal quantities in all. Somatic cell hybridization of cultured cells from patients with different types of G_{M1} gangliosidosis have failed to yield complementary β-galactosidase A in the mixed heterokaryons as a consequence of subunit rearrangements [55] as expected [52]. The finding of complementary activity in heterokaryons derived from fusions between Type 2 cells and those from an unusual β-galactosidase-deficient patient, termed Type 4 [55] has been explained. The Type 4 patient has been shown to have sialidosis (primary α-neuraminidase deficiency) with secondary β-galactosidase deficiency [56].

F. Explanation of the G_{M1} gangliosidosis phenotype

The phenotypic effects of a single gene mutation involving acid β-galactosidase may be explained by a one gene: one polypeptide: many substrates model [52] (Fig. 5). Acid β-galactosidase is heterocatalytic, cleaving β-D linked galactose from several galacto-conjugates (34). Keratosulfate-like glycosaminoglycans (sulfated and sulfate free) galactose-oligosaccharides, glycoproteins with carbohydrate side chains similar to asialofetuin, and G_{MI} ganglioside are a few of the natural substrates known. Many others, yet to be defined, probably exist. A structural mutation giving rise to altered β-galactosidase A with impaired catalytic activity can simultaneously lead to the accumulation of multiple galactoconjugates affecting many organ systems, depending upon the nature, location, degree and consequences of storage.

1) *Multiple phenotypes* - One explanation for the multiplicity of G_{MI} gangliosidosis phenotypes is that different mutations exist which lead to different kinetic alterations of each mutant polypeptide. Two possibilities may be considered: (a) mutant A protein is synthesized which possesses about the same residual activity for each natural substrate; or (b) mutant A protein is synthesized which possesses significantly different residual activities for different natural substrates.

Possibility (a) above may be the case in the most infantile patients in whom storage of gangliosides and galactoproteins is massive, and residual activity for all substrates is less than 1 percent of normal. Many patients with juvenile onset also may belong to category (a), since β-galactosidase activities for G_{MI} and galactoprotein (in this case asialofetuin) are deficient to about the same extent. However, in most later onset patients, residual activity is two to five times higher in liver than in those with Type 1 [43] and five to 30 times higher in cultured skin fibroblasts [57-58]. This difference in residual enzyme activity could account for the phenotypic differences; the rate of storage being slower in Type 2 because of higher residual activity.

Mutations of type (b), in which residual β-galactosidase activity is higher for one type of substrate than for another, might be expected from the differences in molecular size, ionic charge, solubility in water, nature of galactose linkage, and type of penultimate amino sugar in the natural substrates thusfar known to be cleaved by acid β-galactosidase. Unusual phenotypes can be predicted. For example, if significant catalytic activity is retained for galactose-containing glycoconjugates but not for G_{MI}, the phenotype will be that of rapid neurological decline due to G_{MI} storage without significant bony deformities. Patients with this phenotype have been described [59]. Alternatively, if significant catalytic activity is retained for G_{MI} and not for galactose-containing glycoconjugates, the phenotype will be that of severe bony abnormalities due to storage of these compounds with minimal cerebral defect. Several patients with such a phenotype have recently been discovered.

2) *MPS IV B Variant* - Four patients in four different sibships have presented with progressive severe skeletal dysplasia without neurological impairment [60, 62].

The first patient [60] had hip pain at age 8 due to progressive dysplastic

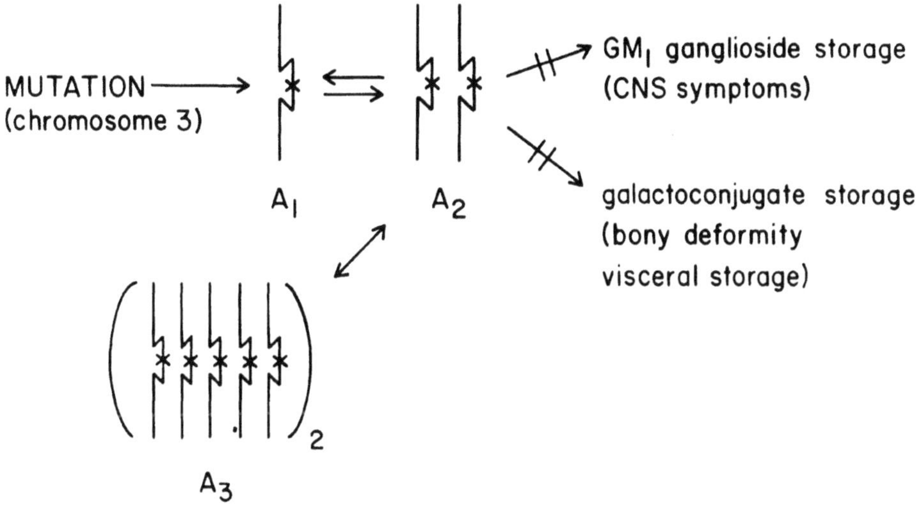

Fig. 6.5 One gene-one polypeptide-many sybstrates model to exlain how a single mutation can give rise to multiple somatic effects in G_{M1} gangliosidosis. The precise number of subunits in enzyme A_3 in unknown.

changes of the pelvis and femoral heads. She also had vertebral platyspondyly and unusual modeling deformities of the vertebral bodies similar to those seen in the hereditary spondyloepiphysial dysplasias. There was no impairment of mental or motor function on careful neurological examination. Both parents had half-normal levels of β-galactosidase activity in leukocytes indicating heterozygosity for a primary mutation of β-galactosidase.

Three other patients with the same phenotype has been reported [61, 62] and an altered Km of the mutant enzyme was evident in two [62]. This condition has been classified as mucopolysaccharidosis IV B. It appears inappropriate to designate such patients as having G_{M1} gangliosidosis since this condition signifies a neurodegenerative disorder. The number of mutations which give rise to β-galactosidase deficiency and clinical disease is larger than previously thought, as expected from the model presented (Fig. 5). It will be of great interest to determine whether the pattern of oligosaccharide excretion differs from other patients with G_{M1} gangliosidosis.

A number of patients with β-galactosidase deficiency who were thought to be unusual G_{M1} gangliosidosis mutants [13] are now known to have sialidosis (primary α-neuraminidase deficiency) [63-64]. In such patients the β-galactosidase deficiency is secondary [63] but the underlying molecular basis for the low levels of activity are still poorly understood.

G. Animal models

G_{M1} gangliosidosis has been reported in Siamese cats in the United States [65] and Japan [66], in a short-hair domestic cat in Cambridge [67], in Friesian cattle in Ireland [68] and in mixed breed beagles from Purdue [69].

The best studied of these animal models are the Siamese cats [70] and beagle dogs [69-71]. A comparison of the general clinical features with the infantile and juvenile forms of the human disease are summarized in table 5. Both the feline and canine diseases resemble the juvenile human form of G_{M1} gangliosidosis since bony abnormalities are not clinically obvious. However, massive amounts of G_{M1} gangliosidoside and galactose-oligosaccharides are stored and excreted; similar to the infantile human form. Thus the dog and cat models fall somewhere in between the infantile and juvenile human disorders.

The hepatic and urinary oligosaccharides present in the affected cats differ in structure from the human oligosaccharides, containing an extra N-acetyl glucosamine residue [72]. Comparison of the mutant cat enzyme with the wild type enzyme has been made [73]. The major differences include an altered Km, increased thermolability and altered molecular weight of the mutant enzyme, suggesting a structural mutation [71]. The mutant and wild type enzyme did not crossreact immunologically.

A similar comparison has been made between mutant and wild type enzyme in G_{M1} begales [74]. A striking deficiency of β-galactosidase (4% of normal) in cultured fibroblasts and in tissues (brain and viscera) accompained by oligosaccharide storage [75] and cerebral G_{M1} ganglioside [69] has been found. A preliminary characterization of the mutant enzyme in frozen liver indicated identity to the wild type enzyme but a deficiency of cross-reactive material [74]. However, more recent studies with fresher sources of tissue (fibroblasts) indicate the presence of normal quantities of immunoreactive mutant enzyme which is very thermolabile [75].

A preliminary study of the oligosaccharides stored in liver in the canine disorder has been made [76]. Examination by thin-layer chromatography (Fig. 6) reveals a difference between the mutant dog and human oligosaccharide patterns. The dog compounds are less polar and contain one extra N-acetylglucosamine residue at the reducing terminus than the human compounds. Recent structural analysis [76] reveals the following structures for the major hepatic oligosaccharides in the G_{M1} dog:

OS 1,2.

Gal β1-4 GlcNAc β1-2 Man α1-6 Man β1-4 GlcNAc β1-4 GlcNAc
Gal β1-4 GlcNAc β1-2 Man α1-3 Man β1-4 GlcNAc β1-4 GlcNAc

and OS 3

Gal β1-4 GlcNAc β1-2 Man α1-6 ╲
                                                      Man β1-4 GlcNAc β1-4 GlcNAc
Gal β1-4 GlcNAc β1-2 Man α1-3 ╱

Except for the extra N-acetylglucosamine residue, these compounds are identical to the two major human oligosaccharides (OS 1 a and b, and OS 3, see Table 3). The biological origin of the additional GlcNAc residue at the reducing terminus is unclear at the present time. However their presence indicates that may be significant differences in glycoconjugate metabolism between closely related mammalian species.

It is important to characterize such animal mutants in detail since they can serve as experimental models for therapy. In the upcoming era of gene therapy, such animals will be especially valuable in determining whether replacement of the cloned gene for β-galactosidase is an efficatious therapeutic approach.

Table 6.5 Comparison of clinical features of animal and human G_{M1} gangliosidosis.

	Human G_{M1} Gangliosidosis		Animal G_{M1} Ganglosidosis	
Clinical Features	Type 1	Type 2	Feline	Canine
Mode of Inheritance	Autosomal recessive	Autosomal recessive	Autosomal recessive	Autosomal recessive
Age at onset	Birth	6-20 months	~4 months	~4 months
Age at death	0.5-2 yr.	3-10 yrs.	~6 months	9 months-1 yr.
Skeletal	Bone and facial abnormalities	slight or none	None apparent grossly	None
Visceromegaly	Hepatosplenomegaly	None	Slight	None
Visual impairment	Early	Late, mild	None	Late
Nervous system	++	+	+	+
Stored G_{M1} and galactose oligosaccharides	++	+	++	++
Enzyme defect	β-galactosidase	β-galactosidase	β-galactosidase	β-galactosidase

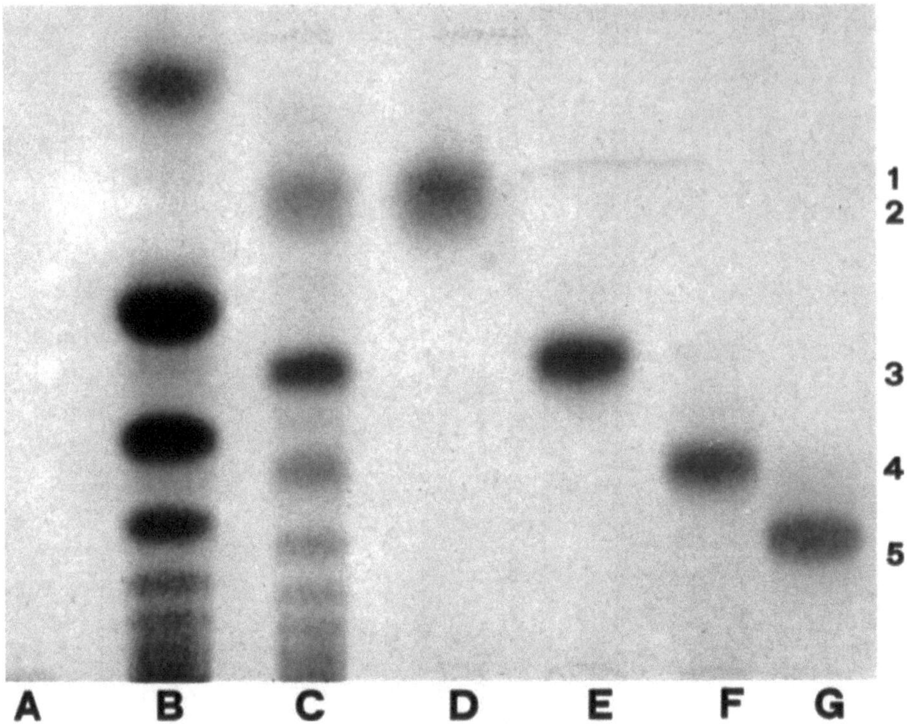

Fig. 6.6 Silicic acid thin-layer chromatography. Lane A, liver extract of a normal dog; lane B, liver extract of G_{M1} gangliosidosis human, Type 1; lane C, liver extract from canine G_{M1} gangliosidosis; lanes D, E, F and G, purified G_{M1} gangliosidosis oligosaccharides OS 1, 2, OS 3, OS 4, and OS 5 respectively.

6.3 SANDHOFF DISEASE

A. Phenotypic description

The clinical and pathologic picture is similar to G_{M2} gangliosidosis type 1, Tay-Sachs disease. Motor weakness begins within the first six months of life. The startle reaction to sound, early blindness, progressive mental and motor deterioration, doll-like facies, cherry-red spots, and macrocephaly are present. Occasional patients have had bony deformities similar to those noted in infantile G_{M1} gangliosidosis. Lymphocyte vacuolation is not prominent. Occasional foamy histiocytes are found in the bone marrow. Death due to bronchopneumonia usually occurs by 3 years of age.

Pathologic changes in the cortex are nearly identical to those seen in Tay-Sachs disease. Lipidosis of cortical, cerebellar, spinal and autonomic neurons is prominent. Vacuolated histiocytes are present in lungs, spleen, lymph nodes and bone marrow, but the degree of histiocytosis is less than that seen in Gaucher's or

Table 6.6 Some properties of human hexosaminidase A and B.

	Hex A	Hex B
Isoelectric pH	5.4	7.9
Molecular weight	102,000	100,000
pH optimum	4.4	4.4
Amino terminal amino acid	blocked	blocked
Carboxy terminal amino acid	serine	aspartic or asparagine
Substrate cleaved	G_{M2}, G_{A2}, globoside, hexosamine oligosaccharides	G_{A2}, globoside, hexosamine oligosaccharides
Subunit (Mol. Wt.)	alpha chain (58,000) beta chain (25,000)	beta chain (25,000)
Proposed structure	$\alpha\beta_2$	$\beta_2\beta_2$

For references, see text.

Niemann-Pick disease. Vacuolation of the tubular epithelial cells of the renal loops of Henle may be prominent, probably due to storage of N-acetyl glucosamine-containing oligosaccharides. Ultrastructural studies reveal intracellular lamellar lipid inclusions and clear vacuolar material in these cells.

B. Enzyme defect in Sandhoff disease

Two hexosaminidase isoenzymes (possessing both β-D-N-acetylgalactosaminidase and β-D-N-acetylglucosaminidase activities) were first demonstrated by Robinson and Stirling in human spleen [74]. They possess similar Michaelis-Menten constants (Table 6) and are both present in the lysosomal fraction, but are readily separated from one another by ion-exchange column chromatography or starch gel electrophoresis.

One of these, hexosaminidase A (hex A) is more negatively charged and is more heat-labile than the other, hexosaminidase B (hex B). Both hex A and B have been found in all normal human tissues (except red cells), such as leucocytes, serum, urine, tears, cultured skin fibroblasts, amniotic fluid, and cultured amniotic cells. A third enzyme, termed hexosaminidase C is nonlysosomal and is not related to hex A or hex B [78].

The activities of both hexosaminidase A and B are deficient in Sandhoff disease [79]. The deficiency is present in all tissues and body fluids that have been examined, including cultured skin fibroblasts and amniotic cells from affected fetuses [80, 81]. The deficiency has been documented using synthetic N-acetylglucosaminides and N-acetylgalactosaminides as well as G_{M2}, G_{A2} and globoside [79] steroid-glucosaminides [82] and a heptasaccharide containing N-acetylgalactosamine [83] as substrates. The diagnosis of homozygotes is conveniently made by the hexosaminidase assay of serum [79] or cultured fibroblasts [84]. Heterozygote detection using serum has been demonstrated; low total hexosaminidase and a low percentage of hex B are used as criteria [85, 86].

Table 6.7 Urinary and hepatic oligosaccharides in Sandhoff disease.

Oligosaccharide	Tissue source	Reference
1. a GlcNAc β1-2 Man α1-3 Man β1-4 GlcNAc b GlcNAc β1-4 Man α1-3 Man β1-4 GlcNAc d GlcNAc β1-2 Man α1-6 Man β1-4 GlcNAc e GlcNAc β1-4 Man α1-6 Man β1-4 GlcNAc	Urine	87b
2. 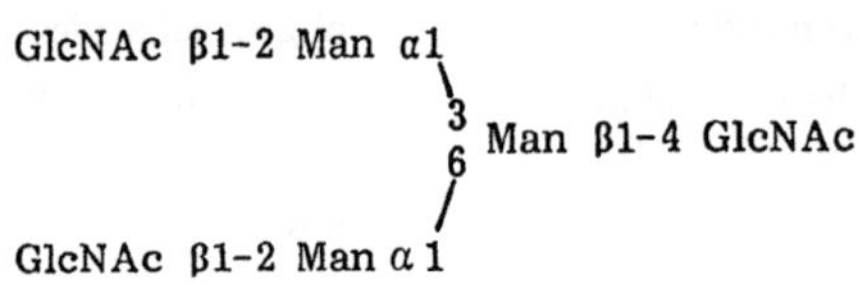	Urine Liver	87a,b
3. 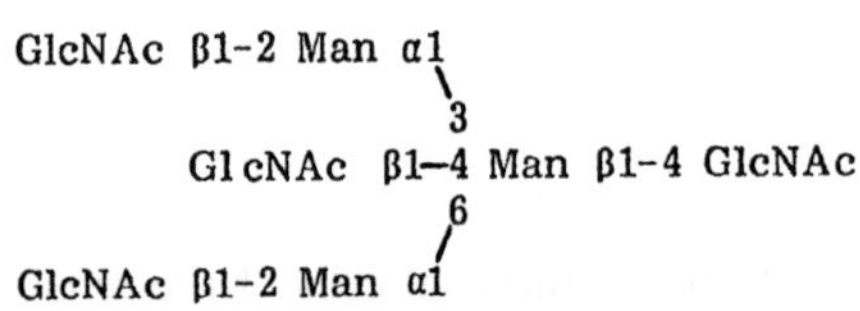	Urine	87b
4. 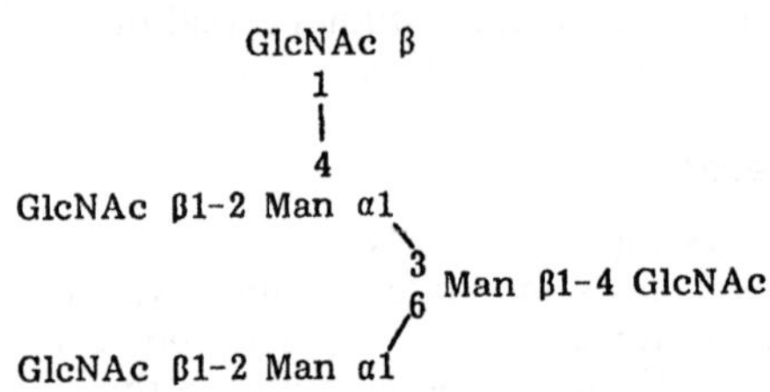	Urine Liver	87a,b
5. 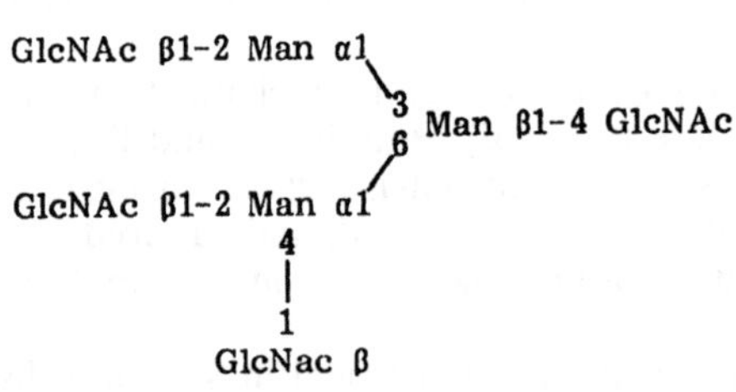	Urine Liver	87a,b

Lowden has demonstrated that hex B in Sandhoff disease heterozygotes is more thermolabile than normal suggesting the presence of a heteropolymer of hex B comparised of both wild type and mutant β-chains rendering the enzyme thermolabile [87]. Sandhoff disease has been correctly diagnosed prenatally in two affected fetuses; both pregnancies were terminated [80, 81].

A patient with a juvenile form of Sandhoff disease has been described. He was normal until age 5 when slurred speech, ataxia and mental decline became apparent. At 10 years he had generalized hypertonia, hyperreflexia and ataxia with normal optic fundi [88]. A deficiency of hex B with preservation of some residual hex A activity in fibroblasts was demonstrated [89].

Several unusual variants of Sandhoff disease have been described, one with considerable hex and B activity in serum but a striking reduction in tissues [90], another who appears to be a genetic compound for two different β-chain mutations [91].

C. Chemistry of the storage products

1) *Lipid storage products* - Cerebral levels of ganglioside G_{M2} are increased 100 to 300 times normal in Tay-Sachs and Sandhoff disease. The asialo-derivative of ganglioside G_{M2}(G_{A2}) accumulates to levels about 20 times normal. The ganglioside content of visceral organs is not significantly increased, but the ganglioside pattern is abnormal; ganglioside G_{M2} comprises a higher proportion of total gangliosides than normal.

Much higher amounts of G_{A2} accumulate in the brain and viscera in Sandhoff disease than in Tay-Sachs disease. Globoside [cer-Glc-Gal-Gal-Gal-(NAc)] accumulates in large quantities in the viscera, especially the kidney and spleen, in Sandhoff disease but not in Tay-Sachs disease. This probably accounts for the lipid inclusions found on ultrastructural examination.

2) *N-Acetylglucosaminyl-oligosaccharides* - A wide variety of N-acetyl glucosaminyl-oligosaccharides have been identified in tissues and urine in Sandhoff disease [87a-87b] (Table 7). Like the galactose oligosaccharides in G_{M1} gangliosidosis these compounds are probably derived from the incomplete degradation of cellular glycoproteins (Figure 3) and they accumulate due to the absence of both hex A and B. The list of compounds identified is not nearly as extensive as those found in G_{M1} gangliosidosis (Table 3); however, further studies will undoubtedly reveal a series of more complex oligosaccharides in these patients as well.

D. Properties of hexosaminidase A and B and their involvement in Sandhoff disease

Hex A and hex B have molecular weights of approximately 100,000 daltons (Table 6). Hex A has a lower isoelectric point than B. The amino terminus of hex A and B appear to be blocked, but the carboxyl-terminal amino acid of hex A appears to be serine and that of hex B aspartic acid (or asparagine). When antibodies are raised against purified hex B, they cross-react with hexosaminidase A indicating that A and B possess a common polypeptide. Conditions favoring subunit disassociation cause disaggregation of hex A and B to smaller units.

Kinectic studies indicate that hex A cleaves ganglioside G_{M2} but hex B does not. Both cleave GA_2 and globoside as well as ρ-nitrophenyl- and 4-methyl-umbelliferyl-β-acetylhexosaminides; when these substrates are used hex A and B have similar pH optimum and kinectics (Table 6).

Fibroblast homogenates from patients with Sandhoff disease fail to hydrolyze N-acetylgalactosamine from a heptasaccharide derived from chondroitin-4 sulfate under conditions in which normal fibroblasts have activity. This suggests that hex A, but not B, cleaves N-acetyl-galactosamine from this compound. Hexosaminidase A and B apparently both cleave N-acetylglucosamine from steroid conjugates such as testosterone 17-β-N-acetylglucosaminide. Studies with a trisaccharide containing a terminal N-acetylglucosamine residue (isolated from hyaluronate) reveal that hex A cleaves the aminosugar at a rate 35 to 40 times that of hex B.

Hexosaminidase A has been reported to cleave synthetic β-actyl-hexosaminide substrates at rates 100 to 6 million times higher than ganglioside G_{M2}, depending on the assay conditions and tissue source. A small molecular weight glycoprotein stimulatory "factor", isolated from human liver [92] and kidney [93] has been shown to stimulate hex A to cleave G_{M2}. Geiger and Arnon [94] and Lee and Yoshida [95] purified hex A and B to homogeneity from human placenta and determined their properties. Hex A contained 1.65 residues of sialic acid per molecule and hex B contained none. Their molecular weights were 100,000 and 108,000, respectively; both could be dissociated into four subunits of 25,000 molecular weight. Hex B contained only one type of polypeptide chain, the β-chain, whereas hex A contained an equal proportion of β-chains and α-chains. This latter were more negatively charged than the former, with the charge difference being apparently due to a high proportion of acidic amino acids as well as sialic acid in the α-chains.

Mahuran and Lowden [96] recently isolated hex A and B in homogeneous form from human placenta and studied their subunit compositions. They confirmed the previous finding that hex B is a tetramer of 2 pairs of β-chains ($\beta_2\beta_2$) but found that hex A contains two β chains and one 50,000 dalton alpha chain. They conclude that the α-chain of hex A is either a 50,000 molecular weight polypeptide or that two 25,000 dalton chains are linked by non-disulfide crosslinkage (e.g. an isopeptide bond), with hex A having the structure ($\alpha\beta_2$).

E. Nature of the mutation in Sandhoff disease

A three locus model [97] coding for two separate polypeptide chains alpha and beta, of hex A and for activator protein appears to explain the molecular genetics of G_{M2} gangliosidosis (Fig. 7). The alpha locus, located on chromosome 15 [98-99] is mutated in the hex A mutants (TSD, juvenile G_{M2} gangliosidosis with hex A deficiency and adult G_{M2} gangliosidosis with hex A deficiency). The beta locus, located on chromosome 5 is mutated in the hex B mutants (Sandhoff disease and juvenile Sandhoff disease). The activator locus (or loci) is involved in the AB(+) mutants.

The model explains how the mutation in Sandhoff disease can lead to simultaneous loss of hex A and B activity [13] and how restoration of hex A activity can occur in fusions of TSD and Sandhoff disease fibroblasts; the alpha subunit derives from the Sandhoff disease fibroblasts, the beta subunit from the TSD fibroblasts [13]. It explains why independent segregation of hex A and B occurs in somatic cell hybrids [96, 99].

F. Animal models

G_{M2} gangliosidosis with deficiency of hex A and B (Sandhoff disease) has been reported in cats (100) and has been used by Rattazzi and co-workers [101-102] to

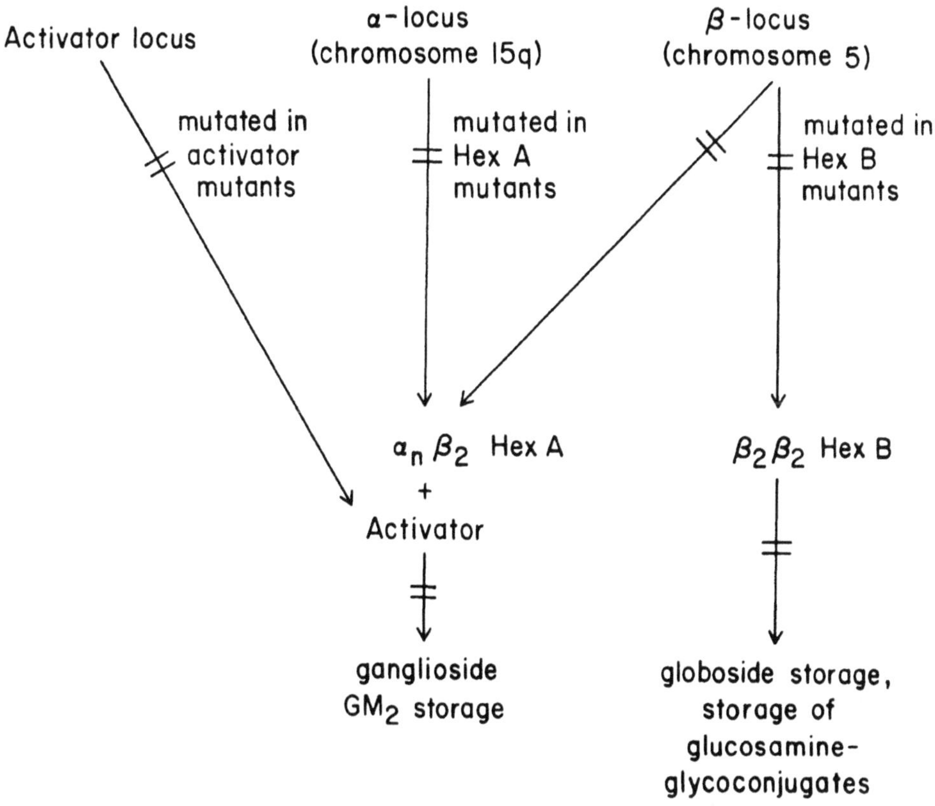

Fig. 6.7 Three-locus model to explain the molecular genetics of Sandhoff disease.

explore enzyme replacement therapy with human hex A. Hepatic uptake of human hex A was enhanced by administration of mannans. Brain uptake was induced by cerebral oxygen embolism without neurological damage. Enzyme levels of hex A which were virtually zero in Sandhoff disease cats were brought to 30% of normal in liver and 10% of normal in brain after enzyme administration. A marked reduction of G_{M2} and globoside storage in liver occurred but no reduction of G_{M2} storage in brain cortex. The excretion of the urinary N-acetylglucosaminyl-oligosaccharides was not determined. Analysis of these compounds would also be helpful in evaluating the effectiveness of enzyme therapy. To date, a structural analysis of the stored and excreted oligosaccharides in the affected cats has not been reported.

This work was supported by grants from the National Institutes of Health, Grant NS 08682, US Public Service, Program Project Grant GM 17702 and the Gould Family Foundation Grant to JSOB.

REFERENCES

1. O'Brien J.S.: *The Gangliosidoses.* In "The Metabolic Basis of Inherited Disease", IV Edition. J.B. Stanbury, J.B. Wyngaarden and D.S. Fredrickson (Eds.), New York, McGraw-Hill, Chapter *40*, 841-865, 1978.
2. Okada S., O'Brien J.S.: *Tay-Sachs disease: generalized absence of a β-D-N-acetylhexosaminidase component.* Science *165*, 698-700, 1968.
3. Sandhoff K., Harzer K.: *Total hexosaminidase deficiency in Tay-Sachs disease (variant O).* In "Lysosomes and Storage Diseases". H.G. Hers and F. Van Hoof (Eds.). New York, Academic Press, 345-356, 1973.
4. Spranger J.W., Langer L.O., Wiedemann H.R.: *G_{M1} gangliosidosis.* In "Bone Displasias: an Atlas of Constitutional Disorders of Skeletal Development". Philadelphia, W.B. Sanders, Co., 171-172, 1974.
5. O'Brien J.S., Bernett J., Veath M.L., Paa D.: *Lysosomal storage disorders: diagnosis by ultrastructural examination of skin biopsies.* Arch. Neurol. *35*, 592-599, 1975.
6. O'Brien J.S.: *Ganglioside storage diseases.* In "Advances in Human Genetics". Vol. 3. K. Hirschhorn and H. Harris (Eds.). New York, Plenum Press, 39-98, 1972.
7. Wolfe L.S., Callahan J. Facwett J.S., Andermann P., Scriver C.: *G_{M1} gangliosidosis without chondrodystrophy or visceromegaly: β-galactosidase deficiency with gangliosidosis and the excessive excretion of a keratan sulfate*" Neurology *20*, 23-43, 1970.
8. Wenger L.S., Sattler M., Mueller O.T., Myers G.G., Schneider R.S., Nixon G.W.: *Adult G_{M1} gangliosidosis: clinical and biochemical studies on two patients and comparison to other patients called variant or adult G_{M1} gangliosidosis.* Clin. Genet. *17*, 323-328, 1980.
9. Stevenson R.E., Taylor H.A., Parks S.E.: *β-Galactosidase deficiency: prolonged survival in three patients following early central nervous deterioration.* Clin. Genet. *13*, 305-309, 1978.
10. Okada S., O'Brien J.S.: *Generalized gangliosidoses: beta-galactosidase deficiency.* Science *160*, 1002-1004, 1968.
11. MacBrinn M.C., Okada S., Ho M.W., Hu C.C., O'Brien J.S.: *Generalized gangliosidosis: impaired cleave of galactose from a mucopolysaccharide and a glycoprotein.* Science *163*, 946-947, 1969.
12. Norden A.G.W., O'Brien J.S.: *Ganglioside G_{M1} β-galactosidase: studies in human liver and brain.* Arch. Biochem. Biophys. *159*, 383-392, 1973.
13. For references see: O'Brien J.S.: *G_{M1} gangliosidosis.* In "The Metabolic Basis of Inherited Disease" IV Edition. J.B. Stanbury, J.B. Wyngaarden and D.S. Fredrickson (Eds.), New York, McGraw-Hill, 841-865, 1978.
14. Wenger D.A., Sattler M., Hiatt W.: *Globoid cell leucodystrophy: deficiency of lactosyl ceramide β-galactosidase.* Proc. Nat. Acad. Sci. USA *71*, 854-858, 1974.
15. Singer H.S., Schafer I.A.: *White cell β-galactosidase activity.* N. Engl. J. Med. *282*, 571-573, 1970.
16. Thomas G.H.: *β-D-Galactosidase in human urine: deficiency in generalized ganglosidosis.* J. Lab. Clin. Med. *74*, 725, 729, 1969.
17. Sloan H.R., Uhlendorf B.W., Jacobson C.B., Fredrickson D.S.: *β-galactosidase in tissue culture derived from human skin and bone marrow: enzyme defect in G_{M1} gangliosidosis.* Pediat. Res. *3*, 532-537, 1969.
18. Lowden J.S., Cutz E., Conen P.E., Rudd N., Doran T.E.: *Prenatal diagnosis of G_{M1} gangliosidosis.* N. Engl. J. Med. *288*, 255-257, 1973.
19 Kaback M.M., Sloan H.R., Sonnerborn M., Herndon R.M., Percy A.K.:*G_{M1} gangliosidosis Type I in utero: detection and fetal manifestations.* J. Pediat. *82*, 1037-1041, 1973.
20. Booth C.W., Gerbie A.G., Nadler H.L.: *Intrauterine diagnosis of G_{M1} gangliosidosis, Type 2.* Pediatrics *52*, 521-524, 1973.
21. Wolfe L.S., Senior R.G., Ng Ying N.M.K.: *The structures of oligosaccharides accumulating in the liver of G_{M1} gangliosidosis, Type 1.* J. Biol. Chem. *249*, 1828-1838,

1974.
22. Lundblad A., Sjoblad S., Svensson S.; *Characterization of a penta and an octasaccharide from urine of a patient with juvenile G_{M1} gangliosidosis*. Arch. Biochem. Biophys. *188*, 130-136, 1978.
23. Yamashita K., Ohkura T., Okada S., Yabuuchi H., Kobata A.: *Urinary oligosaccharides of G_{M1} gangliosidosis*. J. Biol. Chem. *259*, 4789-4793, 1981.
24. Ohkura T., Yamashita K., Kobata A.: (in press). J. Biol. Chem., 1981.
25. Mega T., Lujan E., Yoshida A.: *Studies on the oligosaccharide chains of human α1-protease inhibitor*. J. Biol. Chem. *255*, 4057-4061, 1980.
26. Yamashita K., Liang C., Funakoshi S., Kobata A.: *Structural Studies of asparagine-linked sugar chains of human ceruloplasmin*. J. Biol. Chem. *256*, 1283-1289, 1981.
27. Fournet B., Montreuil J., Strecker G., Dorland K., Haverkamp J., Vliejenthart J.F.G., Binett J.P., Schmid K.: *Determination of the primary structures of 16 asialo-carbohydrate units derived from human plasma α1-acid glycoprotein by 360 MHz NMR spectroscopy and permethylation analysis*. Biochemistry *17*, 5206-5211, 1978.
28. Prasad R., Hudson B.G., Strickland D.K., Ebner K.E.: *The structure of the asparagine-linked carbohydrate unit of rat α-lactalbumin*. J. Biol. Chem. *255*, 1244-1249, 1980.
29. Hayes M.L., Castillino F.J.: *Carbohydrate of the human plasminogen variants*. J. Bioch. Chem. *254*, 8772-8777, 1979.
30. Strecker G., Montreuil J.: *Glycoproteines et glycoproteinoses*. Biochemie *61*, 1196-1231, 1979.
31. Robertson A., Warner T.G., O'Brien J.S.: *Unpublished observations*.
32. Robinson D.: *Multiple forms of glycosidases in the normal and pathological states*. Enzyme *18*, 114-116, 1974.
33. Cheetham P.S., Dance N.E., Robinson D.: *Isoenzymes of human liver β-galactosidase*. Biochem. Soc. Trans, 241-243, 1975.
34. Norden A.G.W., Tennant L.L., O'Brien J.S.: *G_{M1} ganglioside β-galactosidase A: purification and studies of the enzyme from human liver*. J. Biol. Chem. *249*, 7969-7976, 1974.
35. Frost R.G., Holmes E.W., Norden A.G.W., O'Brien J.S.: *Characterization of purified human liver acid β-galactosidases A_2 A_3*. Biochem. J. *175*, 181-188, 1978.
36. Ho M.W., O'Brien J.S.: *Stimulation of acid β-galactosidase activity by chloride ions*. Clin. Chim. Acta *30*, 531-534, 1970.
37. Ho M.W., O'Brien J.S.: *Differential effect of chloride ions on β-galactosidase isoenzymes: a method for separate assay*. Clin. Chim. Acta *32*, 443-450, 1971.
38. Ho M.W., Cheetham P., Robinson D.: *Hydrolysis of G_{M1}-ganglioside by human liver β-galactosidase isoenzymes*. Biochem. J. *136*, 351-359, 1973.
39. Tennant L.L., Veath L.M., O'Brien J.S.: *Unpublished data*, 1980.
40. Norden A.G.W., O'Brien J.S.: *Binding of human liver β-galactosidase to plant lectins insolubilized on agarose*. Biochem. Biophys. Res. Commun. *56*, 193-198, 1974.
41. Suzuki Y., Crocker A., Suzuki K.: *G_{M1} gangliosidosis: correlation of clinical and biochemical data*. Arch. Neurol. *24*, 48-54, 1971.
42. Van Hoof F.: *G_{M1} gangliosidosis*. In "Lysosomes and Storage Diseases". H.G. Hers and F. Van Hoof (Eds.). New York, Academic Press, 305-320, 1973.
43. Norden A.G.W., O'Brien J.S.: *An electrophoretic variant of β-galactosidase with altered catalytic properties in a patient with G_{M1} gangliosidosis*. Proc. Nat. Acad. Sci. USA *72*, 240-244, 1975.
44. Cheetham P.S.J., Dance N.E., Robinson D.A.: *Benign deficiency of Type B β-galactosidase in human liver*. Clin. Chim. Acta *38*, 67-73, 1978.
45. Van Hoof F., Hers H.G.: *The abnormalities of lysosomal enzymes in mucopolysaccharidoses*. Europ. J. Biochem. *7*, 34-39, 1968.
46. Hindman J., Cotlier R.: *Glycosidases in normal human leucocytes and abnormalities in G_{M1} gangliosidosis*. Clin. Chem. *18*, 971-978, 1972.
47. Suzuki K., Suzuki Y.: *Globoid cell leucodystrophy (Krabbe's disease). Deficiency*

of galactocerebroside β-galactosidase. Proc. Nat. Acad. Sci. USA *66*, 302-305, 1970.
48. Wenger D.A., Sattler M., Hiatt W.: *Globoid cell leucodystrophy: deficiency of lactosyl ceramide β-galactosidase*. Proc. Nat. Acad. Sci. USA *71*, 854-858, 1974.
49. Tanaka H., Suzuki K., *Lactosylceramide β-galactosidase in human sphingolipidoses: evidence for two genetically distinct enzymes*. J. Biol. Chem. *250*, 2324-2328, 1975.
50. Lowden J.A., Callahan J.W., Norman M.G., Tahin M., Prichard J.S.: *Juvenile G_{M1} gangliosidosis: occurrence with absence of two β-galactosidase components*. Arch. Neurol.*31*, 200-208, 1974.
51. Naylor S.L., Lalley P.A., Elliott R.W., Brown J.A., Shows T.B.: *Evidence for homologous regions of human chromosome 3 and mouse chromosome 9 predicts location of human genes*. Amer. J. Hum. Genet. *32*, 159a, 1980.
52. O'Brien J.S.: *Molecular genetics of GM_1 β-galactosidase*. Clin. Genet. *8*, 303-313, 1980.
53. Meisler M., Rattazzi M.C.: *Immunological studies of β-galactosidase in normal human liver and in G_{M1} gangliosidosis*. Amer. J. Hum. Genet. *26*, 683-691, 1974.
54. Ben-Yosef Y., Burton B.K., Nadler H.L.: *Quantification of the enzymatically deficient cross-reacting material in G_{M1} gangliosidosis*. Amer. J. Hum. Genet. *29*, 575-580, 1977.
55. Galjaard H., Hoogeveen A., Keijzer W., De Wit-Verbeek H.A., Reuser A.J.J., Ho M.W., Robinson D.: *Genetic heterogeneity in G_{M1} gangliosidosis demonstrated by somatic cell hybridization and enzyme assays in single heterokaryons*. Nature *257*, 60-62, 1978.
56. O'Brien J.S.: *An explanation for complementation in G_{M1} gangliosidosis*. Amer. J. Hum. Genet. *30*, 36a, 1978.
57. O'Brien J.S., Norden A.G.W.: *Nature of the mutation in adult β-galactosidase deficient patients*. Amer. J. Hum. Genet. *29*, 184-190, 1977.
58. Pinsky K., Miller J., Shanfield B., Walters G., Wolfe L.S.: *G_{M1} gangliosidosis in skin fibroblast cultures: enzymatic differences between types 1 and 2 and observations on a third variant*. Amer. J. Genet. *26*, 563-570, 1974.
59. Feldges A., Mueller H.J., Buehler E., Stalder G.: *G_{M1} gangliosidosis. Part I: Clinical aspects and biochemistry*. Helv. Paediatr. Acta *28*, 511-519, 1973.
60. O'Brien J.S., Gugler E., Giedion A., Wiessmann U., Hershkowitz N., Meier C., Leroy J.: *Spondyloepiphyseal dysplasia, corneal clouding, normal intelligence and β-galactosidase deficiency*. Clin. Genet. *9*, 495-504, 1976.
61. Arbisser A.L., Donnelly K.A., Scott C.I.: *Morquio-like syndrome with β-galactosidase deficiency and normal hexosamine sulfatase activity: mucopolysaccharidosis IVB*. Amer. J. Med. Genet. *1*, 195-201, 1977.
62. Groebe H., Krins M., Schmidberger H., Von Figura K., Harzer K., Kresse E., Paschke E., Sewell A., Ullrich K.: *Morquio syndrome (mucopolysaccharidises IVB) associated with β-galactosidase deficiency. Report of two cases*. Amer. J. Hum. Genet. *32*, 258-265, 1980.
63. Wenger D.A., Tarby T.J., Wharton C.: *Macular cherry-red spots and myoclonus with dementia: coexistent neuraminidase and β-galactosidase deficiencies*. Biochem. Biophys. Res. Commun. *82*, 589-595, 1978.
64. Lowden J.A., O'Brien J.S.: *Sialidosis: A review of human neuraminidase deficiency*. Amer. J. Hum. Genet. *31*, 1-18, 1979.
65. Baker H.J., Lindsey J.R.: *Feline G_{M1} gangliosidosis*. Amer. J. Pathol. *74*, 649-657, 1974.
66. Handa S., Yamakawa T.: *Biochemical studies in cat and human gangliosidosis*. J. Neurochem. *18*, 1275-1280, 1971.
67. Blakemore W.F.: *G_{M1} gangliosidosis in a cat*. J. Comp. Pathol. *82*, 179-185, 1972.
68. Donnelly W.J.C., Sheahan B.J., Kelly M.: *Beta-galactosidase deficiency in G_{M1} gangliosidosis of Friesian calves*. Res. Vet. Sci. *15*, 139-144, 1973.
69. Read D.H., Harrington D.D., Keenan T.W., Hinsman E.J.: *Neuronal visceral G_{M1} gangliosidosis in a dog with β-galactosidase deficiency*. Science *194*, 442-445, 1976.
70. Baker H.J., Lindsey J.R., McKhann G.M., Farrell D.M.: *Neuronal G_{M1} gangliosidosis in a Siamese cat with β-galactosidase deficiency*. Science *174*, 838-839, 1971.
71. Holmes E., O'Brien J.S.: *Feline G_{M1} gangliosidosis: characterization of the residual*

liver acid β-*galactosidase*. Amer. J. Hum. Genet. *30*, 505-515, 1978.
72. Holmes E., O'Brien J.S.: *Hepatic storage of oligosaccharides and glycolipids in a cat affected with G_{M1} gangliosidosis*. Biochem. J. *175*, 945-953, 1978.
73. Holmes E., O'Brien J.S.: Purification of acid β-*galactosidase from feline liver*. Biochemistry *18*, 952-958, 1979.
74. Rittmann L.S., Tennant L.L., O'Brien J.S.: *Dog G_{M1} gangliosidosis: characterization of the residual liver acid* β-*galactosidase*. Amer. J. Hum. Genet *32*, 880-889, 1980.
75. Warner T.G., O'Brien J.S.: *Unpublished data*, 1981.
76. Warner T.G., O'Brien J.S.: *Structure analysis of the major oligosaccharides accumulating in canine G_{M1} gangliosidosis liver*. J. Biol. Chem. *257*, 224-232, 1982.
77. Robinson D., Stirling J.L.: *N-acetyl-*β*-glucosaminidases in human spleen*. Biochem. J. *107*, 321-328, 1968.
78. Braidman I., Carroll M., Dance N., Robinson D., Poenam L., Weber A., Dreyfus J.S., Overdijk B., Hoochwinkel G.J.M.: *Characterization of human hexosaminidase C*. FEBS Lettr. *41*, 181-186, 1974.
79. Sandhoff K., Harzer K.: *Total hexosaminidase deficiency in Tay-Sachs disease (variant O)*. In "Lysosomes and Storage Disease" H.G. Hers and F. Van Hoof (Eds.). New York, Academic Press, 345-367, 1973.
80. Desnick R.J., Krivit W., Sharp H.R.: *In utero diagnosis of Sandhoff disease*. Biochem. Biophys. Res. Commun. *51*, 20-23, 1973.
81. Harzer K., Stengel-Rutkowski S., Gley E.O., Albert A., Murken J.D., Zahr V., Henkel K.P.: *Prenatale diagnose der G_{M2} gangliosidose type 2*. Sandhoff-Jatzkewitz Krankeit Dtsch. Med. Wschr. *100*, 106-110, 1975.
82. Tomasi L.G., Fukushima D.K., Kolodny E.H.: *Steroid hexosaminidase activity in Tay-Sachs and Sandhoff-Jatzkewitz disease*. Neurology *24*, 1158-1162, 1974.
83. Thompson J. N., Stoolmiller A.G., Matalon R., Dorfman A.: *N-acetyl-*β*-hexosaminidase: role in the degradation of glycosaminoglycans*. Science *181*, 866-868, 1973.
84. Okada S. McCrea M., O'Brien J.S.: *Sadhoff disease (G_{M2} gangliosidosis type 2): clinical, chemical and enzymes studies in five patients*. Ped. Res. *6*, 606-615, 1972.
85. Kolodny E.H.: *Sandhoff disease: studies on the enzyme defect in homozygotes and detection of heterozygotes*. In "Sphingolipidoses and Allied Disorders" B.W. Volk and S.M. Aronson (Eds.). New York, Plenum Press, 321-337, 1972.
86. Lowden J.A., Ives E.J., Keene D.L., Burton A.L., Skomorowski M.A., Howard F.: *Carrier detection in Sandhoff disease*. Am. J. Hum. Genet. *31*, 281-288, 1979.
87. Lowden J.A.: *Evidence for a hybrid hexosaminidase isoenzyme in heterozygotes for Sandhoff disease*. Am. J. Hum. Genet. *31*, 281-288, 1979.
87a. Ng Ying Kin N.M.K., Wolfe L.S.: *Oligosaccharides accumulating in the liver from a patient with G_{M2} gangliosidosis variant O*. Biochem. Biophys. Res. Comm. *59*, 837-841, 1974.
87b. Strecker G., Herlant-Peers M., Fournet B., Montreuil J., Dorland L., Haverkamp J., Vliejenthart F.G.: *Structure of seven oligosaccharides excreted in urine of a patient with Sandhoff disease (G_{M2} gangliosidosis-variant O)*. Eur. J. Biochem. *81*, 165-172, 1977.
88. Mc Leod P.M., Wood S., Jan J.E., Applegarth D.A., Dolman C.L.: *Progressive cerebellar ataxia, psychomotor retardation, and hexosaminidase deficiency in a 20 year old child: juvenile Sandhoff disease*. Neurology *27*, 571-577, 1977.
89. Wood S., Mac Dougall B.G.: *Juvenile Sandhoff disease: some properties of the residual hexosaminidase in cultured fibroblasts*. Am. J. Hum. Genet. *28*, 489-494, 1976.
90. Spence M.W., Ribley B.A., Embil J.A., Tibbles J.A.: *A new variant of Sandhoff disease*. Pediat. Res. *8*, 628-636, 1974.
91. Lane A.B., Jenkins T.: *Two variant hexosaminidase* β*-chain alleles segregating in a South African family*. Clin. Chim. Acta *38*, 219-224, 1978.
92. Li S.C., Hirabayashi Y., Li Y.T.: *A protein activator for the enzyme hydrolysis of G_{M2} ganglioside*. J. Biol. Chem. (in press).

93. Conzelmann E., Sandhoff K.: *AB Variant of infantile G_{M2} gangliosidosis: deficiency of ganglioside G_{M2} and glycolipid G_{A2}.* Proc. Nat. Acad. Sci *75*, 3979-3982, 1978.
94. Geiger B., Arnon R.: *Chemical characterization and subunit structure of human N-acetylhexosaminidase A and B.* Biochemistry *15*, 3484-3490, 1976.
95. Lee J.E.S., Yoshida A.: *Purification and chemical characterization of human hexosaminidase A and B.* Biochem. J *159*, 535-541, 1976.
96. Mahuran D., Lowden J.A.: *The subunit and polypeptide structure of hexosaminidases from human placenta.* Can. J. Biochem. *58*, 287-293, 1980.
97. Beutler E.: *The biochemical genetics of the hexosaminidase system in man.* Am. J. Hum. Genet. *32*, 95-106, 1979.
98. Gilbert F., Kucherlapati R., Creagan R.P., Murnane J.M., Darlington G.I., Ruddle F.H.: *Tay-Sachs and Sandhoff diseases: The assignment of genes for hexosaminidases A and B to individual human chromosomes.* Proc. Nat. Acad. Sci. USA *72*, 263-269, 1975.
99. Lalley P.A., Rattazzi M.C., Shows T.B.: *Human β-D-N-acetylhexosaminidase A and B: expression and linkage relationships in somatic cell hybrids.* Proc. Nat. Acad. Sci. USA *71*, 1569-1574, 1974.
100. Cork L.C., Munnel J.F., Lorenz M.D., Murphy J.N., Baker H.J., Rattazzi M.C.: *G_{M2} ganglioside storage disease in cats with β-hexosaminidase deficiency.* Science *196*, 1014-1018, 1977.
101. Rattazzi M.C., Lanse S.B., Mc Cullough R.A., Nester J.A., Jacobs E.A.: *Towards enzyme replacement in G_{M2} gangliosidosis: organ disposition and induced central nervous system uptake of human hexosaminidase in the cat.* Birth Defects. Original Article Series *16*, 179-184, 1980.
102. Rattazzi M.C., Appel A.M., Baker H.J.: *Enzyme replacement in G_{M2} gangliosidosis: a reduction of glycolipid storage in visceral organs.* Am. J. Hum. Genet. *32*, 51A, 1980.

Pertti Aula
Seppo Autio
Kari O. Raivio
Juhani Rapola
Martin Renlund

7
Salla disease

NAME OF DISEASE	Salla disease
PREVALENCE	Finnish people
MODE OF INHERITANCE	Autosomal recessive
MUTANT GENE PRODUCT	Unknown
CHROMOSOMAL LOCATION	Unknown
EXPRESSION	
1. Altered DNA structure	?
2. Disturbed protein function	?
3. Disrupted cell and organ function	Accumulation of free sialic acid in liver cells and possibly in neurons.

7.1 INTRODUCTION

During the last few years a group of severely retarded patients with ataxia and other neurological signs have been shown to represent a new inherited lysosomal storage disorder [1]. The eponym Salla disease has been adopted according to the geographical area of northern Finland where most of the early patients originated from. The basic biochemical defect of Salla disease is still unknown but the patients consistently excrete increased amounts of free sialic acid in the urine [2]. This chapter reviews the clinical and biochemical data so far available from 34 Finnish patients. No patients outside this country have yet been reported.

7.2 CLINICAL FEATURES

The patient material consists of 20 male and 14 female individuals ranging from 2 to 63 years in age. They belong to 23 families with a total of 133 sibs.

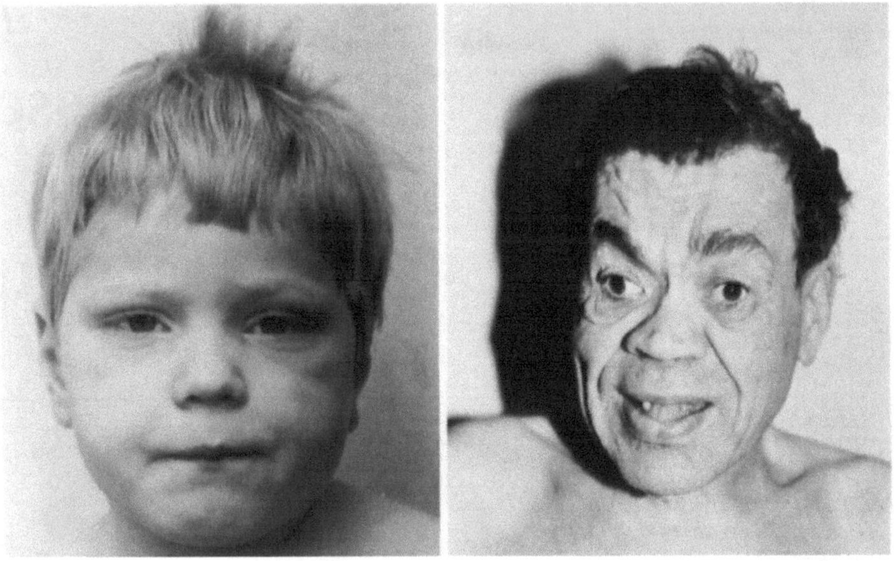

Fig. 7.1 Two patients with Salla disease. Left: a 3-year-old boy with normal facial features. He has delayed psychomotor development, severe ataxia and no speech. Right: a-63-year-old severely retarded male patient has right external strabismus but no other abnormal external features.

A. Early hystory

The fetal and neonatal history has been unremarkable in all cases. The first signs of the disease, hypotonia and/or mild delay in psychomotor development have been noticed by the parents usually between 4 and 12 months of age. At the end of the first year of life most of the patients have been developmentally retarded with motor clumsiness and truncal ataxia. By the end of the second year all have been obviously retarded. Twenty six of the patients started walking at the age of 1.9 to 12 years while eight never did, and another eight later lost the ability to walk. Most patients learned to speak single words or short sentences at the age of 2 to 5 years. No patient has been able to attend normal school.

B. Physical findings

External features of the patients have not shown any consistent abnormalities (Fig. 1). Some patients have growth retardation; 18 patients were more than 2 SD below the expected height. Seventeen patients have divergent strabismus but no other ocular manifestations of Salla disease have been detected. X-ray studies have shown a thickening of the calvarium in 19 of 28 patients as the only radiological abnormality in Salla disease.

C. Mental retardation

All patients above four years of age with Salla disease have been severely or profoundly mentally retarded. Patients with repeated IQ-testing have shown an impairment of the mental performance after the age of 15 to 20 years. They require constant care and assistance in daily life and most of them live in institutions for mentally retarded. Life expectancy of Salla disease patients seems not to be much reduced as the oldest patient today is already 63 years old. Salla disease patients are characterized by good-humoured nature in contrast to patients with AGU. Severe speech impairment presenting with dysarthria and dyspraxia is a common finding in Salla disease. The majority of the patients are or have been able to express single words or short sentences.

D. Neurological findings

The main motor abnormalities have been ataxia, athetosis and rigidity or spasticity (Table 1). Ataxia is more prominent in the younger patients whereas athetosis, rigidity, and spasticity are found more frequently in adult patients. As mentioned earlier several patients have been hypotonic during the first year of life when no other signs of the disease were yet detectable.

Table 7.1 Clinical findings in 34 patients with Salla disease.

Neurological signs	
Ataxia	34/34
Athetosis	22/34
Abnormal reflexes	28/34
brisk tendon reflexes/clonus	24/34
positive Babinski sign	17/34
Rigidity/spasticity (lower limb)	28/34
Speech abnormality	34/34
dysarthric/dyspractic	34/34
single words/short sentences	24/34
no speech	10/34
Unable to walk	16/34
Convulsions	8/34
Mental retardation	
IQ 30-40	2/34
20-30	5/34
below 20	27/34
Physical findings	
Growth more than 2 SD below expected height	18/34
Exophoria	17/34
Thick calvarium	19/34

Convulsive seizures have not been a major problem in Salla disease. Two patients have had generalized convulsions since childhood and six other patients have had absence attacks since second to fifth decade or life.

E. Neurophysiological findings

A gradual decrease in the amplitude of the EEG recording has been a consistent finding in Salla disease. In cases where EEG was taken during the first two years of life no abnormalities have been seen, but no patient older than 10 years has had a normal EEG. In addition to low-voltage recordings no other abnormalities have been found. Nerve conduction velocities and needle electromyogram have shown normal results in patients studied for these parameters.

7.3 BIOCHEMISTRY

The basic defect in Salla disease is still unresolved but increased urinary free sialic acid excretion suggests an abnormality in sialic acid metabolism.

All patients with Salla disease have shown to have a 10 to 15 fold increase in the amount of free sialic acid in urine [2]. Detailed analyses of carbohydrate and sialic acid derivatives in the urine showed that the sialic acid containing material is N-acetylneuraminic acid and that no other abnormal metabolites are present.

Increased urinary free sialic acid can conveniently be demonstrated with thin layer chromatography [3]. Urine samples from Salla disease patients show a typical strong blue band with the same mobility as authentic N-acetylneuraminic acid (see Fig. 5.6). In order to demonstrate the anticipated enzyme deficiency several lysosomal hydrolases, including neuraminidase, have been assayed both in peripheral blood lymphocytes and cultured skin fibroblasts, but no enzyme abnormality has been detected.

7.4 LYSOSOME MORPHOLOGY

Evidence for lysosomal storage in Salla disease stems from several findings: vacuoles in peripheral blood lymphocytes, enlarged lysosomes in skin and liver biopsies and staining with a lectin specific for sialic acid in cultured fibroblasts [4]. The percentage of vacuolated lymphocytes has ranged from 1 to 14% and only one patient, the youngest in the series has not shown vacuoles. Electron microscopy (EM) of a skin biopsy specimen has been done in 23 patients and in all cases typical enlarged electron-lucent lysosomal bodies have been seen (Fig. 2). Liver biopsy has been studied by EM in five patients and similar large lysosomal vacuoles with fibrillo-granular material have been detected in hepatocytes and Kupffer cells. The fine structure of the storage lysosomes in Salla disease has been identical to that seen in several other lysosomal storage disorders, particularly in aspartylglucosaminuria and mannosidosis (see Figs. 5.10, 4.3, 4.4). Histochemical stainings of skin biopsy sections have not yielded any information on the biochemical nature of the storage substance but fluorescence microscopic studies with oligosaccharide specific lectins in cultured fibroblasts have indicated the presence of increased amounts of sialic acid in the lysosomal vacuoles. In contrast to normal and mutant cell lines from other lysosomal storage diseases cultured

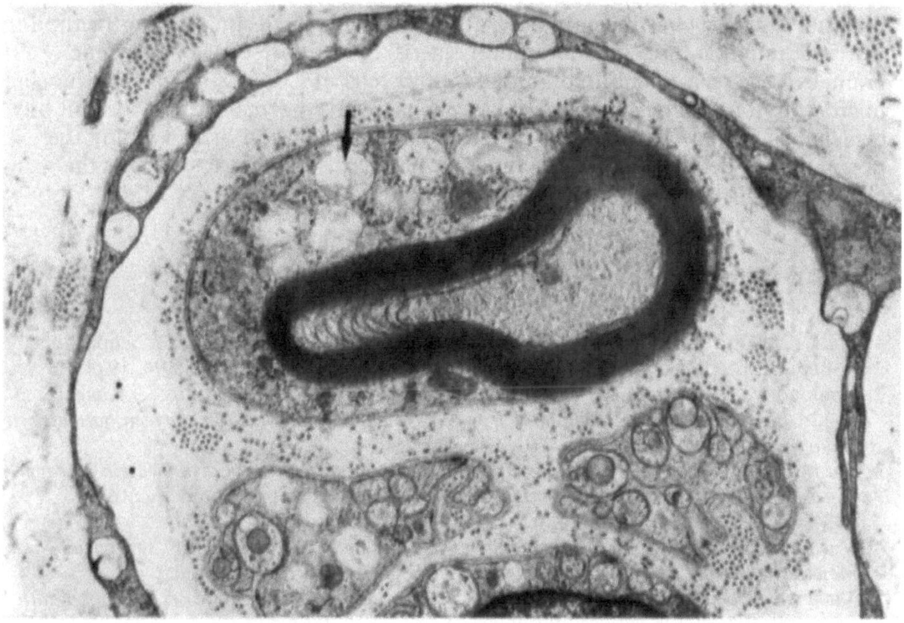

Fig. 7.2 Electron microscopic picture of the skin biopsy in Salla disease. A Schwann cell encircling myelinated nerve contains several storage lysosomes (one marked with arrow). Note also storage lysosomes in the perineural fibrocytes (x 5,000).

skin fibroblasts from Salla disease patients showed fluorescent cytoplasmic granules when stained with LPA- (limulus polyphemus) lectin coupled to a fluorescent stain [5]. This finding is in accordance with the recent biochemical results suggesting that cultured fibroblasts and liver tissue of Salla disease patients contain more free sialic acid than appropriate controls (Renlund et al. unpublished results). No autopsy studies have so far been performed in Salla disease, and therefore the extent of the storage phenomenon in other organs in unknown.

7.5 GENETICS OF SALLA DISEASE

The proportion of affected sibs in the families of the 34 patients with verified Salla disease is in favor of autosomal recessive mode of inheritance. In addition to the 34 patients with a confirmed diagnosis the 23 families have had eight deceased siblings who very likely have suffered from Salla disease. The proportion of affected sibs including the deceased siblings is 0.24 according to correction for complete truncate ascertainment by a priori method. Genealogical data also are in agreement with autosomal recessive inheritance. 21 families originate from the same restricted area in northern Finland and the parents are known to be related in at least five families.

At present there are no tests for heterozygote detection in Salla disease. Prenatal diagnosis has recently been attempted with the lectin staining method in cultured amniotic fluid cells in a family at risk. No fluorescent cytoplasmic granules were seen in cultured amniotic fluid cells in LPA-staining. Electron microscopy of cultured amniotic fluid cells also gave normal results without any evidence of storage lysosomes. The family made the option of continuing the pregnancy. A definite prenatal diagnosis of Salla disease has to wait for the elucidation of the basic biochemical defect in Salla disease.

REFERENCES

1. Aula P., Autio S., Raivio K.O., Rapola J., Thoden C.-J., Koskela S.-L., Yamashina I.: *"Salla disease" a new lysosomal storage disorder.* Arch. Neurol. *36*, 88-94, 1979.
2. Renlund M., Chester M.A., Lundblad A., Aula P., Raivio K.O., Autio S., Koskela S.-L.: *Increased urinary excretion of free N-Acetylneuraminic acid in thirteen patients with Salla disease.* Eur. J. Biochem. *101*, 245-250, 1979.
3. Humbel R., Collart M.: *Oligosaccharides in urine of patients with glycoprotein storage diseases. 1. Rapid detection by thin-layer chromatography.* Clin. Chim. Acta *60*, 143-145, 1975.
4. Virtanen I., Ekblom P., Laurila P.: *Subcellular compartmentalization of saccharide moieties in cultured normal and malignant cells.* J. Cell. Biol. (in press).
5. Virtanen I., Ekblom P., Laurila P., Nordling S., Raivio K.O., Aula P.: *Characterization of storage material in cultured fibroblast by specific lectin binding in lysosomal storage diseases.* Pediatr. Res. *14*, 1199-1203, 1980.

8

I-cell disease and pseudo-Hurler polydystrophy

Paolo Durand
Aldo Pelizza
Carla Borrone
Rosanna Gatti

NAME OF DISEASES	I-cell disease and pseudo-Hurler polydystrophy
PREVALENCE	?
MODE OF INHERITANCE	Autosomal recessive
MUTANT GENE PRODUCT	UDP-N-acetylglucosamine: glycoprotein-N-acetylglucosaminylphosphotransferase
CHROMOSOMAL LOCATION	?
EXPRESSION	
1. Altered DNA structure	?
2. Disturbed protein function	Enzyme activity defect
3. Disrupted cell and organ function	Accumulation of mucopolysaccharides, glycolipids, free and bound sialic acid and cystine in mesenchymal tissues, leads to somatic changes and brain dysfunction.

8.1 BACKGROUND

I-cell disease and pseudo-Hurler polydystrophy are considered here because they have been useful models for studies on the recognition and the uptake of the enzymes into lysosomes and on glycoprotein processing, secretion and packaging.

Both diseases result from a genetic defect which affects activity of a specific phosphotransferase which acts during posttranslational processing of glycoproteins. The enzyme phosphorylates a mannose residue in the oligosaccharide chain of acid hydrolases. This phosphomannosyl residue acts as a recognition marker for receptor-mediated uptake of the enzyme into lysosomes of mesenchymal cells. Failure to phosphorylate this mannosyl residue in I-cell disease and pseudo-Hurler polydystrophy results in excessive loss of lysosomal enzymes into the extracellular fluid. These two disorders will be reviewed together. They have many clinical and biochemical features in common which suggests a defect in the same pathway. No attempt is made to review the literature in a comprehensive manner and only that information which we consider to be essential will be presented here.

8.2 I-CELL DISEASE

I-cell disease was recognized as a separate disease by Leroy and De Mars in 1967 [1]. They found, using a phase contrast microscope, unusual cytoplasmic inclusions in the cultured fibroblasts of a girl who had been considered to have Hurler's syndrome, despite the fact that she did not have cloudy corneas or excessive acid mucopolysaccharides in the urine. As a result of this observation, Leroy et al. proposed the term "Inclusion cell" or "I-cell" disease for this new storage disorder [2]. In 1970, Spranger and Wiedemann classified this disorder as mucolipidosis II [3].

Shortly after, the finding that the multiple lysosomal enzyme deficiency in fibroblasts from I-cell patients is accompanied by excessive excretion of the same enzymes in the culture media as well as in extracellular fluids [4, 5] provided a test for diagnosis and generated considerable fundamental research on lysosomal enzyme turnover, uptake by cells and packaging within lysosomes. The precise nature of the genetic defect has recently been uncovered as a consequence of innovative studies in the laboratories of Elizabeth Neufeld, William Sly and Stuart Kornfeld.

The clinical features of I-cell disease are reminiscent of those in the Hurler syndrome. Symptoms are present from birth and the course is rapidly progressive with a fatal outcome usually occurring before 4 years.

Table 8.1 I-cell disease: clinical symptoms.

- — clinical symptoms expressed before or from birth; fatal outcome usually before 4 years
- — mental and motor retardation
- — severe growth failure
- — coarse facial features
- — protruding eyes, shallow orbits, puffy eyelids, rarely clouded cornea
- — small nose, flat bridge, anteverted nostrils, excessive discharge
- — prominent mouth, gingival hypertrophy, enlarged tongue, hoarse voice
- — thick ears, frequent otitis
- — short neck
- — sistolic murmur, ECG abnormalities
- — thoracic deformities, recurrent respiratory infections
- — dorsal lumbar gibbus
- — thick skin, hirsutism
- — protuberant abdomen, umbilical and inguinal hernias
- — progressive stiffness of all joints

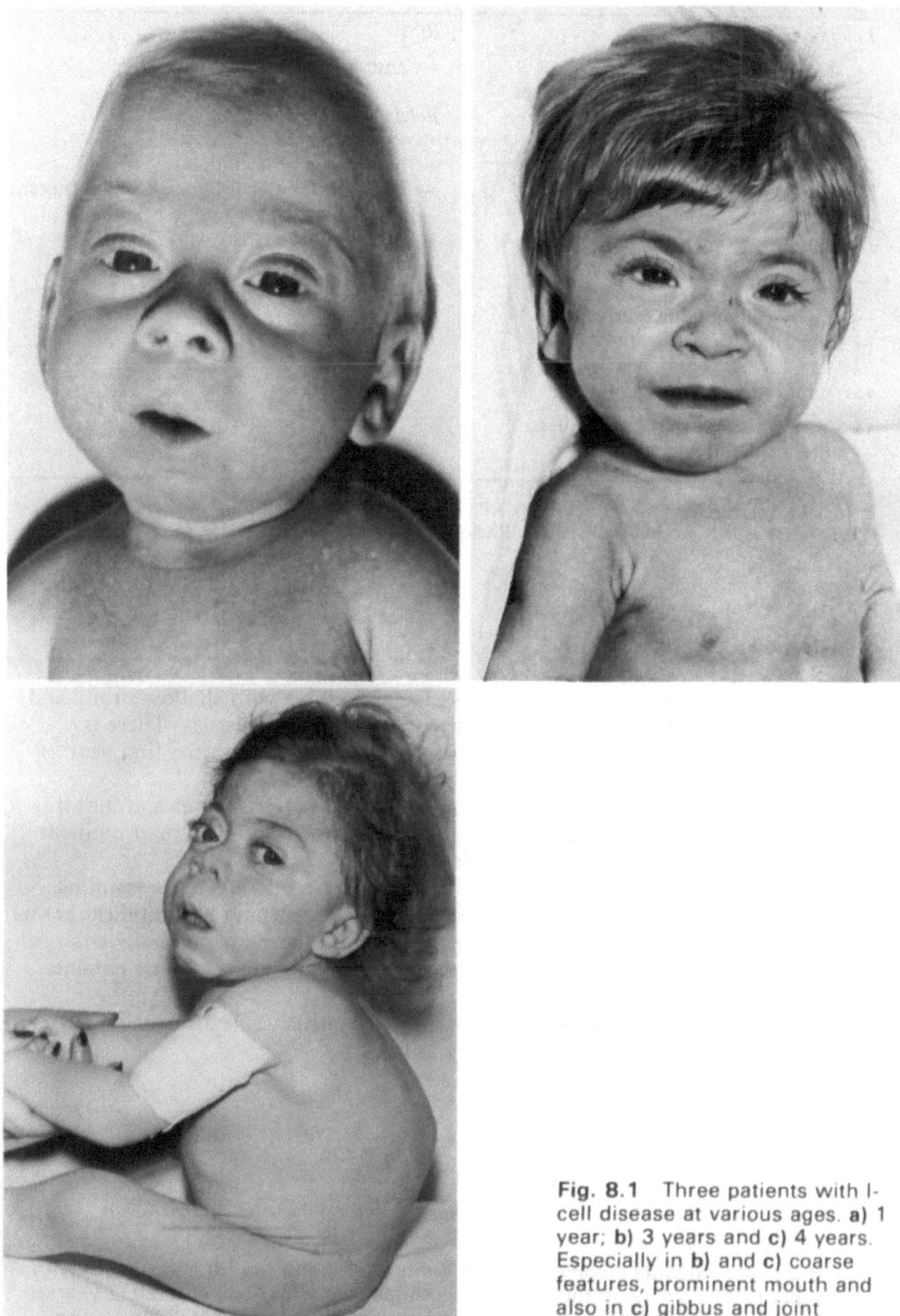

a b c

Fig. 8.1 Three patients with I-cell disease at various ages. **a)** 1 year; **b)** 3 years and **c)** 4 years. Especially in **b)** and **c)** coarse features, prominent mouth and also in **c)** gibbus and joint stiffness.

Table 8.2 Radiologic signs of I-cell-disease*.

Long bones	*Ribs*
— shortening	— anterior arches enlarged
— wide and undermodelled diaphyses	
— epiphyseal hypoplasia	*Pelvis*
— periostal apposition	
— deformations	— hypoplasia of acetabula
— fracture (early stage)	— supra-acetabular iliac constriction (notch)
	— flared iliac wings
Hands/feet	— pelvic-pubic dysostosis
	— coxa valga
— shortening	— hip dislocation
— epiphyseal stippling (early stage)	
— retardation of ossification	
Vertebral column	
— short, biconvex vertebral bodies	
— anterior, inferior beaking	

* See also: Lemaitre L., Remy J., Farriaux J.P., Dhondt J.L., Walbaum R.: *Radiologic Signs of Mucolipidosis II or I-cell disease*. Pediat. Radiol. 7, 97-105, 1979.

The main clinical symptoms are coarse facial features with shallow orbits and prominent mouth due to the early hypertrophy of gingival tissues. There is marked growth deficiency and joint stiffness already evident in the first year of life (Table 1 and Figs. 1 a, b, c).

The majority of the patients have progressive and severe mental and motor deterioration. Generalized and progressive "dysostosis multiplex" is prominent and evident from the earliest age (Table 2 and Figs. 2, 3, 4, 5).

Additional features include thick and tight skin, systolic murmurs resulting from valvular heart disease. The abdomen is protuberant and an umbilical hernia may be present but the liver and spleen are normal in size or minimally enlarged. Head size is usually normal. Corneal clouding is usually absent; some patients have had glaucoma.

Upper respiratory infections and otitis recur frequently. Death usually occurs in early childhood due to cardiorespiratory failure.

Fig. 8.2 Marked wide diaphyses of the femura **(a)**, humerus **(b)** and ribs **(c)** with dense periostal cloaking in the femora (a) of a patient with I-cell disease (2 years). Hypoplasia of the capital epiphyses. The iliac wings flare immediately above the acetabula. The acetabular roofs blend into the iliac bone. Osteoporosis.

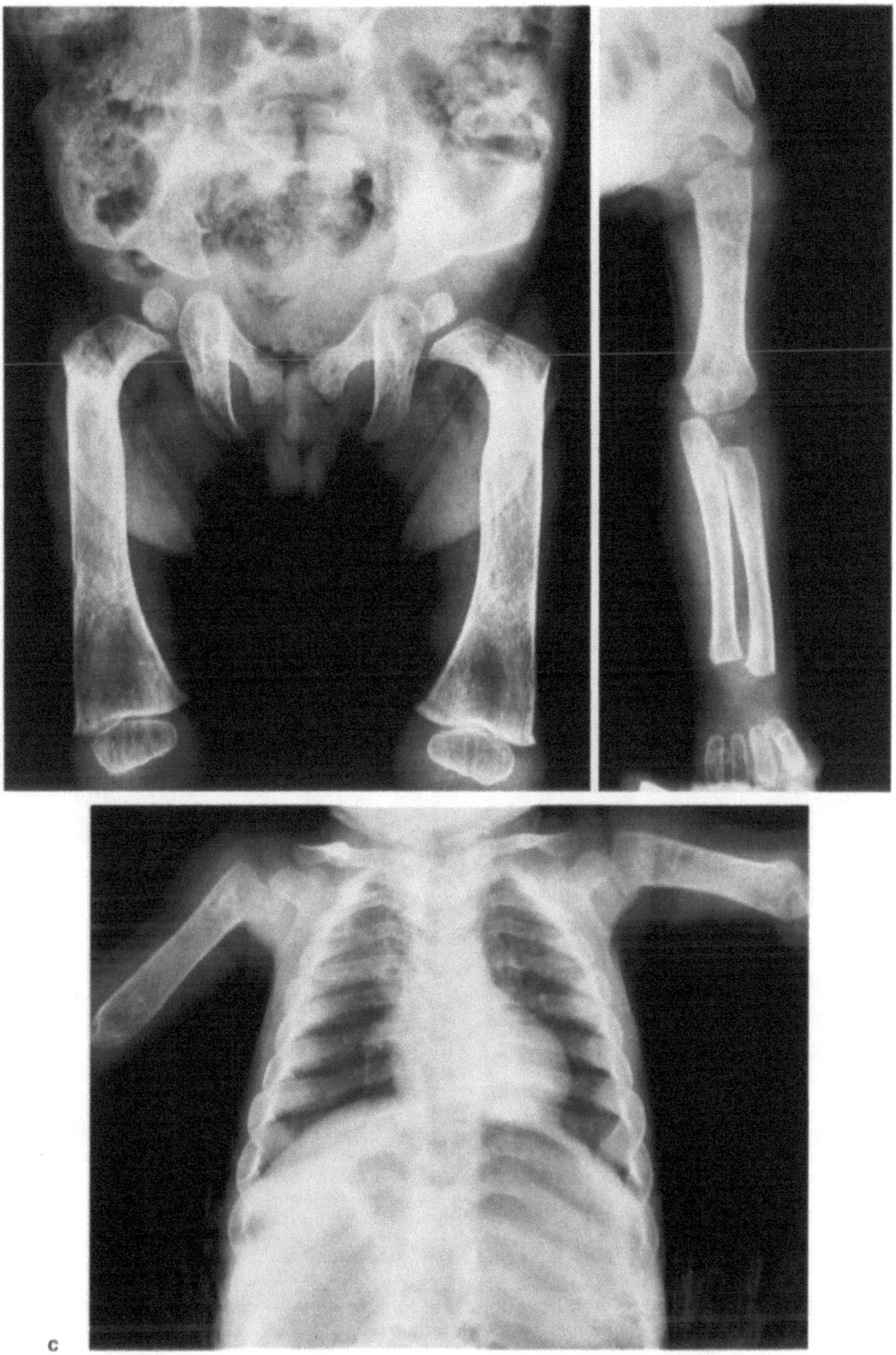

a b c

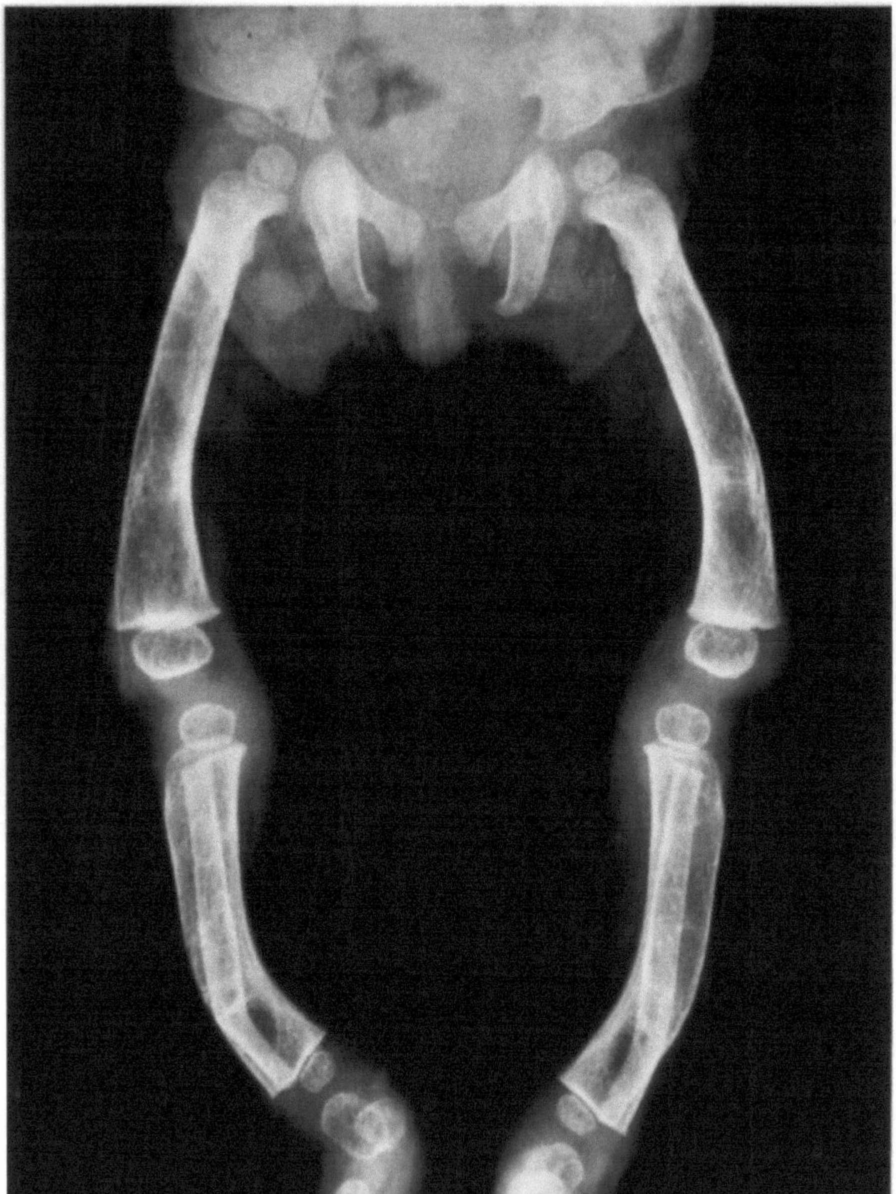

Fig. 8.3 Poor modelling and bowing of the femora as well as tibiae and fibulae in another case of I-cell disease (1 year). Osteoporosis.

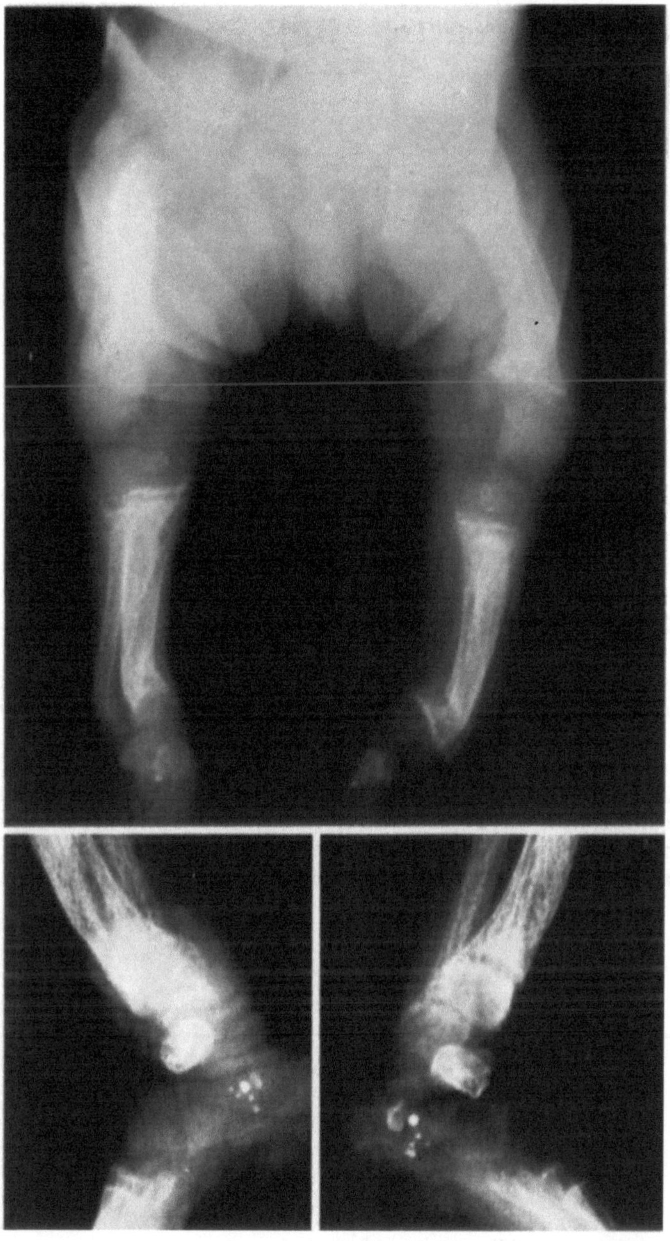

Fig. 8.4 Abnormalities of bone modelling, absence of diaphyseal constriction and tubulation, enlarged metaphyses and lateral beaks, fractures of the tibiae and stippling of the calcaneum in a patient with I-cell disease (2 months). Fibrillar trabeculation with no differentiation between the cortical and medullary zones.

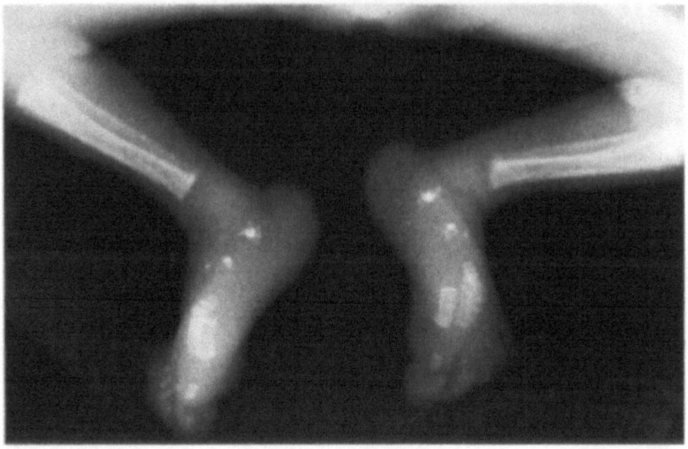

Fig. 8.5 Stippling of calcaneum in a fetus of 22 weeks with I-cell disease.

8.3 PSEUDO HURLER POLYDYSTROPHY

The first patients, described by Maroteaux and Lamy in 1966 [6], were four subjects with some features of the Hurler syndrome but a much slower clinical evolution and no excessive mucopolysacchariduria. As suggested by Spranger and Wiedeman this disorder was classified as Mucolipidosis III [3]. In 1973 Thomas et al. [7] found that the I-cell phenomenon, with both its morphologic and biochemical features, was apparent in fibroblasts from patients with pseudo-Hurler polydystrophy. An extensive report on the clinical phenotype of this disease has been provided by Kelly et al. [8].

Table 8.3 Pseudo-Hurler polydystrophy: clinical symptoms.

— clinical symptoms expressed at 2-4 years; slow progressive course
— growth failure
— short neck and trunk
— joint stiffness (shoulders, elbows, hips, hands)
— slow mental development
— mildly coarsened features
— fine corneal clouding
— cardiac murmur (aortic and/or mitral valve disease)
— carpal tunnel syndrome
— dysostosis multiplex

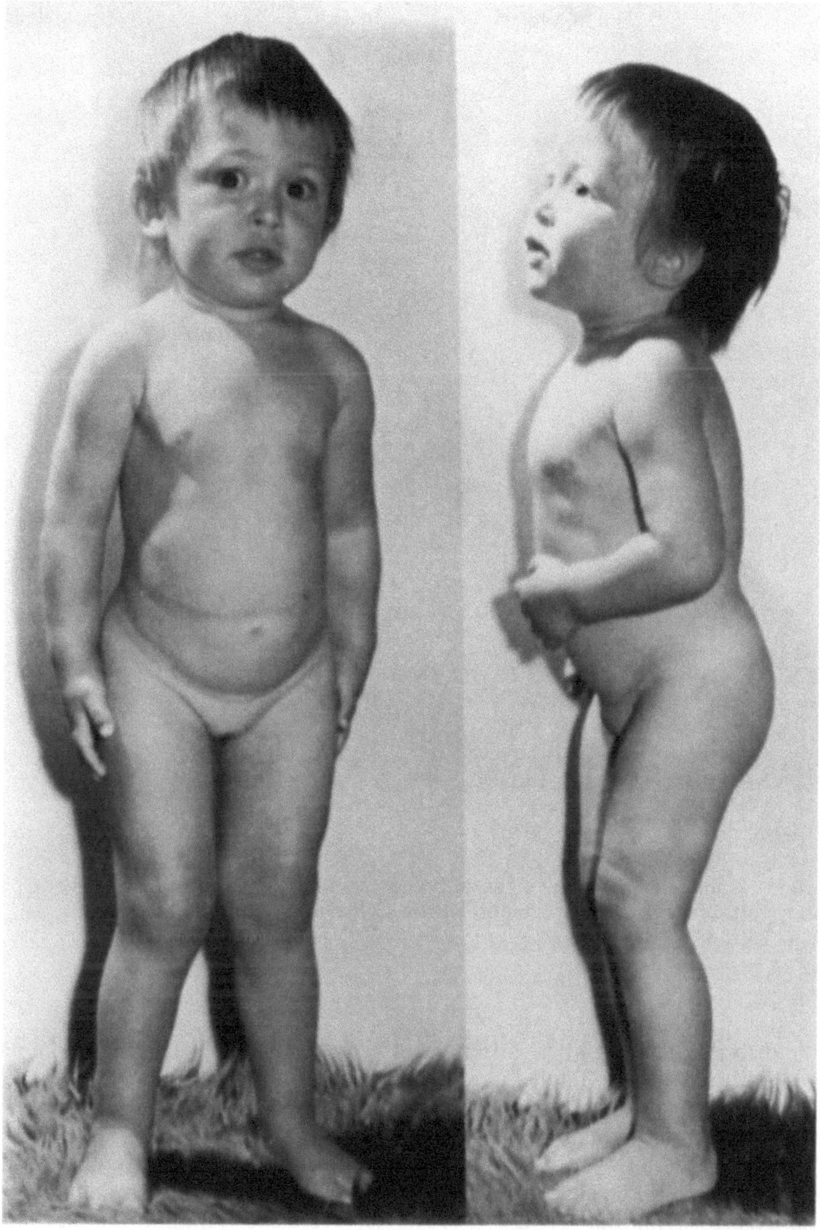

Fig. 8.6 Patient with pseudo-Hurler polydystrophy at 3 years of age. She shows plump facies, short neck, short trunk; stiff joints are most apparent in the hand. Today, at 9 years, growth is severely retarded and she shows stiff joints, claw hands, the murmur of aortic regurgitation and extensive radiographic skeletal changes. She performed at a "below age" level in an ungraded school.

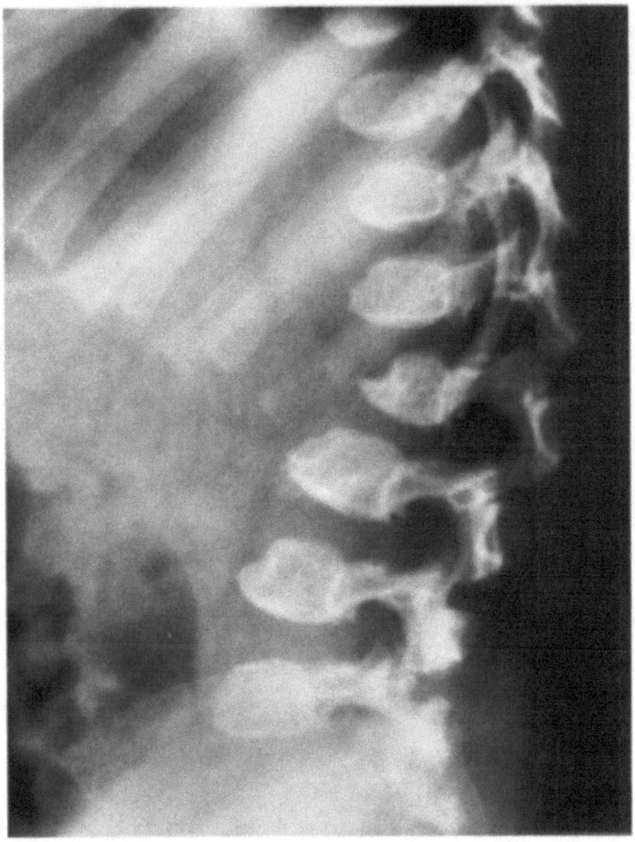

Fig. 8.7 Short and beaking vertebral bodies, especially evident in the second lumbar vertebra in a case of pseudo-Hurler polydystrophy (4 years).

The disorder has its clinical onset at 1-4 years of age with the initial complaint of joint stiffness (Table 3 and Fig. 6). By 5 to 8 years linear growth becomes retarded, the facial features are mildly coarsened, the neck is short and dorsolumbar kyphosis may be present. Physical disability is most often evident in the hands, hips, elbows and shoulders resulting in progressive restriction of mobility. At this age skeletal deformities become manifest and roentgenograms reveal dysostosis multiplex (Fig. 7). However severe skeletal abnormalities have been reported before one year of life. Carpal tunnel syndrome often occurs. Slit-lamp examination frequently reveals fine peripheral corneal opacities which increase with time. Murmur of aortic and/or mitral valve disease have been heard. Many patients have mild mental retardation but some have a near normal intelligence even late in adult life.

The course of the disorder is slowly progressive and the oldest patients are

Table 8.4 Radiologic signs of pseudo-Hurler polydystrophy*.

Long bones

— shortening
— undermodelling diaphyses
— irregular epiphyses
— oblique distal radius/ulna
— radial bowing

Pelvis

— narrow iliac body
— acetabular hypoplasia
— coxa valga

Spine

— anterior inferior beaking
— oval vertebral bodies
— short odontoid

Thorax

— short, thick clavicles
— wide, oar-shaped ribs

Hands-Feet

— shortening
— claw hand
— small, irregular carpal-tarsal bones

Skull

— premature suture closure

* See also:
Melhem R., Dorst J.P., Scott C.I., McKusick V.A.: *Roentgen findings in mucolipidosis III (Pseudo-Hurler polydystrophy).* Radiology *106*, 153-160, 1973.
Nolte K., Spranger J.: *Early skeletal changes in mucolipidosis III.* Ann. Radiol. *19*, 151-159, 1976.

more than 32 years old. It is frequently found that male patients are more severely affected than females.

Average life expectancy is probably reduced but data on longevity is insufficient to be sure.

A clinical dividing line between I-cell disease and pseudo-Hurler polydystrophy may be difficult because of the existence of some patients with an intermediate clinical appearance and evolution.

8.4 LABORATORY FINDINGS

No excessive mucopolysaccharides are excreted in urine. Sialyloligosaccharides are found in the urine of patients with I-cell disease and pseudo-Hurler polydystrophy, with the largest amounts being found in I-cell disease.

Both diseases are characterized by a deficiency of lysosomal hydrolase activities within cultured skin fibroblasts, and greatly elevated levels of the same enzymes in body fluids. Except for β-galactosidase activity, which is often low, activities of other lysosomal enzymes are usually normal in liver [4, 5, 8-10]. Examination of cultured fibroblasts obtained from patients with I-cell disease and pseudo-Hurler polydystrophy reveal severe deficiencies of β-D-galactosidase, α-L-fucosidase and sialidase activities as compared to normal and partially reduced activities of

N-acetyl-β-D-hexosaminidase, β-D-glucuronidase, α-L-iduronidase, arylsulfatase A, α-D-galactosidase, α-D-mannosidase, cholesterol esterase, sphingomyelinase, α-L-arabinosidase and galactosylceramide-β-D-galactosidase; activities of acid phosphatase, α-D-glucosidase and β-D-glucosidase are usually normal.

Both types of sialidase activity (sialoglycoprotein sialidase and sialoganglioside sialidase) are reduced within fibroblasts.

In addition lysosomal hydrolases from I-cell disease fibroblasts have altered electrophoretic mobilities.

A nearly complete lack of sialoglycoprotein sialidase activity is found in leukocytes from patients with I-cell disease and a partial deficiency of the same enzyme is found in leukocytes from patients with pseudo-Hurler polydystrophy; activities of the other hydrolases in leukocytes are normal.

In the media from cultured fibroblasts and in extracellular fluids (urine, plasma, tears, cerebrospinal fluid) the activity of a number of lysosomal enzymes including β-D-glucuronidase, β-D-galactosidase, α-D-galactosidase, α-D-mannosidase, α-L-iduronidase, N-acetyl-β-D-hexosaminidase, α-L-fucosidase, arylsulfatase A and sialidase is considerably increased (2-4 times normal).

The decrease in lysosomal enzyme activity within cultured fibroblasts and the elevated levels of the same enzymes in surrounding medium and in body fluids is evident in all cases of I-cell disease. The somewhat higher residual activity of lysosomal enzyme activity in cultured fibroblasts and the somewhat lower activity in culture media and in body fluids found in pseudo-Hurler polydystrophy may be correlated with the milder phenotype of this disorder.

The simplest test to confirm the diagnosis of these diseases is the demonstration of an elevated serum level of one or more lysosomal enzymes (usually hexosaminidase or arylsulfatase A).

It is preferable not to use α-L-fucosidase as the tested enzyme since two of our patients with proven pseudo-Hurler polydystrophy were "low activity" variants of α-L-fucosidase (see Chapter 3); both had normal serum activity, obscuring the diagnosis.

8.5 PATHOLOGY

The pathological manifestations of I-cell disease are confined to mesenchymal tissues such as skin and bone and to a few other types of cells, mainly Schwann cells, glomerular epithelial cells, vascular perithelial cells, and peripheral neurons [11-13].

Fibroblasts from patients with I-cell disease and from some patients with pseudo-Hurler polydystrophy are especially sensitive to freezing.

Cultured fibroblasts contain a large number of perinuclear cytoplasmic inclusions seen by phase contrast microscopy. These inclusions appear by electron microscopy to be enlarged lysosomes containing principally a fibrillo-granular reticulum and some lamellar images.

Conjunctival biopsy shows normal epithelium contrasting with very severely affected fibroblasts, together with dense inclusions of the blood capillary walls.

Examination of bone reveals alterations of enchondral ossification and involvement of proliferating cartilage.

The cytoplasm of chondrocytes is filled with numerous inclusions containing a fine fibrillo-granular reticulum and juxtaposed some lamellar rings.

Hepatocytes are filled with pleomorphic inclusions. The most frequent types contain osmiophilic material organized in spherules or as lamellae.

Others contain large neutral lipid droplets not bounded by a unit membrane or inclusions consisting of rectangular crystals within filamentous material limited by a unit membrane.

The involvement of Kupffer's cells is variable.

In the fibrocytes of the cardiac valves and in sarcolemma of the myocardial fibers pleomorphic cytoplasmic inclusions are present. In the kidney the glomerular epithelial cells contain clear cytoplasmic inclusions.

No important abnormalities are found at the light microscopic level in the brain or in peripheral nerves.

The electron microscopic examination of these tissues reveals the presence, in neurons and astrocytes, of membrane-bound vacuoles containing fibrillo-granular reticulum, few lamellar structures and dense osmiophilic inclusions [12, 13]. The Schwann cells of the sensory nerves also present several similar vacuoles.

No autopsy studies of pseudo-Hurler polydystrophy have been reported in the literature. However the ultrastructural examination of tissue biopsies reveals changes similar to those in I-cell disease.

8.6 STORAGE SUBSTANCES

The deficiency of multiple acid hydrolase in fibroblasts appears to account for the accumulation of sulfate-containing acid mucopolysaccharides (dermatan sulfate 32%, herparan sulfate 32% and chondroitin sulfate 16%), glycolipids and free and bound sialic acid [14, 16]. Fibroblasts from I-cell disease patients also accumulate lysosomal cystine, in a manner similar to that seen in cystinosis [15].

No significant storage of glycolipids, glycosaminoglycans or glycopeptides is found in brain and visceral organs. Only ganglioside GT1b is slightly increased.

The amount of protein-bound sialic acid is normal in I-cell disease brain and a general decrease of the total sialic acid content in the brain is mainly due to a decrease in the amount of ganglioside [17, 18].

Therefore neither the morphological changes nor the storage products are severe enough to explain the severe mental retardation that occurs in I-cell disease. Further investigation of the biochemical basis for the mental retardation needs to be conducted.

8.7 INCIDENCE AND GENETICS

I-cell disease and pseudo-Hurler polydystrophy have a widespread ethnic occurence and there does not appear to be any ethnic predominance. Evidence that these diseases are transmitted as autosomal recessive traits comes from pedigree analyses in which there is approximately equal involvement of boys and girls, increased parental consanguinity, absence of symptoms in parents and, most important, a partial expression of the biochemical I-cell phenomenon in obligate heterozygotes.

Intermediate phenotypes between I-cell disease and pseudo-Hurler polydystrophy have been reported in the literature [19], indicating genetic heterogeneity or the existence of genetic compounds.

The observation that the same acid hydrolases from various I-cell patients have differing electrophoretic mobilities [20] and normal to altered lectin profiles also suggests the existence of biochemical heterogeneity [21, 51].

Somatic cell hybridization studies performed with fibroblasts of various I-cell strains derived from unrelated patients or from I-cell disease and pseudo-Hurler polydystrophy patients show the existence of at least two complementation groups. Those patients with I-cell disease and pseudo-Hurler polydystrophy which do not complement are probably allelic variants. The fact that some do complement suggests that at least two different biochemical defects are involved [21, 51]. Perhaps different reaction steps in the sequence of post-translational modification of the carbohydrate side-chain of lysosomal enzymes are involved, or the enzyme in question is a heteropolymer comprised of genetically distinct subunits.

8.8 CARRIER DETECTION

In obligate heterozygotes for I-cell disease and pseudo-Hurler polydystrophy the average activity of lysosomal enzymes is not marked increased in their serum and it is not significantly reduced in their cultured fibroblasts; detection of carriers is not possible by these tests.

Several laboratories have recently reported the usefulness of serum β-hexosaminidase assays to identify carriers for both diseases [22, 23].

This involves quantitation of the four β-hexosaminidase isoenzymes in plasma and serum separated by chromatography on DEAE cellulose. In plasma and in serum of obligate heterozygotes and presumptive heterozygotes there is an intermediate increase of hexosaminidases I_1 and I_2 as a percentage of total β-hexosaminidase.

However, in almost 10% of heterozygotes, values overlap with controls rendering certain identification of all heterozygotes impossible [23].

8.9 PRENATAL DIAGNOSIS

Prenatal diagnosis of I-cell disease and pseudo-Hurler polydystrophy is possible by the demonstration of increased activity of lysosomal hydrolases in amniotic fluid supernatants, and of multiple lysosomal enzyme deficiencies in cultured amniotic cells.

An additional test, which does not always need to be performed, is the demonstration of intracellular accumulation of glycosaminoglycans using $^{35}SO_4$ incorporation.

About 60 at-risk pregnancies for I-cell disease have been prenatally monitored following the first reports of Huijing et al. [24] and Warren et al. [25]. The prenatal diagnosis of pseudo-Hurler polydystrophy has been performed in only a few cases.

The fetal pathology of I-cell disease has been investigated in some pregnancies that were interrupted because of an affected fetus [26; 27]. Membrane-bound vacuoles with clear contents or with fibrillo-granular material were observed in fibroblasts and in a few other cells (macrophages, endothelial cells, podocytes) of the organs investigated. Roentgenograms revealed stippling of the calcaneum in one of our affected fetuses (Fig. 5). Examination of the placenta also revealed the storage process and this organ can be used for pathological studies [27].

8.10 THE NATURE OF THE GENETIC DEFECT

A large number of investigations preceded the discovery of the primary defect. Hickman and Neufeld [28], Neufeld et al. [29] in a series of elegant experiments, reported that cultured skin fibroblasts from I-cell disease patients secrete acid hydrolases into the medium which cannot be taken up by fibroblasts.

Thus, neither I-cell nor normal fibroblasts take up I-cell derived lysosomal enzymes. For this reason, I-cell secretion enzymes which are specific for the mucopolysaccharidoses, fail to function as corrective factors in mucopolysaccharidoses. However, I-cell fibroblasts maintain the ability to pinocytose lysosomal enzymes from normal cells.

These findings led Hickmann and Neufeld [28] to postulate that there is a recognition marker common to a number of lysosomal enzymes, which is defective in I-cell and pseudo-Hurler polydystrophy fibroblasts.

Periodate oxidation of several lysosomal enzymes causes a reduction in uptake of the same enzymes suggesting that specific pinocytosis of lysosomal enzymes by fibroblast is dependent on a carbohydrate residue.

This residue acts as a recognition marker common to a number of enzymes, which appears to be lacking from enzymes elaborated by fibroblasts from patients with I-cell disease [30].

Using mannans and phosphates as competitive inhibitors of uptake, and alkaline phosphatase to destroy uptake, Kaplan et al. [31] and Natowicz et al. [32] deduced that a phosphomannosyl residue is the recognition marker for the uptake of β-glucuronidase.

These findings were extended to the uptake of α-L-iduronidase [33], β-hexosaminidase, and several other enzymes [34, 35] and led to the conclusion that mannose-6-phosphate is the recognition marker on lysosomal hydrolases. Its absence in lysosomal enzymes from I-cells was reported by Hasilik et al. [36] and Hasilik and Neufeld [37], who found that I-cell fibroblasts (in contrast to normal) fail to incorporate phosphate into newly synthesized β-hexosaminidase, cathepsin D and α-glucosidase.

Considerable information has been gleaned concerning the abnormalities of I-cell acid hydrolases from other research.

It was found, for instance, that several of the hydrolases show aberrant concanavalin A binding properties [38].

Also the I-cell secreted N-acetyl-β-D-hexosaminidase is quantitatively retained on Ricinus communis-Sepharose columns while only 10-30% of N-acetyl-β-D-hexosaminidase secreted by normal fibroblasts is retained on this galactose-recognizing lectin [39].

Moreover it was found that several of the I-cell hydrolases are electrophoretic abnormal. The intracellular enzymes are more electropositive and the excreted enzymes are more electronegative than normal. After neuraminidase treatment the I-cell excreted enzymes resemble the I-cell intracellular pattern [40, 41].

The results of this research suggest that I-cell secretion enzymes contain excess sialic acid and have abnormal carbohydrate composition found predominantly in complex-type oligosaccharide chains characteristic of secretory glycoproteins [42].

As discussed above newly synthesized acid hydrolases, destined for transport to lysosomes, acquire a phosphomannosyl targeting signal in two steps. The first step involves the transfer of N-acetylglucosamine-1-phosphate from uridin 5′-diphosphate (UDP)-N-acetylglucosamine to a mannose residue of the acid hydrolase. The second step is removal of the outer phosphodiester-linked

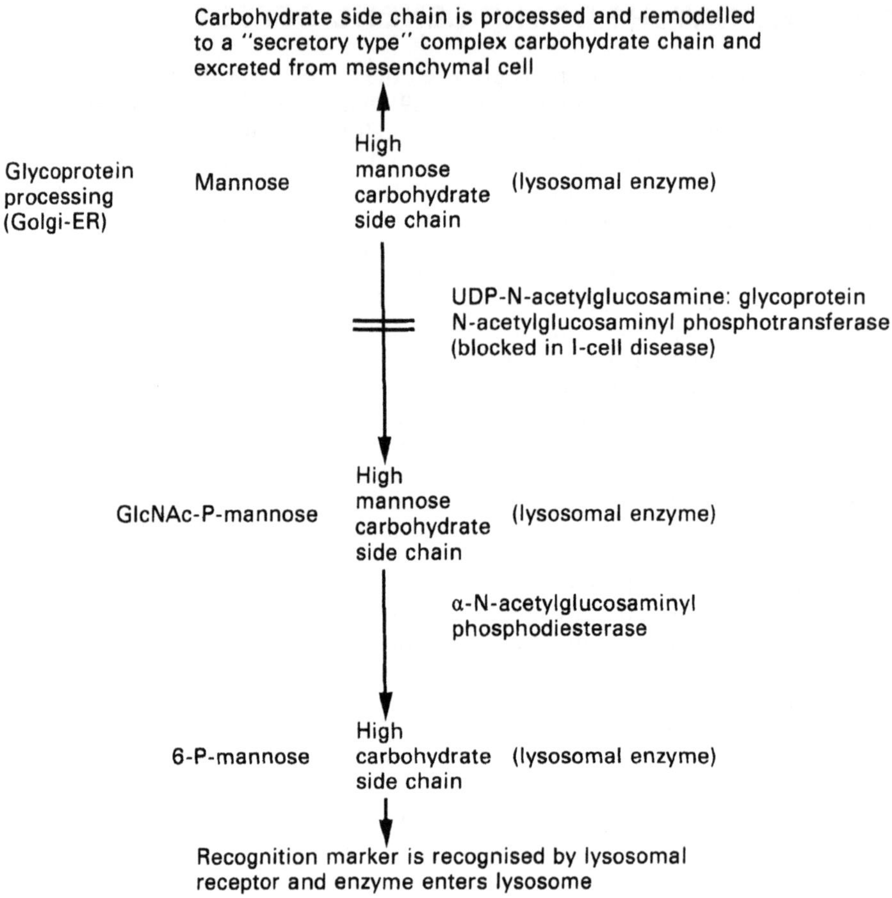

Fig. 8.8 Schematic representation of putative defect in lysosomal enzyme processing in I-cell disease and pseudo-Hurler polydystrophy.
GlcNAc: N-acetylglucosamine; ER: endoplasmic reticulum.

N-acetylglucosamine to expose the mannose-6-phosphate recognition marker [43, 44].

Reitman and Kornfeld [45], Varki and Kornfeld [46] have recently developed assays for UDP-N-acetylglucosamine: glycoprotein N-acetylglucosaminyl phosphotransferase and for α-N-acetylglucosaminyl phosphodiesterase, the enzymes catalyzing the two reactions in the aforementioned sequence.

Reitman et al. [47] have also demonstrated that fibroblasts from five patients with I-cell disease and five with pseudo-Hurler polydystrophy are severely deficient in N-acetylglucosaminylphosphotransferase activity, thus identifying the enzyme defect in these diseases (Fig. 8).

The decrease in the phosphotransferase activity explains the lack of

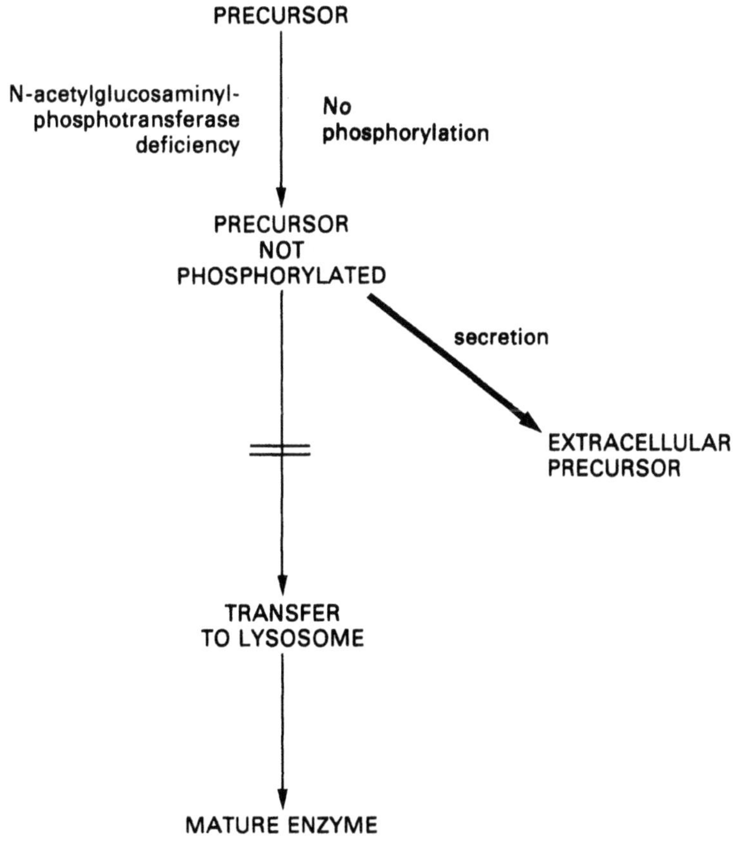

Fig. 8.9 Schematic representation of the consequences of enzyme defect in I-cell disease and pseudo-Hurler polydystrophy. The hydrolitic enzymes produced by I-cells are not phosphorylated. The absence of the recognition marker prevents the receptor-mediated transfer of the hydrolytic enzymes from endoplasmic reticulum to lysosomes and causes the immature enzymes to be treated as a secretory proteins (see for comparison figure 1.3).

phosphorylation of acid hydrolases synthesized by I-cell and pseudo-Hurler polydystrophy fibroblasts.

Failure to add the recognition marker leads to failure to transfer the enzymes to lysosomes and they are secreted instead. This explains the abnormally high levels of enzyme secretion (Fig. 9). The abnormal biochemical composition of these enzymes is the result of further processing of their high mannose-type oligosaccharide chains to complex-type chains by terminal glycosyltransferases in the Golgi apparatus.

As yet no patients have been found to be deficient in activity of the α-N-acetylglucosaminyl phosphodiesterase; in theory this could give rise to a similar defect since the mannose-6-phosphate is blocked (see Fig. 8). Furthermore, no explanation is available as yet for the normal activity of some enzymes (α-glucosidase, β-glucosidase and acid phosphatase) in I-cell fibroblasts: whether these three enzymes have different recognition markers for uptake is unclear. At least one of them, α-glucosidase, is not properly phosphorylated in I-cell fibroblasts indicating it may also carry the same recognition marker as those enzymes which are excreted in excess [37].

A recent study on transfer of lysosomal β-hexosaminidase by cultured hepatocytes showed that neither the secreted nor intracellular hexosaminidase from these cells were taken up by fibroblasts [48].

This study questioned the importance of the phosphomannosyl-recognition marker in transport of lysosomal enzymes in liver. On the other hand Fischer et al. [49] reported the finding that the phosphomannosyl enzyme receptor exists on intracellular membranes of every organ examined including liver. However, hepatocytes were also reported to express receptors for galactose-terminal [50] and mannose-terminal glycoproteins on their surface which may explain the situation in I-cell disease where low intracellular enzyme levels are found in fibroblasts, high levels in serum and above normal levels in liver.

Presumably these alternate receptors provide an alternative pathway for pinocytosis of circulating I-cell enzymes which lack the phosphomannosyl recognition marker.

Genetic heterogeneity within I-cell disease is demonstrated by complementation of lysosomal enzyme processing mutants [21].

Examination by Miller and Kornfeld's group of three of our patients with pseudo-Hurler polydystrophy indicates that these patients do not have a deficiency of phosphotransferase activity.

These patients, one from an Italian and the other two from a Spanish family, may have a defect in another step in the lysosomal processing pathway [51].

It is apparent that considerable further research has to be carried out before answers are available for the important questions raised above.

REFERENCES

1. Leroy J.G., De Mars R.I.: *Mutant enzymatic and cytological phenotypes in cultured human fibroblasts.* Science *157*, 804-806, 1967.
2. Leroy J.G., Spranger J.W., Feingold M., Opitz J. M., Crocker A.C.: *I-cell disease: a clinical picture.* J. Pediat. *79*, 360-365, 1971.
3. Spranger J.W., Wiedemann H.R.: *The genetic mucolipidosis. Diagnosis and differential diagnosis.* Humangenetik *9*, 113-139, 1970.
4. Wiesmann U.N., Lightbody J., Vassella F., Herschkowitz N.N.: *Multiple lysosomal enzyme deficiency due to enzyme leakage?* New Engl. J. Med. *284*, 109-110, 1971.
5. Wiesmann U.N., Vassella F., Herschkowitz N.N.: *I-cell disease: leakage of lysosomal enzymes into extracellular fluids.* New Engl. J. Med. *285*, 1090-1091, 1971.
6. Maroteaux P., Lamy M.: *La pseudo-polydystrophie de Hurler.* Presse Méd. *74*, 2889-2892, 1966.
7. Thomas G.H., Taylor H.A., Reynolds L.W., Miller C.S.: *Mucolipidosis III (pseudo-Hurler polydystrophy). Multiple lysosomal enzyme abnormalities in serum and cultured fibroblast cells.* Pediat. Res. *7*, 751-756, 1973.
8. Kelly T.E., Thomas G.H., Taylor H.A., Mc Kusick V.A., Sly W.S., Glaser J.H., Robinow M., Luzzati L., Espiritu C., Feingold M., Bull M.J., Ashenhurst E.M., Ives E.J.: *Mucolipidosis III (pseudo-Hurler polydystrophy): clinical and laboratory studies in a series of 12 patients.* Johns Hopkins Med. J. *137*, 156-175, 1975.
9. Leroy J.G., Martin J.J.: *Mucolipidosis II (I-cell disease): present status of knowledge.* Birth Defects. Orig. Art. Series *11*, 283-293, 1975.

10. Gatti R., Borrone C., Torreblanca J., Cavalieri S., De Martini I., Filocamo M., Antelo M.C.: *Caracteristicas biologicas de las mucolipidosis II y III*. An. Esp. Pediat. *12*, 563-574, 1979.
11. Tondeur M., Vamos-Hurwitz E., Mockel-Pohl S., Dereume J.P., Cremer N., Loeb H.: *Clinical, biochemical and ultrastructural studies in a case of chondrodystrophy presenting the I-cell phenomenon in culture*. J. Pediat. *79*, 366-378, 1971.
12. Martin J.J., Leroy J.G., Farriaux J.P., Fontaine G., Desnick R.J., Cabelo A.: *I-cell disease (mucolipidosis II). A report on its pathology*. Acta Neuropath. *33*, 285-305, 1975.
13. Nagashima K., Sakakibara K., Endo H., Konishi Y., Nakamura H., Suzuki Y., Abe T.: *I-cell disease (mucolipidosis II) pathological and biochemical studies of an autopsy case*. Acta Path. Jap. *27*, 255-264, 1977.
14. Schmickel R.D., Distler J., Jourdian G.W.: *Accumulation of sulfate-containing acid mucopolysaccharides in I-cell fibroblasts*. J. Lab. Clin. Med. *86*, 672-682, 1975.
15. Tietze F., Debrohun Butler J.: *Elevated cystine levels in cultured skin fibroblasts from patients with I-cell disease*. Pediatr. Res. *13*, 1350-1355, 1979.
16. Leroy J.G., Ho M.W., MacBrinn M.C., Zielke K., Jacob J., O'Brien J.S.: *I-cell disease: biochemical studies*. Pediatr. Res. *6*, 752-757, 1972.
17. Dacremont G., Kint J.A., Cocquyt G.: *Brain sphingolipids in I-cell disease (mucolipidosis II)*. J. Neurochem. *22*, 599-602, 1974.
18. Berra B., Di Palma S., Lindi C., Sandhoff K.: *Content of gangliosides and protein-bound sialic acid in post-mortem brain patients with mucolipidosis I and II (ML I and II)*. Cell. Molec. Biol. *25*, 281-284, 1979.
19. Erickson R.P., Sandman R., Langer L.O. Jr.: *A lysosomal storage disease with severe dwarfism, severe joint contractures, mild-moderate mental retardation, corneal opacities and retinal abnormalities*. In "Disorders of Connective Tissues". D. Bergsma (Ed.), Birth Defects. Orig. Art. Ser. XI, *6*, 317-324, National Foundation-March of Dimes, New York, 1975.
20. Champion M.J., Shows T.B.: *Electrophoretic abnormalities of lysosomal enzymes in mucolipidosis fibroblasts lines*. Am. J. Hum. Genet. *29*, 149-163, 1977.
21. Miller A.L., Freeze H.H., Kress B.C.: I Cell disease. In "Lysosomes and lysosomal storage diseases" J.W. Callahan and J. A. Lowden (Eds.), 271-287, Raven Press, 1981.
22. Van Elsen A.F., Leroy J.G., Vanneuville F.J., Vercruyssen A.L.: *Isoenzymes of serum N-acetyl-β-D-glucosaminidase in the I-cell disease heterozygote*. Hum. Genet. *31*, 75-81, 1976.
23. Jolly R.D., Desnick R.J.: *Inborn errors of lysosomal catabolism. Principles of heterozygotes detection*. Am. J. Med. Genet. *4*, 293-307, 1979.
24. Huijing F., Warren R.J., Mc Leod A.G.W.: *Elevated activity of lysosomal enzymes in amniotic fluid of a fetus with mucolipidosis II*. Clin. Chim. Acta *44*, 453-455, 1973.
25. Warren R.J., Condron C.J., Hollister D., Huijing F., Neufeld E.F., Hall C.W., Mc Leod A.G.W., Lorincz A.E.: *Antenatal diagnosis of mucolipidosis II (I-cell disease)*. Pediatr. Res. *7*, 343-349, 1973.
26. Aula P., Rapola J., Autio S., Raivio K., Karijalainen O.: *Prenatal diagnosis and fetal pathology of I-cell disease*. J. Pediat. *87*, 221-226, 1975.
27. Erickson R.P., Pflenger O.H., Sandman R., Hall B.O.: *Placental pathology in Mucolipidosis II*. In "Disorders of Connective Tissue" D. Bergsma (Ed.). Birth Defects. Orig. Art. Series XI, 365-370, National Foundation-March of Dimes, New York, 1975.
28. Hickman S., Neufeld E.F.: *A hypothesis for I-cell disease: Defective hydrolases that do not enter lysosomes*. Biochem. Biophys. Res. Commun. *49*, 992-999, 1972.
29. Neufeld E.F., Sando G.N., Garvin A.J., Rome L.H.: *The transport of lysosomal enzymes*. J. Supramol. Struct. *6*, 95-101, 1977.
30. Hickman S., Shapiro L.J., Neufeld E.F.: *A recognition marker required for uptake of a lysosomal enzyme by cultured fibroblasts*. Biochem. Biophys. Res. Commun. *57*, 55-61, 1974.
31. Kaplan A., Achord D.T., Sly W.S.: *Phosphoexosyl components of a lysosomal enzyme*

are recognized by pinocytosis receptors on human fibroblasts. Proc. Natl. Acad. Sci. USA *74*, 2026-2030, 1977.
32. Natowicz M.R., Chi M.M.Y., Lowry O.H., Sly W.S.: *Enzymatic identification of mannose-6-phosphate on the recognition marker for receptor-mediated pinocytosis of β-glucuronidase by human fibroblasts.* Proc. Natl. Acad.Sci. USA *76*, 4322-4326, 1979.
33. Sando G.N., Neufeld E.F.: *Recognition and receptor-mediated uptake of a lysosomal enzyme, α-L-iduronidase, by cultured human fibroblasts.* Cell. *12*, 619-627, 1977.
34. Kaplan A., Fischer D., Achord D.T., Sly W.S.: *Phosphoexosyl recognition is a general characteristic of pinocytosis of lysosomal glycosidases by human fibroblasts.* J. Clin. Invest. *60*, 1088-1093, 1977.
35. Ullrich R., Mersmann G., Weber E., Von Figura K.: *Evidence for lysosomal enzyme recognition by human fibroblasts via a phosphorylated carbohydrate moiety.* Biochem. J. *170*, 643-650, 1978.
36. Hasilik A., Rome L.H., Neufeld E.F.: *Processing of lysosomal enzymes in human skin fibroblasts.* Fed. Proc. *38*, 467, 1979 (Abstract).
37. Hasilik A., Neufeld E.F.: *Biosynthesis of lysosomal enzymes in fibroblasts: II Phosphorylation of mannose residue.* J. Biol. Chem. *255*, 4946-4950, 1980.
38. Miller A.L.: *I-cell disease: Isoelectric focusing, Concanavalin A-Sepharose 4 B binding and kinetic properties of human liver acid β-D-galactosidases.* Biochim. Biophys. Acta *522*, 174-186, 1978.
39. Gonzales-Noriega A.: *Receptor-mediated transport of acid hydrolases in fibroblasts.* Thesis, Washington University, St. Louis, Missouri, 1980.
40. Vladutiu G.D., Rattazzi M.C.: *Abnormal lysosomal hydrolases excreted by cultured fibroblasts in I-cell disease (mucolipidosis II).* Biochem. Biophys. Res. Commun. *67*, 956-964, 1975.
41. Vladutiu G.D., Rattazzi M.C.: *I-cell disease. Desialylation of β-hexosaminidase and its effect on uptake by fibroblasts.* Biochim. Biophys. Acta *539*, 31-36, 1978.
42. Kress B.C., Freeze H.H., Herd J.K., Alhadeff J.A., Miller A.L.: *Purification and characterization of I-cell disease α-L-fucosidase.* J. Biol. Chem. *255*, 955-961, 1980.
43. Tabas I., Kornfeld S.: *Biosynthetic intermediates of β-glucuronidase contain high mannose oligosaccharides with blocked phosphate residues.* J. Biol. Chem. *255*, 6633-6639, 1980.
44. Hasilik A., Klein U., Waheed A., Strecker G., Von Figura K.: *Phosphorylated oligosaccharides in lysosomal enzymes: identification of α-N-acetylglucosamine (1) phospho (6) mannose diester groups.* Proc. Natl. Acad. Sci. USA *77*, 7074-7078, 1980.
45. Reitman M.L., Kornfeld S.: *UDP-N-acetylglucosamine: glycoprotein N-acetylglucosamine-1-phosphotransferase-proposed enzyme for the phosphorylation of the high mannose oligosaccharide units of lysosomal enzymes.* J. Biol. Chem. *256*, 4275-4281, 1981.
46. Varki A., Kornfeld S.: *Identification of a rat liver α-N-acetylglucosaminyl phosphodiesterase capable of removing "blocking" α-N-acetylglucosamine residues from phosphorylated high mannose oligosaccharides of lysosomal enzymes.* J. Biol. Chem. *255*, 8389-8401, 1980.
47. Reitman M.L., Varki A., Kornfeld S.: *Fibroblasts from patients with I-cell disease and pseudo-Hurler polydystrophy are deficient in UDP-N-acetylglucosamine: glycoprotein-N-acetylglucosaminylphosphotransferase activity.* J. Clin. Invest. *67*, 1574-1579, 1981.
48. Ullrich K., Voss V., Von Figura K.: *Transfer of lysosomal β-hexosaminidase by cultivated hepatocytes.* Pediatr. Res. *15*, 1182, 1981 (abstract).
49. Fischer H.D., Gonzales-Noriega A., Sly W.S., Morrè P.J.: *Phosphomannosyl-enzyme receptors in rat liver.* J. Biol. Chem. *255*, 9608-9615, 1980.
50. Ashwell G., Morrel H.G.: *The role of surface carbohydrates in the hepatic recognition and transport of circulating glycoproteins.* Adv. Enzymol. *41*, 99-128, 1974.
51. Miller A.L.: Personal communication.

John O'Brien
Paolo Durand

Conclusions and speculations

This volume is a collection of both human and animal studies on glycoprotein storage diseases.

These studies have proved to be informative as to:

1) Identification of previously unrecognized diseases.
2) Identification of the metabolic products of the mutant genes.
3) Classification within a group.
4) Demonstration of genetic principles.

In each of these separate diseases in man there is a relatively characteristic combination of clinical lesions, enzyme deficiency and storage of glycoconjugates which forms the basis for classification.

For most of these disorders, demonstration of the enzymatic deficiency has led to accurate diagnosis of affected individuals, recognition of variant and atypical forms, detection of carriers and prenatal diagnosis of affected fetuses.

Many of the misteries of glycoprotein metabolism have been unravelled in the past decade. However, the relationship between enzyme deficiency and the observed pathology, is poorly understood.

For example, the precise mechanism by which mutations of one class of enzymes can produce a variety of clinical syndromes is not clear. This suggests that other factors in addition to the primary enzyme deficiency may interact *in vivo* to produce or alleviate pathology. Detailed knowledge of the structure-function of the mutated enzyme and of possible cofactor interactions may allow recognition of the mechanism by which clinical variants occur. Now, for instance, deficiencies in specific endogenous protein activators have been demonstrated to occur in known cases of sphingolipidoses with apparently normal activity of the specific hydrolases and it is anticipated that also some glycoproteinoses may be mimicked by deficiencies of activators.

It is likely that new glycoprotein storage diseases will be discovered upon future research. For example, β-mannosidosis has been discovered in goats but has not yet been described in man. Other disorders with sialic acid storage and a distinct form of I-cell disease are in the process of being defined.

Progress in unravelling these disorders depends on defining the metabolic derangements as precisely as possible, and examining any available member of each family in which findings suggestive of such metabolic disorders are present.

Other aims of this volume have been:

1) To review the methodologies used for diagnosis (including technical limitations).

2) To point out potential diagnostic hazards for clinicians.

3) To explore the "state of the art" of the molecular pathology of each disease.

Extreme care must be taken in the performance, interpretation and reporting of the results using the various assays described in this book.

Although there are technical problems to overcome for the different methodologies used for diagnosic purposes, they appear to be surmountable.

Sensitive and reproducible enzyme assay methods are available for each disorder.

How can one be confident that the clinical findings are due to the biochemical abnormalities, rather than being coincidental or that both are not due to some other primary defect?

The strategy of approach, for obviating misdiagnoses, especially of atypical forms is:

a) to use, when possible, naturally occurring glycoconjugates as substrates;

b) to measure the enzyme activity in terms of release of the sugar involved;

c) to include the identification of the stored substances;

d) to identify the carrier state of the parents.

Therefore without a pathophysiologic mechanism and without clear patterns of inheritance for the biochemical and clinical findings, it is difficult to deduce a causative relationship.

Although basic studies have not been informative on the gene defects of the diseases considered in this book, they have increased our understanding of the biology of lysosomal hydrolases, the molecular recognition system which controls the subcellular and extracellular fates of these enzymes and the processing of glycoproteins in human metabolism.

One of the most exciting areas for the future will be to determine the structures of each hydrolase gene using recombinant DNA technology and ascertain the coding sequence abnormalities in the mutant genes.

It is likely that many different types of mutations will be found and that some may be amenable to gene therapy.

In spite of the achievements in diagnosis, prognosis and prevention of glycoprotein storage diseases, therapeutic advances have been minimal. Enzyme therapy has been disappointing since no lasting clinical improvements have resulted. In many cases not even metabolic improvement has been achieved.

One of the major problems in the therapy of these diseases is to deliver the deficient enzyme to target tissues.

The existence of organ specific carbohydrate recognition (lectin) systems may be exploited in future attempts at enzyme replacement in these and other lysosomal storage diseases. Efforts directed towards enzyme therapy are probably doomed to failure until more is known about targeting, dosage and avoidance of immune responses.

The ability to insert normal human glycosidase genes into the appropriate mutant cells, producing stable transformants with correction of the metabolic defect, will open up a whole new area of therapeutics aimed not at compensating a metabolic defect, but at actually abolishing the defect.

Obviously a great deal of fundamental research on cloning of such genes, their delivery to mammalian cells in appropriate vectors, then targeting to specific organs, their regulation, stability and effect upon metabolism and on the potentially deleterious consequences of such therapy, will have to be carried out with animal models before trials in humans can be considered. For the glycoprotein storage diseases animal models (for α-mannosidosis beef and cats, for β-mannosidosis goats, for G_{M1} gangliosidosis dogs, cats and beef and for Sandhoff disease cats) are available.

Looking forward to this new era, the editors hope that the data presented in this volume will be useful in identification of an increasing number of patients, in prevention of these diseases and in providing inspiration for investigators and physicians in their attempts at understanding more about the multifaced area of glycoprotein storage diseases.

Index

MIX
Papier aus verantwortungsvollen Quellen
Paper from responsible sources
FSC® C105338

If you have any concerns about our products,
you can contact us on
ProductSafety@springernature.com

In case Publisher is established outside the EU,
the EU authorized representative is:
Springer Nature Customer Service Center GmbH
Europaplatz 3, 69115 Heidelberg, Germany

Printed by Libri Plureos GmbH
in Hamburg, Germany